T0300433

CHAPMAN & HALL/CRC FINANCIAL MATHEMATICS SERIES

American-Style Derivatives

Valuation and Computation

CHAPMAN & HALL/CRC
Financial Mathematics Series

Aims and scope:
The field of financial mathematics forms an ever-expanding slice of the financial sector. This series aims to capture new developments and summarize what is known over the whole spectrum of this field. It will include a broad range of textbooks, reference works and handbooks that are meant to appeal to both academics and practitioners. The inclusion of numerical code and concrete real-world examples is highly encouraged.

Proposals for the series should be submitted to one of the series editors above or directly to:
CRC Press, Taylor and Francis Group
24-25 Blades Court
Deodar Road
London SW15 2NU
UK

CHAPMAN & HALL/CRC FINANCIAL MATHEMATICS SERIES

American-Style Derivatives

Valuation and Computation

Jérôme Detemple

Published in 2006 by
Chapman & Hall/CRC
Taylor & Francis Group
6000 Broken Sound Parkway NW, Suite 300
Boca Raton, FL 33487-2742

International Standard Book Number-10: 1-58488-567-X (Hardcover)
International Standard Book Number-13: 978-1-58488-567-2 (Hardcover)
Library of Congress Card Number 2005052861

Library of Congress Cataloging-in-Publication Data

Detemple, Jérôme.
American-style derivatives : valuation and computation / Jerome Detemple.
p. cm. -- (Chapman & Hall/CRC financial mathematics series)
Includes bibliographical references and index.
ISBN 1-58488-567-X (alk. paper)
1. Derivative securities--United States. 2. Derivative securities--Valuation. I. Title. II. Series.

HG6024.U6D49 2005
332.64'57--dc22 2005052861

Visit the Taylor & Francis Web site at
http://www.taylorandfrancis.com

and the CRC Press Web site at
http://www.crcpress.com

Dedication

To Pamela, Michael and Alfie, for their love and support.

Acknowledgments

I benefitted from the comments of several generations of students at MIT, McGill and Boston University to whom I taught material included in this book. The remarks of Weidong Tian, Zhiguo He and Jing Zhou were especially appreciated. Special thanks are due to Mark Broadie. Much of this book is based on joint work that we undertook during the past decade. I have benefitted from our discussions on various aspects associated with the valuation of derivatives. Most of all I am indebted to him for introducing me to some of the subtleties arising in the computation of derivative securities' prices. Thanks are also due to Ioannis Karatzas, who taught me a great deal about stochastic processes and their application to financial models. His insights and contributions have shaped my own thinking and permeate several chapters of this book. Finally, I bear an obvious intellectual debt to the founding fathers and early pioneers of the field: Fisher Black, Myron Scholes, Bob Merton, Steve Ross and John Cox.

Biography

Jérôme Detemple is Professor of Finance and Economics at Boston University School of Management and an Associate Fellow at CIRANO. He holds a Ph.D. in Finance from the Wharton School, University of Pennsylvania, and a Doctorat D'État ès Sciences Économiques from Université Louis Pasteur in Strasbourg. His scholarly work is primarily in the fields of mathematical finance and financial economics. He is the author of 40 articles in leading academic journals and is widely known for his contributions to derivative securities valuation, risk management and asset allocation. Professor Detemple is currently on the editorial boards of *Mathematical Finance*, *Management Science* and the *Review of Derivatives Research*. He is a past associate editor of the *Review of Financial Studies*.

Contents

Chapter 1

Introduction

Derivative securities have become common financial instruments that appear in multiple areas of economic activity. Contracts of this type have been exchanged for several centuries among economic agents. But it is only during the past three decades, since the creation of the first organized options market, the Chicago Board of Options Exchange (CBOE), that the derivatives industry has experienced rapid growth. Since the opening of this market, the number and types of contracts have substantially increased. Today's investor can trade foreign exchange options, futures contracts, index options, bond options, energy derivatives and weather derivatives in organized markets. Additionally, advances in pricing theory and in technology during the past ten years have made it possible to engineer contracts with new provisions designed to meet all kinds of specific investment or financing needs. Capped options, Asian options, shout options and other types of exotic securities can now be purchased in the over-the-counter market or are issued by firms with particular objectives in mind. Option-like payoffs also permeate the activities of firms and the reward structures granted to managers and employees. Economic agents are becoming increasingly aware of the benefits and pitfalls associated with the use of derivative products.

This book provides an extensive treatment of theoretical and computational aspects of derivative securities pricing, with a particular emphasis on the valuation of American-style derivatives. The cross-disciplinary nature of this topic has been the trigger for a multitude of contributions dealing with various aspects of the valuation process. Inevitably, a selection of themes had to be made. Every effort has been made to present the most important aspects of the field. Part of this selection, nevertheless, reflects my own perceptions as well as personal research interests. Effort has also been made to provide appropriate references to the literature dealing with the results presented. Apologies are extended to those whose contributions may have been inadvertently overlooked.

The remainder of this introduction provides perspective on the various themes addressed. Important milestones in the development of a pricing theory for American-style contracts are surveyed. Part of this overview parallels the order in which the various topics are dealt with in the subsequent chapters. In certain places, however, chronological order takes precedence, in order to provide a better understanding of the evolution of the field.

The valuation of derivative securities has been the object of a long quest. A model describing the random behavior of speculative asset prices was proposed very early on, by Bachelier [1900]. The development of a rigorous theory of option pricing, however, only dates back to the 1970s, with the seminal papers of Black and Scholes [1973] and Merton [1973a]. The major contribution of this work can be viewed as a valuation approach consistent with the absence of arbitrage opportunities in the financial market. Black and Scholes used this methodology to price European options and related contingent claims in a simple model, where the underlying price follows a geometric Brownian motion and the interest rate is constant (the standard market model). Merton refined and extended the approach to more general settings. An equivalent methodology, based on an appropriately chosen "risk-neutral" valuation operator, was pioneered, a few years later, by Cox and Ross [1976]. The foundations and principles underlying this valuation method were identified and characterized in the fundamental papers of Harrison and Kreps [1979] and Harrison and Pliska [1981].

The valuation of American options also has a long history. Samuelson [1965] and McKean [1965] were the first to treat this problem as a stopping time problem involving the option's payoff and the underlying price distribution. McKean, furthermore, showed that the stopping problem could be converted into a free boundary problem. It is only more recently, however, that the optimal stopping problem has been posed relative to an appropriate measure, the Cox-Ross risk neutral measure, that correctly prices American options (Bensoussan [1984] and Karatzas [1988]). Karatzas [1988], in particular, shows that the American option payoff can be replicated by a carefully chosen policy of investment in the primary assets in the model. The value of the American option must then be equal to the value of the replicating portfolio in order to avoid arbitrage opportunities and be consistent with equilibrium in the capital market.

While the stopping time approach to American option valuation is instructive, it does not provide immediate insights into the properties of the optimal exercise boundary, nor does it lead to efficient numerical procedures. Nevertheless, building on this approach, Kim [1990], Jacka [1991] and Carr, Jarrow and Myneni [1992] were able to derive, in the context of the standard market model, a useful decomposition of the American option value, that emphasizes the premium attached to early exercise, i.e., exercise before the maturity date. This early exercise premium (EEP) representation expresses the value of the American option as the corresponding European option value plus the gain from early exercise. The gain from early exercise is the present value of the dividend benefits in the exercise region net of the interest losses on the payments incurred upon exercise.

In fact, the EEP formula ties back to the general theory of stochastic processes and more specifically to the Riesz decomposition of a supermartingale. This connection materializes because the option valuation problem, which is a stopping time problem involving a particular payoff function, is solved by taking the smallest supermartingale majorant of the payoff. This particular supermartingale is called the *Snell envelope* of the payoff function. The Riesz decomposition breaks the Snell envelope down into two parts, a martingale and a potential, and the latter corresponds to the EEP component of the option value. The Riesz decomposition of the value function associated with a stopping time problem was established by El Karoui and Karatzas [1991]. Their result was subsequently applied by Myneni [1992], to price an American put in the standard market model. The decomposition has since been extended, by Rutkowski [1994], to a more general class of market models with semimartingale prices.

The early exercise premium representation has a parametric flavor as it expresses the American option value in terms of the unknown optimal exercise boundary. But, by using the fact that immediate exercise is optimal when the asset price reaches the boundary, the EEP formula also produces a recursive integral equation for the optimal exercise boundary. This integral equation is useful for several purposes. For one it can serve as a starting point for a study of the exercise boundary and its properties. It can also be used for computational purposes. To calculate the American option value it suffices to solve the integral equation for the boundary and then to substitute this solution back in the EEP formula. In recent years substantial research efforts have been devoted to this two-stage procedure. Several computational schemes have been proposed and various modifications of the integral equation, designed to simplify calculations, have been developed.

While the valuation of standard American option contracts has now achieved a fair degree of maturity, much work remains to be done regarding the new contractual forms that are constantly emerging in response to evolving economic conditions and regulations. Innovations that have received attention include barrier options and capped options. Both types of contracts contain provisions designed to limit the payoffs under certain conditions. Barrier options, typically, condition the payoffs on the occurrence of events involving first passage times. For instance, a *knock-out* barrier option automatically expires if the underlying asset price reaches a prespecified threshold. A *knock-in* barrier option, on the contrary, comes into existence if such an event materializes. Capped options have a ceiling on their payoff that limits the potential gains from early exercise. Barrier and capped options are attractive from the perspective of issuers because they limit potential liabilities. Yet, they also remain attractive for purchasers as they retain an upside potential and are less costly than their plain vanilla counterparts. As a result, payoffs of this sort have appeared as components of securities issued by firms to cover certain financing needs. The valuation of these instruments can be quite involved depending on the exact nature of the contractual provisions. Thorough treatments of American-style versions of these contracts, in the standard market model, can be found in Broadie and Detemple [1995] and Gao, Huang and Subrahmanyam [2000]. Valuation formulas and principles for capped options also prove useful for the valuation of American-style vanilla options. Properly designed capped options, indeed, can provide accurate approximations of standard American option prices.

Our understanding of contracts written on multiple underlying assets is also limited, in comparison to the single asset case. Recent papers have made progress on that front, identifying the structure of immediate exercise regions associated with a variety of multiasset derivatives such as max-options, min-options and spread options (Tan and Vetzal [1994], Riddiough, Geltner and Stojanovic [1996], Gerber and Shiu [1997], Broadie and Detemple [1997a], Villeneuve [1997], Detemple, Feng and Tian [2003]). Some of the surprising properties documented in this literature show that optimal exercise decisions are not simple extrapolations of those prevailing in the single asset case. Proper identification of these policies becomes critical, in some cases, for capturing the full benefits associated with the early exercise provision.

Occupation time derivatives are among the latest innovations to appear in derivatives markets. Contracts of this sort have the common property of depending on the time spent by the underlying asset price in certain regions of the state space. Simple examples are payoffs depending on the amount of time above (or below) a fixed barrier. An option that comes into existence when the underlying price spends more than a prespecified amount of time below a given barrier is a more specific illustration. This contract, called a *cumulative knock-in barrier* option, is a natural generalization of the standard (one-touch) knock-in option that comes into existence at the first touch of the barrier. A variety of contracts with provisions based on occupation times have been examined. Quantile options, Parisian options and step options are perhaps the better known examples. European-style contracts of these types have been thoroughly studied and pricing formulas have been derived. Fundamental contributions in the area include Akahori [1995], Dassios [1995], Chesney, Jeanblanc-Picque and Yor [1997], Hugonnier [1999] and Linetsky [1999]. Little attention has been devoted to American-style versions of these claims: some results are provided herein.

Computational aspects of derivatives valuation have also been the subject of much research. The set of numerical techniques available for valuing American options is now vast and has increased substantially since the early developments of the 70s. Major approaches include the binomial method initiated by Cox, Ross and Rubinstein [1979], techniques based on partial differential equations developed by Schwarz [1977] and Brennan and Schwarz [1977], [1978], methods based on the integral equation of Kim [1990], Jacka [1991] and Carr, Jarrow and Myneni [1992], variational inequalities techniques proposed by Jaillet, Lamberton and Lapeyre [1990] and Monte Carlo simulation methods initiated by Boyle [1977]. A number of new approaches or modifications of existing methods have been proposed recently. Recent techniques based on approximations of stopping times (Broadie and Detemple [1996]), of partial differential equations (Carr and Faguet [1995]) or of integral equations (Ju [1998]) yield substantial improvements over traditional approaches. Modifications of the binomial method such as BBS (Binomial with Black-Scholes modification) also exhibit improved performance. An extensive study comparing the performances of various methods available appears in Broadie and Detemple [1996]. The new developments, that have taken place since that study, concern mainly the numerical valuation of American multiasset derivatives. But even though several computational approaches have been proposed to handle these complex securities (Broadie and Glasserman [1997a], [1997b], Rogers [2002], Haugh and Kogan [2003], Andersen and Broadie [2004]), their efficient numerical evaluation still remains a challenge.

The layout of the book is as follows. The second chapter reviews valuation principles for European contingent claims in a financial market in which the underlying asset price follows an Ito process and the interest rate is stochastic. Chapter 3 extends the analysis to American contingent claims. In this context the basic valuation principles for American options are laid out. Two instructive representation formulas, the EEP decomposition and the delayed exercise premium (DEP) decomposition, are also described. These results are applied in chapter 4 to American option valuation in the Black-Scholes market setting (i.e., the model with constant coefficients). A reduction of the integral equation as well as studies of related contractual forms are also presented in this context. American barrier and capped options are analyzed in chapter 5. For capped options, valuation formulas are first provided in the standard model with dividend-paying assets. Results are also presented for non-dividend-paying assets when the underlying asset price follows an Itô process with stochastic volatility and the cap's growth rate is an adapted stochastic process. Chapter 6 examines options written on multiple underlying assets. Immediate exercise regions are identified and valuation formulas provided, for a variety of popular multiasset option contracts. Chapter 7 provides an introduction to occupation time derivatives. Standard contractual forms are described and analyzed. Valuation formulas are reported for European-style and some American-style claims. Chapter 8 is devoted to numerical methods. Focus is on the computation of American-style options and emphasis is placed on recent developments. An efficiency study is carried out to evaluate the performance of some of these methods. All proofs are collected in appendices, found at the end of each chapter.

Chapter 2

European Contingent Claims

This chapter deals with the valuation of European-style financial claims. We first define the class of contingent claims that is the focus of our analysis (section 2.1) and describe the financial market setting (section 2.2). We then value European contingent claims (sections 2.3-2.7), examine option and forward contracts (section 2.8) and specialize the results to markets with deterministic coefficients (section 2.9). Lastly, we outline an extension to a market with multiple assets (section 2.10).

2.1 Definitions

Broadly speaking a *contingent claim* is a financial contract whose payoff is contingent on the state of nature. In its most general form a contingent claim generates a flow of payments over periods of time as well as cash payments at specific dates. The cash flows involved need not be paid at fixed points in time or during fixed periods of time. Some claims involve payments at prespecified random times or even at (random) times that are chosen by the holder of the contract. A contingent claim is said to be *European-style* if the timing of the contractual payments is specified at the outset. It is *American-style* if the timing of the payments is chosen by the owner of the claim.

Derivative securities are examples of contingent claims. They are typically defined as financial contracts whose payoffs depend on the price(s) of some *underlying* or *primary* asset(s). In this case the state dependence in the payoff of the contract arises through the market price(s) of some other asset(s).

The standard example of a derivative security is an *option* contract. An option gives the holder of the contract the right, but not the obligation, to buy (or sell) a given asset, at a predetermined price (the *exercise* or *strike* price), at or before some prespecified future date (the *maturity* date). The option to buy (sell) is a *call* (*put*) option. A European-style option contract can only be exercised at the fixed maturity date T. At that time, exercise is optimal if and only if the option produces a non-negative cash flow; in that event the option is said to be *in the money*. The payoff of a European call option is therefore $(S_T - K)^+ \equiv \max(S_T - K, 0)$, where S_T is the underlying asset price at the specified maturity date and $K > 0$ is the exercise price of the contract. The payoff of a European put option is $(K - S_T)^+$. American-style options, which can be exercised as desired, at or before the maturity date, have similar payoff functions upon exercise.

2.2 The Economy

The economy has the following characteristics. Uncertainty is represented by a complete probability space $(\Omega, \mathcal{F}, P)$ where Ω is the set of elementary events or "states of nature" with generic element ω, $\mathcal{F}$ is a σ-algebra representing the collection of observable events and P is a probability measure defined on $(\Omega, \mathcal{F})$. The time period is the finite interval $[0, T]$. A standard Brownian motion process z is defined on $(\Omega, \mathcal{F}, P)$ with values in $\mathbb{R}$. The flow of information is given by the natural filtration $\mathcal{F}_{(.)}$, i.e., the P-augmentation of the filtration generated by the Brownian motion. Thus, information at any time t consists in the observed trajectory of the Brownian motion process up to time t. Without loss of generality we set $\mathcal{F}_T = \mathcal{F}$ so that all the observable events are eventually known. Our model for information and beliefs is $(\Omega, \mathcal{F}, \mathcal{F}_{(\cdot)}, P)$.

Two types of investment opportunities, riskless and risky, are available in the asset market. The riskless asset (the bond) can be interpreted as a bank account, bearing interest at the locally riskless rate r. An investment in this account equal to B_0, at time $t = 0$, evolves according to

$$dB_t = r_t B_t dt, \qquad B_0 \text{ given.} \tag{2.1}$$

The interest rate r is a stochastic process that is bounded, strictly positive and progressively measurable.[1] For later use, define the discount factor $R_{s,t} = \exp\left(-\int_s^t r_v dv\right)$ for $s, t \in [0, T]$. The risky asset (the stock) has a price S satisfying the stochastic differential equation

$$dS_t = S_t \left[(\mu_t - \delta_t)\, dt + \sigma_t dz_t \right], \quad S_0 \text{ given.} \tag{2.2}$$

The process δ represents the dividend yield on the stock; μ and σ are, respectively, the drift and the volatility coefficients of the stock's total rate of return (the return from capital gains plus the dividend yield: $dS/S + \delta dt$). The coefficients δ, μ, and σ are bounded and progressively measurable processes of the filtration. The dividend rate is non-negative, $\delta \geq 0$; the volatility σ is bounded above and bounded away from zero (P-a.s.). The price that solves (2.2) is the

[1] Given the uncertainty setting adopted here, a process X is said to be *progressively measurable* if it depends on the trajectories of the Brownian motion z and on time. Formally, X is progressively measurable if X_t is $\mathcal{F}_t \times \mathcal{B}([0, t])$ measurable for all $t \in [0, T]$, where $\mathcal{B}([0, t])$ denotes the Borel sigma-field on $[0, t]$.

ex-dividend stock price. The gains process G is composed of capital gains and dividends ($dG = dS + S\delta dt$). The (total) rate of return is the gains process divided by the price ($dS/S + \delta dt = dG/S$).

The assumption that the return volatility be bounded away from zero ensures that an investor can always trade in the Brownian motion by trading the stock. As a result any financial exposure that is affected by the random fluctuations in the Brownian motion can be hedged. When this is the case the financial market is said to be *complete.*

Remark 1 *Alternative definitions of market completeness can be formulated. A natural one is to say that the financial market is complete when any state contingent claim with an $\mathcal{F}_{(.)}$-progressively measurable payoff process (i.e., cash flows that depend on the realized trajectories of the Brownian motion process z) can be manufactured by selecting an appropriate portfolio of traded financial assets and investing a suitable initial amount. Modulo some technical conditions on the class of claims involved, this definition is equivalent to the one above. When the volatility coefficient σ is bounded away from zero, the stochastic shocks affecting the financial market (the Brownian motion z) can be hedged, at all times, by investing in the stock. The ability to pursue unconstrained investment policies in the stock and in the locally riskfree asset, then, ensures the attainability of these contingent claims (Karatzas and Shreve [1998, Section 1.6]. See also Harrison and Kreps [1979], Harrison and Pliska [1981], Duffie [1986]). A formalization of some of these ideas is provided in the next section.*

It has become standard practice to use stochastic processes of the form (2.2) to model the behavior of stock prices. For instance the geometric Brownian motion process, obtained by taking constant coefficients (μ,σ,δ), is used as a basis for the Black and Scholes [1973] analysis. Typical Markovian models with price-dependent volatility, interest rate or dividend rate can also be embedded in this structure (see section 4.6 in chapter 4). The specification in (2.2) is more general to the extent that it allows for general dependencies between the coefficients and the trajectories of the underlying Brownian motion.[2]

In order to determine the prices of contingent claims one needs to characterize the set of random variables (payoffs) that can be manufactured (replicated) by trading the assets available, namely the stock and the bond. To this end a description of the stochastic process representing the value of an account associated with a portfolio policy is needed. Let X be the value of this account, i.e., the wealth process associated with an investment policy in the financial assets (2.1)–(2.2). We first define the notions of consumption (i.e., withdrawal) and portfolio policies, and the set of "feasible" or "admissible" consumption-portfolio policies.

A *portfolio* policy π is a progressively measurable, $\mathbb{R}$-valued stochastic process such that $\int_0^T \pi_t^2 \sigma_t^2 dt < \infty$, ($P$-a.s.). Here π_t denotes the (dollar) investment in

[2] Alternative formulations that have received attention include processes with jumps (Merton [1973a], Cox and Ross [1976]).

the stock at date t; the amount invested in the riskless asset is $X_t - \pi_t$. A cumulative *consumption* policy C is a progressively measurable, non-decreasing, right-continuous process with values in $\mathbb{R}$, initial value $C_0 = 0$ and $C_T < \infty$ $(P - a.s.)$. Consumption amounts to a withdrawal of funds from the account; injections are not permitted at this stage. When cumulative consumption is null at all times the portfolio is said to be *self-financing*: it involves neither infusions nor withdrawals of funds but only rebalancing of the positions held in the different assets.

An investment amounting to π_t in the stock at date t produces a gain (capital gains plus dividends) equal to $\pi_t\left[(dS_t/S_t) + \delta_t dt\right]$. An investment of $X_t - \pi_t$ in the bond gives a gain of $(X_t - \pi_t)r_t dt$. The activity of consumption reduces wealth by the corresponding amount dC_t. A consumption-portfolio policy (C, π) therefore leads to a wealth process X that is described by the stochastic differential equation

$$\begin{aligned} dX_t &= (X_t - \pi_t)r_t dt + \pi_t[(dS_t/S_t) + \delta_t dt] - dC_t, \quad X_0 = x, \\ &= r_t X_t dt + \pi_t[(\mu_t - r_t)dt + \sigma_t dz_t] - dC_t, \quad X_0 = x, \end{aligned} \tag{2.3}$$

where x denotes the initial amount invested. Given an initial investment $x > 0$, a consumption-portfolio policy (C, π) is *admissible*, if the associated wealth process X solving (2.3) satisfies the non-negativity constraint

$$X_t \geq 0, t \in [0, T], (P - a.s.). \tag{2.4}$$

This condition is a no-bankruptcy condition mandating that wealth cannot be negative during the trading period. Non-negativity of wealth does not preclude policies that involve short sales of the stock ($\pi < 0$) or borrowing at the riskfree rate ($X - \pi < 0$). Let $\mathcal{A}(x)$ be the set of admissible policies.

A *European-style contingent claim* (f, Y) is composed of a cumulative payment process f and a terminal cash flow Y at date T. The cumulative payment process is a finite variation process that is non-decreasing, progressively measurable, right-continuous and null at zero. The terminal cash flow is a non-negative $\mathcal{F}_T$-measurable random variable. A consumption-portfolio policy (C, π) *replicates* a European contingent claim (f, Y) at initial cost x if (C, π) is admissible, $dC_t = df_t$, and $X_T = Y$. We also say that (C, π) *generates* the claim (f, Y). The claim (f, Y) is *attainable* from an initial investment x if there exists an admissible consumption-portfolio policy such that $dC_t \geq df_t$ for all $t \in [0, T]$ and $X_T \geq Y$ (P-a.s.). Such a consumption-portfolio policy is said to *attain* or *super-replicate* (f, Y) from x.

2.3 Attainable Contingent Claims

The pricing of contingent claims amounts to the identification of an appropriate valuation operator that maps future payoffs into current prices. Given that the processes satisfying (2.1) and (2.2) represent the values of traded assets, the valuation operator must be consistent with these existing prices. In fact, as will become clear below, the price processes (2.1) and (2.2) completely determine the valuation operator in our economic setting.

The market model (2.1)-(2.2) implies a unique *market price of risk* (MPR) θ defined as $\theta_t = \sigma_t^{-1}(\mu_t - r_t)$. This one-dimensional process is progressively measurable and bounded, because σ is bounded away from zero. It is also uniquely defined. This is a typical implication of the ability to trade, at all times, in the underlying source of uncertainty, the Brownian motion. That is, it is a direct implication of market completeness. The market price of risk represents the expected excess return (over the riskfree rate) implicitly assigned by the model (2.1)-(2.2) to the stochastic shocks z affecting the financial market. It is also known as the *Sharpe ratio.*

Consider now the exponential process η defined by

$$\eta_t = \exp\left(-\int_0^t \theta_s dz_s - \frac{1}{2}\int_0^t \theta_s^2 ds\right). \tag{2.5}$$

This process is progressively measurable and positive. An application of Ito's lemma shows that η is a local martingale ($d\eta_t = -\eta_t\theta_t dz_t$).[3] Given that any non-negative local martingale is a supermartingale we see that the process η is actually a supermartingale (see Karatzas and Shreve [1988, Chapter 1, Problem 5.19]).[4] Moreover, boundedness of the market price of risk implies that the Novikov condition is satisfied,[5] i.e., for some constant K

$$E\left[\exp\left(\frac{1}{2}\int_0^T \theta_s^2 ds\right)\right] \le \exp\left(\frac{1}{2}K^2 T\right) < \infty.$$

It then follows that η is a martingale (Karatzas and Shreve [1988, Chapter 3, Corollary 5.13]) with initial value $\eta_0 = 1$. As a result, the new measure, $Q(A) \equiv E[\eta_T 1_A]$, $A \in \mathcal{F}_T$, where $E[\cdot]$ denotes the expectation under P, can be defined. It is easily verified that Q is a probability measure that is equivalent to P. For reasons that will become clear shortly, Q is called the *equivalent martingale measure* or the *risk neutral measure.* Additionally, by the Girsanov Theorem (Karatzas and Shreve [1988, Chapter 3, Theorem 5.1]), the process

$$\tilde{z}_t = z_t + \int_0^t \theta_s ds \tag{2.6}$$

for $t \in [0, T]$, is a standard Q-Brownian motion process. It is often useful to express it in its differential form $d\tilde{z}_t = dz_t + \theta_t dt$.

Under Q the discounted ex-dividend price augmented by the cumulative discounted dividends, $\widetilde{S}_t \equiv R_{0,t}S_t + \int_0^t R_{0,v}\delta_v S_v dv$, is a Q-martingale. Indeed,

[3] A process X is a *local* martingale if the stopped process $\{X(t \wedge \tau_n) : t \ge 0\}$ is a martingale for any sequence of stopping times $\{\tau_n : n = 1, ...\}$ such that $\lim_{n\to\infty} \tau_n = \infty$.

[4] A process X is a supermartingale if $E_t X_s \le X_t$ for any $s \ge t$.

[5] A process X is said to satisfy the Novikov condition if $E\left[\exp\left(\frac{1}{2}\int_0^T X_s^2 ds\right)\right] < \infty$.

recalling that $R_{0,t} \equiv \exp\left(-\int_0^t r_v dv\right)$, using Ito's lemma and the definitions (2.2) and (2.6) yields

$$\begin{aligned} d\widetilde{S}_t &= R_{0,t}dS_t + S_t dR_{0,t} + R_{0,t}\delta_t S_t dt \\ &= R_{0,t}S_t\left[(\mu_t - r_t)\,dt + \sigma_t dz_t\right] \\ &= R_{0,t}S_t\sigma_t d\tilde{z}_t, \end{aligned} \tag{2.7}$$

with initial condition $\widetilde{S}_0 = S_0$. The martingale property of $\widetilde{S}$ follows from the boundedness of r, σ, the square-integrability property $S \in \mathcal{L}^2[z]$ and the fact that $\tilde{z}$ is a Q-Brownian motion, hence a Q-martingale.[6] Thus, $\widetilde{S}_t = \widetilde{E}_t\left[\widetilde{S}_T\right]$ for all $t \leq T$, where $\widetilde{E}_t[\cdot] \equiv \widetilde{E}[\cdot \mid \mathcal{F}_t]$ is the conditional expectation relative to Q given the information $\mathcal{F}_t$. Substituting the definition of $\widetilde{S}$ on both sides of this equality and rearranging enables us to conclude that the valuation formula

$$S_t = \widetilde{E}_t\left[R_{t,T}S_T + \int_t^T R_{t,v}\delta_v S_v dv\right] \tag{2.8}$$

holds. The stock price can also be written in terms of the P-expectation as

$$S_t = E_t\left[R_{t,T}\eta_{t,T}S_T + \int_t^T R_{t,v}\eta_{t,v}\delta_v S_v dv\right] \tag{2.9}$$

where $\eta_{t,v} \equiv \eta_v/\eta_t$ and $E_t[\cdot] \equiv E_t[\cdot \mid \mathcal{F}_t]$ is the conditional expectation under P. This formula follows by applying Bayes' law to (2.8) (Karatzas and Shreve [1988, Chapter 3, Lemma 5.3]). Expression (2.8) also shows that the discounted ex-dividend price process $R_{0,t}S_t$ is a supermartingale under Q: given that dividends are non-negative (2.8) implies $R_{0,t}S_t \geq \widetilde{E}_t\left[R_{0,T}S_T\right]$, for all $t \leq T$.

The martingale property of the discounted stock price process (resp. of the process $\widetilde{S}$) in the absence (resp. presence) of dividends (2.7) motivates the terminology "equivalent martingale measure" used to describe the measure Q. Because the MPR θ is uniquely determined, the measure Q that yields the martingale property of $\widetilde{S}$ is unique. This feature of the model is a consequence of market completeness. The valuation formula (2.8) motivates the alternative terminology "risk neutral" measure. Indeed, as dividends are discounted at the locally riskfree rate the economy appears risk neutral. One should bear in mind, however, that a suitable risk adjustment is accounted for in the density η of the measure Q, through the MPR θ. This risk correction must always be taken into account when valuing contingent claims.

The formulas (2.8) and (2.9) are alternative representations of the stock price. The risk neutral valuation formula (2.8) calculates the stock price by

[6] The bracket $[z]$ is used to represent the quadratic variation of the process z (i.e., $d[z]$ is the local variance of z). For Brownian motion $d[z] = dt$. See Karatzas and Shreve [1988, Section 1.5] for formal definitions and properties. A process X belongs to $\mathcal{L}^2[z]$ if and only if $E\left[\int_0^T X_v^2 d[z]_v\right] = E\left[\int_0^T X_v^2 dv\right] < \infty$.

taking the expectation under the risk neutral measure of the discounted future dividends, where discounting is at the riskfree rate. The alternative expression (2.9) calculates the price as the expectation under the original measure of the discounted future dividends, where discounting is at a risk-adjusted rate implicit in the deflator $R\eta$. This latter formula reflects the standard notion that the price of an asset (here the stock) is the present value of its future cash flows.

In (2.8) the discount rate is locally riskless (conditional on contemporaneous information) but risky relative to the information available strictly prior to current time. Hence, the discount factor $R_{t,T}$ is an $\mathcal{F}_T$-measurable random variable that cannot be factored out of the conditional expectation operator $\widetilde{E}_t[\cdot]$. The same applies in the case of (2.9).

As we shall demonstrate below and in the next chapter the valuation operator for the stock in (2.8) also prices arbitrary contingent claims introduced in the financial market, as long as their payoffs depend on the same source of uncertainty that affects the stock price and the interest rate. With this interpretation in mind note that the system of Arrow-Debreu prices implied by the price system (2.1)-(2.2) is given by $R_{0,t}\eta_t dP$: each of these prices represents the value attributed by the market at date 0 to one dollar paid in a state (t, ω). The state price density (SPD) is defined as $\xi_t \equiv R_{0,t}\eta_t$, i.e., it represents Arrow-Debreu prices normalized by probabilities. With these definitions, the stock price formula can be written in the form

$$S_t = E_t \left[\xi_{t,T} S_T + \int_t^T \xi_{t,v} \delta_v S_v dv \right]$$

where $\xi_{t,v} = \xi_v / \xi_t$ represents the conditional state price density as of time t.

Consider European contingent claims (f, Y) with finite "values", i.e., claims that satisfy the integrability condition[7]

$$\widetilde{E} \left[R_{0,T} Y + \int_0^T R_{0,s} df_s \right] = E \left[\xi_T Y + \int_0^T \xi_s df_s \right] < \infty. \tag{2.10}$$

Let $\mathcal{I}_+$ be the class of European claims satisfying this condition (recall that the definition of a European claim adopted above mandates $df \geq 0, Y \geq 0$).

Our first theorem provides a characterization of the set of attainable contingent claims.

Theorem 1 *(Karatzas and Shreve [1988]). Consider a contingent claim (f, Y) $\in \mathcal{I}_+$. If (f, Y) is attainable from an initial investment x then*

$$\widetilde{E} \left[R_{0,T} Y + \int_0^T R_{0,s} df_s \right] \leq x \tag{2.11}$$

[7] Condition (2.10) says that the random variable $D_T \equiv \xi_T Y + \int_0^T \xi_s df_s$, representing the deflated cash-flows generated by the claim, is in $\mathcal{L}^1(P)$ (i.e., is finite in expectation).

(equivalently, $E\left[\xi_T Y + \int_0^T \xi_s df_s\right] \leq x$, where the expectation is taken relative to the measure P). Conversely, suppose that (2.11) holds. Then there exists a consumption-portfolio policy (C, π) that attains (f, Y) from the initial wealth x. Furthermore, if (2.11) holds with equality then there exists a consumption-portfolio policy (C, π) that replicates (f, Y) at initial cost x. If $\xi_T Y + \int_0^T \xi_s df_s \in \mathcal{L}^2$ and (2.11) holds with equality the replicating consumption-portfolio policy is unique.

In Proposition 4 below we show that $\widetilde{E}\left[R_{0,T} Y + \int_0^T R_{0,s} df_s\right]$ represents the market value at date 0 of the contingent claim (f, Y). Condition (2.11) then states that the value (i.e., the cost) of a contingent claim (f, Y) cannot exceed the value of any initial wealth level x from which the claim can be attained.

Remark 2 *Suppose that (C, π) replicates (f, Y) at initial cost x. The sufficiency part of the proof of Theorem 1 in the appendix (see (2.30)) shows that the wealth process associated with (C, π) is*

$$X_t = \widetilde{E}_t\left[R_{t,T} Y + \int_t^T R_{t,s} df_s\right], \qquad X_0 = x.$$

As f is non-decreasing (and null at 0) and Y is non-negative we conclude that wealth is non-negative at all times. The wealth process is the present value of the future cash flows generated by the policy (C, π) from an initial outlay of funds equal to x.

2.4 Valuation of Attainable Claims

With the characterization of an attainable contingent claim in Theorem 1 it is now easy to deduce its market value. To this end, we introduce the notions of an arbitrage opportunity and of the rational price of a contingent claim. Suppose that the claim (f, Y) is marketed at some price $V = V(f, Y)$ such that

$$dV_t = V_t[\alpha_t dt + \rho_t dz_t] - df_t, \tag{2.12}$$

where α, ρ are progressively measurable processes with $\int_0^T (\alpha_t \vee \rho_t^2) dt < \infty, (P-a.s.)$ and $\alpha_t \vee \rho_t^2 \equiv \max(\alpha_t, \rho_t^2)$. Individuals can then invest in the stock, the riskless asset, as well as in the contingent claim. Let $\widehat{\pi}^v = n^v V$ be the amount invested in the claim, where n^v is the number of claims held (long or short). Suppose that $\widehat{\pi}^v$ is a progressively measurable, $\mathbb{R}$-valued process such that $\int_0^T \left(\widehat{\pi}_t^v \alpha_t \vee (\widehat{\pi}_t^v \rho_t)^2\right) dt < \infty$, $(P-a.s.)$.

Applying the arguments underlying the derivation of (2.3), we see that a consumption-portfolio policy $(\widehat{C}, \widehat{\pi}, \widehat{\pi}^v)$ leads to a wealth process $\widehat{X}$ that solves the stochastic differential equation

$$d\widehat{X}_t = r_t \widehat{X}_t dt + \widehat{\pi}_t [dS_t/S_t - r_t dt] + n_t^v [dV_t - r_t V_t dt] + \widehat{\pi}_t \delta_t dt + n_t^v df_t - d\widehat{C}_t$$

subject to the initial condition $\widehat{X}_0 = \widehat{x}$.

Definition 2 *A consumption-portfolio policy* $(\widehat{C}, \widehat{\pi}, \widehat{\pi}^v)$ *is an arbitrage opportunity if and only if* $\widehat{x} = 0$, $d\widehat{C}_t \geq 0$, $(P - a.s.)$, $P(\widehat{X}_T \geq 0) = 1$ *and* $P(\widehat{X}_T > 0) > 0$.

An arbitrage opportunity is a consumption-portfolio policy that has zero initial cost, requires no intermediate cash infusions (but allows for intermediate withdrawals) and has a positive probability of positive wealth at time T (along with a null probability of negative wealth). An arbitrage opportunity need not be admissible in the sense discussed before: no restrictions are placed on intermediate values of wealth. Terminal wealth (the liquidation value of the account at the final date), however, must be non-negative. An arbitrage policy is said to satisfy a *solvency* constraint at the terminal date.

Definition 3 *A rational price process for the claim* (f, Y) *is a price process* V *that is consistent with the absence of arbitrage opportunities in the financial market.*

A rational price is a no-arbitrage price. The set of rational prices for the contingent claim (f, Y) must contain the market *value* of the claim. Indeed, deviations between the market price and the set of rationals prices would lead to the existence of an arbitrage opportunity, a situation that is inconsistent with equilibrium in the financial market. Completeness of the financial market ensures that the set of rational prices associated with a given attainable claim is a singleton: the rational price of an attainable claim is unique.

The absence of arbitrage opportunities and the structure of the claim, characterized by non-negative cash flows $df \geq 0$, $Y \geq 0$, impose an immediate restriction on the price process (2.12). Indeed, it is clear that V must be non-negative (in fact strictly positive if $(f, Y) \neq (0, 0)$). In the opposite event one could simply buy the claim, thus pocketing an immediate positive amount and collecting additional non-negative cash flows in the future. This strategy entails no outlay of funds, but generates non-negative inflows over time. Positive inflows could be systematically reinvested at the riskfree rate until the maturity date: the resulting policy has positive terminal value with positive probability. A similar argument also establishes that $V_T \geq Y$. As the symmetric argument ensures that $V_T \leq Y$, we conclude that $V_T = Y$.

With these definitions we are now ready to provide a valuation formula for the contingent claim.

Proposition 4 *The rational price at time* t *of the European contingent claim* $(f, Y) \in \mathcal{I}_+$ *is uniquely given by*

$$V_t(f, Y) = \widetilde{E}_t \left[R_{t,T} Y + \int_t^T R_{t,s} df_s \right] = E_t \left[\xi_{t,T} Y + \int_t^T \xi_{t,s} df_s \right] \tag{2.13}$$

for $t \in [0, T]$.

Proposition 4 provides the most general pricing formulas, for claims in the class under consideration, in the context of the financial market model with stochastic coefficients (2.1)–(2.2). It shows that the value of any European-style contingent claim involving payments over $[0, T]$ is given by the expected value of the discounted cash flows. Discounting is at the locally riskfree interest rate when the expectation is taken under the equivalent martingale measure implicit in the market model (2.1)–(2.2). It is at a risk-adjusted rate when the expectation is calculated under the original probability measure. Note that these pricing formulas are valid even though the riskfree rate as well as the drift and volatility of the stock price process are progressively measurable processes of the Brownian filtration, i.e., even though these coefficients may depend on the history of the Brownian motion. Note also that the valuation operator in Proposition 4 is identical to the valuation operator for the stock in (2.8). Thus, all the properties satisfied by the stock price process are also satisfied by the prices of contingent claims. In particular the Q-martingale property, of the process composed of the discounted price of the claim augmented by its cumulative discounted cash flows, holds. The justification for the pricing formulas draws on the no-arbitrage principle: when the market price of the contingent claim deviates from the rational price prescribed in (2.13), it is possible to construct a portfolio policy involving the claim, the stock and the bond that constitutes an arbitrage opportunity. The universality of the pricing operator is an implication of market completeness.

2.5 Claims Involving Negative Payoffs

Claims involving negative cash flows or combinations of negative and positive cash flows can be handled using a modification of the arguments developed in the last two sections.

Consider a claim (f, Y) composed of a cumulative cash flow process f whose increments are paid over time and a terminal payment Y at date T. Suppose that f is a progressively measurable, right-continuous, finite variation process with null initial value and that Y is an $\mathcal{F}_T$-measurable random variable. Also assume that f and Y are uniformly bounded from below. Thus, cash flows can take negative values but cannot become unboundedly negative. Examples of claims satisfying these conditions include forward and futures contracts, interest rate swaps with caps on the flexible rate, break forward options (also known as Boston options) and other exotic contracts.

To incorporate this class of claims in the analysis we modify the notion of attainability as follows. A consumption-portfolio policy (C, π) is said to be K-*admissible*, if C is bounded from below and the associated wealth process X solving (2.3) satisfies a uniform lower bound

$$X_t \geq -K, t \in [0, T], (P - a.s.) \tag{2.14}$$

where $K > 0$ is some arbitrary, but finite constant. Let $\mathcal{A}(x; K)$ denote the set of K-admissible policies. Negative values of consumption, representing infusions of funds, are now permitted. The value of the account can also take negative values, as long as these remain bounded from below.

A consumption-portfolio policy (C, π) *K-replicates* or *K-generates* a European contingent claim (f, Y) at initial cost x if (C, π) is K-admissible, $dC_t = df_t$, and $X_T = Y$. The claim (f, Y) can be *K-attained* or *K-superreplicated* from an initial investment x if there exists a K-admissible consumption-portfolio policy such that $dC_t \geq df_t$ for all $t \in [0, T]$ and $X_T \geq Y$ (P-a.s.).

Let $\mathcal{I}_{ubb}$ be the set of claims satisfying the integrability condition (2.10) and such that the random pair (f, Y) is uniformly bounded from below. A characterization of K-attainable claims is as follows.

Theorem 5 *Consider a contingent claim $(f, Y) \in \mathcal{I}_{ubb}$. If (f, Y) is K-attainable from an initial investment x then*

$$\widetilde{E}\left[R_{0,T}Y + \int_0^T R_{0,s} df_s\right] \leq x \tag{2.15}$$

(equivalently, $E\left[\xi_T Y + \int_0^T \xi_s df_s\right] \leq x$, where the expectation is taken relative to the measure P). Conversely, suppose that (2.15) holds. Then there exists a consumption-portfolio policy (C, π) that K-attains (f, Y) from the initial wealth x. Furthermore, if (2.15) holds with equality then there exists a consumption-portfolio policy (C, π) that K-replicates (f, Y) at initial cost x. If $\xi_T Y + \int_0^T \xi_s df_s \in \mathcal{L}^2$ and (2.15) holds with equality the K-replicating consumption-portfolio policy is unique.

Theorem 5 shows that the earlier characterization of attainable claims remains valid for K-attainable claims. The reason has to do with the properties of the consumption-portfolio policies used to K-replicate or K-superreplicate a given contingent claim (f, Y). By definition the wealth process associated with a K-admissible policy satisfies a uniform lower bound. This ensures that the process $R_{0,t}X_t + \int_0^t R_{0,s} dC_s$ representing the sum of discounted wealth plus cumulative discounted consumption is a supermartingale under the risk neutral measure. The budgetary restriction (2.15) follows from this supermartingale property and the relations $dC_t \geq df_t$ for all $t \in [0, T]$ and $X_T \geq Y$ (see the proof for further details).

The existence of a K-replicating policy ensures that an arbitrage portfolio can be constructed whenever the market price deviates from the rational price of a claim. This guarantees that our previous valuation formulas extend to the claims under consideration.

Proposition 6 *The rational price at time t of the European contingent claim $(f, Y) \in \mathcal{I}_{ubb}$ is uniquely given by*

$$V_t(f, Y) = \widetilde{E}_t\left[R_{t,T}Y + \int_t^T R_{t,s} df_s\right] = E_t\left[\xi_{t,T}Y + \int_t^T \xi_{t,s} df_s\right]$$

for $t \in [0, T]$.

2.6 The Structure of Contingent Claims' Prices

The price of the contingent claim, in earlier sections, was represented as a local semimartingale with unspecified drift coefficient α (see (2.12)). Intuitive arguments suggest that the drift of this price process ought to be restricted if the financial market is complete. We formalize this intuition next.

A self-financing consumption-portfolio policy $(0, \widehat{\pi}, \widehat{\pi}^v)$ induces a wealth process satisfying

$$d\widehat{X}_t = r_t\widehat{X}_t dt + (\widehat{\pi}_t(\mu_t - r_t) + \widehat{\pi}_t^v(\alpha_t - r_t))\,dt + (\widehat{\pi}_t\sigma_t + \widehat{\pi}_t^v\rho_t)\,dz_t$$

subject to the initial condition $\widehat{X}_0 = x$. Moreover, selecting the particular policy $\widehat{\pi}_t = -\sigma_t^{-1}\widehat{\pi}_t^v\rho_t$ yields the locally riskless account value

$$d\widehat{X}_t = r_t\widehat{X}_t dt + \widehat{\pi}_t^v\left(-\sigma_t^{-1}\rho_t(\mu_t - r_t) + (\alpha_t - r_t)\right)dt$$

with same initial condition. Define the excess portfolio return as $e_t \equiv -\sigma_t^{-1}\rho_t(\mu_t - r_t) + (\alpha_t - r_t)$ and let $A_t \equiv \{\omega : e_t \geq 0\}$ represent the set of date t states in which $e_t \geq 0$. Consider the portfolio policy $\widehat{\pi}_t^v = \widehat{X}_t\left(2 \times 1_{\{A_t\}} - 1\right)$. This policy invests all the resources available in the claim if $e \geq 0$ and shorts a similar amount in the opposite event. The associated wealth process satisfies $d\widehat{X}_t = r_t\widehat{X}_t dt + \widehat{X}_t\,|e_t|\,dt$ or, equivalently,

$$\widehat{X}_t = x\exp\left(\int_0^t (r_s + |e_s|)\,ds\right).$$

A potential arbitrage portfolio can then be constructed by financing this policy at the locally riskfree rate. The initial cost of this new strategy is null and the terminal cash flow equals

$$x\exp\left(\int_0^T r_s ds\right)\left(\exp\left(\int_0^T |e_s|\,ds\right) - 1\right).$$

Clearly, this is an arbitrage portfolio if $\{(t,\omega) : e_t \neq 0\}$ has positive $l \times P$-measure (where l is Lebesgue measure): in this situation an investor would be able to profit for sure without disbursing funds. The absence of arbitrage opportunities, in equilibrium, implies that $\{(t,\omega) : e_t = 0\}$ must have full measure. We summarize the implication of this condition in our next theorem.

Theorem 7 *The return of the attainable contingent claim (f, Y) satisfies the spanning relation*

$$\alpha_t - r_t = \rho_t\theta_t = \beta_t(\mu_t - r_t)$$

where $\beta_t \equiv \rho_t/\sigma_t$. If the underlying asset corresponds to the market portfolio of risky securities this relation coincides with the Capital Asset Pricing Model (CAPM) which states that the expected excess return on any security is proportional to the expected excess return of the market portfolio. The proportionality *factor β is the beta of the claim. If the underlying asset does not correspond to the market portfolio, but satisfies the CAPM, then a simple transformation establishes that the CAPM still holds for the claim, i.e., $\alpha_t - r_t = \beta_t^m(\mu_t^m - r_t)$ where μ_t^m is the drift of the market portfolio and β^m the beta of the claim with respect to the market portfolio.*

This theorem records two results. First, it states that the market price of risk implied by the price of any contingent claim is identical to the market price of risk implied by the underlying asset. This is an intuitive relation given the completeness of asset markets (risks ought to be priced identically across securities). Second, it also states that the price of the claim is consistent with general pricing principles characterizing equilibrium in frictionless markets. In that regard the theorem shows that the risk premium of the claim satisfies the CAPM, in that it is proportional to the risk premium of the market portfolio of risky securities. Further insights regarding this equilibrium relationship can be found in the original articles of Sharpe [1964], Lintner [1965] and Mossin [1969], who developed the static version of the CAPM. The dynamic version, for markets with diffusion price processes, can be found in Merton [1973b].

2.7 Changes of Numeraire and Valuation

The present value formulas in Propositions 4 and 6 can be restated in the form,

$$R_t V_t(f, Y) + \int_0^t R_s df_s = \widetilde{E}_t \left[R_T Y + \int_0^T R_s df_s \right],$$

to emphasize that the process on the left hand side, i.e., the discounted price augmented by the cumulative discounted cash flow, is a Q-martingale. This relation can be reinterpreted in the following manner. Suppose that we choose, as a numeraire, the value of an account continuously reinvested at the short rate r. With an initial investment equal to 1 this gives an account value $b_t \equiv R_t^{-1} = \exp\left(\int_0^t r_v dv\right)$ at date t. Let us call this value the "bond" numeraire. In this unit of account the claim is worth

$$V_t^b(f, Y) \equiv \frac{V_t(f, Y)}{b_t} = R_t V_t(f, Y)$$

at $t \in [0, T]$ and generates a cumulative payment process f^b, such that $df_v^b = R_v df_v$ at time $v \in [0, T]$, as well as a terminal payment $Y^b = R_T Y$. The present value formula now states that the process $V_t^b(f, Y) + \int_0^t df_s^b$, expressed in the bond numeraire, is a Q-martingale. Alternatively, if cash flows are non-negative $(df^b \geq 0, Y^b \geq 0)$, we can also say that the claim's price in the bond numeraire, $V_t^b(f, Y)$, is a Q-supermartingale.

As the choice of a numeraire is arbitrary one may naturally wonder about the properties of prices when alternative "currencies" prevail. Of particular interest is whether one can identify an equivalent measure that preserves the martingale property in the new numeraire. The remainder of this section is devoted to these issues, first addressed in an insightful paper by Geman, El Karoui and Rochet [1995].

Consider a consumption-portfolio policy $(C', \pi') \in \mathcal{A}(x)$ with $C'_t = \int_0^T X_v dC_v$ and $\pi'_t = X_t \pi_t$ for $t \in [0,T]$, and where $\pi\sigma$ satisfies the Novikov condition. Under the risk neutral measure, the associated wealth process satisfies the equation $dX_t = X_t [r_t dt - dC_t + \pi_t \sigma_t d\widetilde{z}_t]$ subject to the initial condition $X_0 = x$. The solution is

$$X_t = x \exp\left(\int_0^t (r_v dv - dC_v)\right) \eta_t^{\pi\sigma}$$

where $\eta_t^{\pi\sigma} \equiv \exp\left(-\frac{1}{2}\int_0^t (\pi_v \sigma_v)^2 dv + \int_0^t \pi_v \sigma_v d\widetilde{z}_v\right)$ is a Q-martingale. Suppose that we select the consumption-adjusted wealth process

$$X_t^{ca} \equiv \exp\left(\int_0^t dC_v\right) X_t \tag{2.16}$$

as our new numeraire. In this new "currency" the claim pays $f_t^{ca} = \int_0^t (1/X_v^{ca}) df_v$, for $t \in [0,T]$ and $Y^{ca} = Y/X_T^{ca}$ at maturity; its price is $V_t^{ca}(f,Y) \equiv V_t(f,Y)/X_t^{ca}$ at $t \in [0,T]$. Substituting these definitions in the present value formula gives

$$R_t X_t^{ca} V_t^{ca}(f,Y) + \int_0^t R_s X_s^{ca} df_s^{ca} = \widetilde{E}_t \left[R_T X_T^{ca} Y^{ca} + \int_0^T R_s X_s^{ca} df_s^{ca} \right]$$

or, equivalently,

$$V_t^{ca}(f,Y) = \widetilde{E}_t \left[\frac{R_T X_T^{ca}}{R_t X_t^{ca}} Y^{ca} + \int_t^T \frac{R_s X_s^{ca}}{R_t X_t^{ca}} df_s^{ca} \right].$$

As $R_s X_s^{ca}/(R_t X_t^{ca}) = \eta_{t,s}^{\pi\sigma}$ for all $s \in [t,T]$ we can also pass to the new measure $Q^{\pi\sigma}$ defined by $dQ^{\pi\sigma} \equiv \eta_{0,T}^{\pi\sigma} dQ$, to conclude that

$$V_t^{ca}(f,Y) = E_t^{\pi\sigma}\left[Y^{ca} + \int_t^T df_s^{ca} \right], \tag{2.17}$$

where $E_t^{\pi\sigma}[\cdot]$ is the expectation under $Q^{\pi\sigma}$. In other words the process $V_t^{ca}(f,Y) + \int_0^t df_s^{ca}$, expressing prices and payments in the new numeraire, is a $Q^{\pi\sigma}$-martingale. The answer to our first question is indeed affirmative. But it is important to realize that the equivalent martingale measure had to be constructed in a very specific way in order to preserve the martingale property in the new currency system. In this operation changes in units of account are intimately related to changes of measure (and conversely). Moreover, we also uncover the interpretation of the measure $Q^{\pi\sigma}$. It represents the unique measure under which the price augmented by the cumulative payments, all expressed in the numeraire X^{ca}, enjoy the martingale property.

A pricing relation, whose importance will become clear in the next section, can also be retrieved modulo an additional transformation. Fix $t \in [0, T]$. Define, for $s \geq t$, the cash flow $f_s^* = X_t \int_0^s (1/X_v) df_v$ and let $Y^* = Y X_t / X_T$. As $X_v / X_v^{ca} = \exp\left(-\int_0^v dC_s\right)$ it follows from (2.17) that

$$V_t(f, Y) = E_t^{\pi\sigma} \left[\exp\left(-\int_t^T dC_v \right) Y^* + \int_t^T \exp\left(-\int_t^s dC_v \right) df_s^* \right]. \quad (2.18)$$

This new expression shows that the value of the claim (f, Y) in the original economy with interest rate r and risk neutral measure Q is identical to the value of the claim (f^*, Y^*) in a new economy with cumulative interest process C and risk neutral measure $Q^{\pi\sigma}$. The result suggests a form of *symmetry* between the claims (f, Y) and (f^*, Y^*) when paired with their respective economies.

We can collect these results in the following theorem,

Theorem 8 *Consider a European contingent claim $(f, Y) \in \mathcal{I}_{ubb}$ and suppose that we adopt the new numeraire X^{ca} defined in (2.16). Expressed in this new currency, the price of the claim, $V_t^{ca}(f, Y)$, augmented by the cumulative cash flow, $f_t^{ca} = \int_0^t df_s^{ca}$, is a $Q^{\pi\sigma}$-martingale where $Q^{\pi\sigma}$ is an equivalent measure with Q-density equal to $\eta_{0,T}^{\pi\sigma} = R_T X_T^{ca} / (R_0 X_0^{ca})$ (see (2.17)). The value of the claim $V_t(f, Y)$ in the original economy with interest rate r and pricing measure Q is the same as the value of a symmetric claim (f^*, Y^*), in a new economy with cumulative interest process C and pricing measure (i.e., equivalent martingale measure) $Q^{\pi\sigma}$.*

The symmetry property described in this proposition holds for any contingent claim and, in particular, for the stock price (2.2). As we shall see in the next sections and chapters, the result covers particular cases that have been extensively studied in the options literature.

A discussion of general symmetry properties can be found in Kholodnyi and Price [1998]. Their analysis focuses on (2.17) and uses concepts from operator and group theories. They show, in particular, that symmetry can be expressed in terms of Kelvin transforms. In their foreign exchange setting the symmetry relations have a natural interpretation of payoffs found on "opposite" sides of a market (i.e., payoffs quoted in different currencies). Schroder [1999] applies the methodology of Geman, El Karoui and Rochet [1995] by selecting the dividend-adjusted asset price as the new numeraire. This choice leads to a special case of (2.18), valid for general claims. Detemple [2001, Section 8] provides further perspective on the change of measure and the symmetry property between seemingly unrelated claims in different economies.

2.8 Option and Forward Contracts

Standard European option contracts involve a payment at the maturity date T only. For a call option the cumulative payment is $f = 0$ and the terminal payoff is $Y = (S_T - K)^+$ where K is the strike price (or exercise price); for a put option $f = 0$ and $Y = (K - S_T)^+$. For these contracts the pricing formula of Proposition 4 specializes as follows.

Corollary 9 *In the financial market model (2.1)–(2.2) the rational price of a European call option with maturity date T and exercise price K is given by*

$$c_t = \widetilde{E}_t \left[R_{t,T}(S_T - K)^+ \right]$$

for $t \in [0,T]$. The price of a European put option is $p_t = \widetilde{E}_t \left[R_{t,T}(K - S_T)^+ \right]$, for $t \in [0,T]$.

Inspection of option payoffs reveals that the European put payoff is equal to the payoff of a call with matching characteristics minus the stock price plus the strike price. That is, $(K - S_T)^+ = (S_T - K)^+ + K - S_T$. The put value is then given by

$$p_t = c_t + K\widetilde{E}_t[R_{t,T}] - \left[S_t - \widetilde{E}_t \left[\int_t^T R_{t,v} S_v \delta_v dv \right] \right]. \tag{2.19}$$

This *put-call parity* (PCP) relationship determines the relative prices of options with identical characteristics written on the same underlying asset.

Another important property of options is the property of *put-call symmetry* that relates the price of a put to the price of a call in an auxiliary financial market with modified characteristics. To state this relation consider a financial market with interest rate δ, in which the underlying asset price satisfies

$$dS_t^* = S_t^*[(\delta_t - r_t)dt + \sigma_t dz_t^*], \quad S_0^* \text{ given} \tag{2.20}$$

under some risk neutral measure Q^*. In this market the asset has dividend rate r and volatility coefficient σ. The process z^* is a Brownian motion under the pricing measure Q^*. Both z^* and Q^* are specified below. As before, the coefficients (δ, r, σ) are adapted to the filtration $\mathcal{F}_{(\cdot)}$ generated by the Brownian motion z, which represents the information available to investors.

Schroder [1999] demonstrates the following general European put-call symmetry (PCS) property.

Theorem 10 *(European PCS). Consider a European put option with characteristics K and T written on an asset with price S given by (2.2) in the market with stochastic interest rate r. Let $p(S,K,r,\delta;\mathcal{F}_t)$ denote the put price process. Then*

$$p(S,K,r,\delta;\mathcal{F}_t) = c(K,S,\delta,r;\mathcal{F}_t) \tag{2.21}$$

where $c(K,S,\delta,r;\mathcal{F}_t)$ is the value of a call with strike price S and maturity date T in a financial market with interest rate δ and in which the underlying asset price follows the Ito process (2.20) with initial value K and with z^* defined by

$$z_t^* = -\widetilde{z}_t + \int_0^t \sigma_v dv \tag{2.22}$$

for $t \in [0,T]$. The process z^* is a Brownian motion under the measure Q^* given by

$$dQ^* = \exp\left(-\frac{1}{2}\int_0^T \sigma_v^2 dv + \int_0^T \sigma_v d\tilde{z}_v\right) dQ \equiv \eta_T^* dQ. \tag{2.23}$$

The key idea behind put-call symmetry is a change of measure converting a put option in the original economy into a call option with symmetric characteristics in the auxiliary economy. In effect this change of measure amounts to a change of numeraire, as outlined in the previous section, in which the dividend-adjusted underlying asset is taken as the new unit of account (the numeraire is $X_T^{ca} = S_T \exp\left(\int_0^T \delta_v dv\right)$ and $X_T = \exp\left(-\int_0^T \delta_v dv\right) X_T^{ca} = S_T$). Expressing the payoff in this new numeraire and correcting for dividends gives $Y^* = YX_t/X_T = (K - S_T)^+ S_t/S_T = (S_T^* - S_t)^+$ where $S_T^* = KS_t/S_T$. This is the payoff of a call option with strike S_t and written on a new asset whose price S^* is the inverse of the original asset price adjusted by a multiplicative factor depending only on the initial conditions. Note also that this equivalence is obtained by switching from (S, K, r, δ) to (K, S, δ, r), but keeping the trajectories of the Brownian motion the same, i.e., the filtration that is used to compute the value of the call is the one generated by the original Brownian motion z. In other words the information used to compute the call value in the auxiliary financial market is the same as in the original market. As the property holds for general environments in which the coefficients of the price process are themselves adapted processes, it will hold, in particular, in the context of diffusion models: the separation between the information filtration and the change of measure is important for proving the property in these models without imposing "symmetry" restrictions directly on the volatility coefficient.

A slightly stronger version of the preceding result is obtained if the coefficients of the model are adapted to the subfiltration generated by the process z^* (see Detemple [2001]). Let $\mathcal{F}_{(\cdot)}^*$ denote the filtration generated by the Q^*-Brownian motion process z^*.

Corollary 11 *Suppose that the coefficients (r, δ, σ) are adapted to the filtration $\mathcal{F}_{(\cdot)}^*$. Then*

$$p(S, K, r, \delta; \mathcal{F}_t) = c(K, S, \delta, r; \mathcal{F}_t^*)$$

where $c(K, S, \delta, r; \mathcal{F}_t^*)$ is the value of a call with strike price S and maturity date T in a financial market with information filtration $\mathcal{F}_{(\cdot)}^*$ generated by the Q^*-Brownian motion process (2.22), interest rate δ and in which the underlying asset price follows the Ito process (2.20) with initial value K.

In the context of this corollary part of the information embedded in the original information filtration generated by the Brownian motion z may be irrelevant for pricing the put option. As all the coefficients are adapted to the subfiltration generated by z^* this is the only information that matters in computing the expectation under Q^* which determines the put value (see (2.34) in the appendix). Note that European PCS in the standard model with constant coefficients (the Black-Scholes setting) is a subcase of this corollary. Indeed, for this setting, it can be verified that the filtrations $\mathcal{F}^*_{(\cdot)}$ and $\mathcal{F}_{(\cdot)}$ coincide and that direct integration over z^* leads to the call value in the auxiliary financial market and the put value in the original economy.

Before turning our attention to forward contracts, note that we can also use the measure Q^* in order to restate the European PCP relation as

$$p_t = c_t + K\widetilde{E}_t\left[R_{t,T}\right] - S_t E^*_t\left[\exp\left(-\int_t^T \delta_v dv\right)\right].$$

Special cases of this formula (for instance in the setting with constant coefficients) have been well documented in the literature.

A forward contract is another example of a derivative security, commonly employed to hedge financial exposures. This financial arrangement involves a terminal payment $Y = S_T - K$ where K is the delivery price and has no intermediate cash flows ($f = 0$). Given that the final payment takes negative values, in the event $S_T < K$, and is bounded below by $-K$ the contract fits in the framework of section 2.5.

Corollary 12 *In the financial market model (2.1)–(2.2) the rational price of a forward contract with maturity date T, delivery price K and written on the asset price (2.2) is given by*

$$\begin{aligned} V_t &= \widetilde{E}_t\left[R_{t,T}(S_T - K)\right] = S_t - \widetilde{E}_t\left[\int_t^T R_{t,v}\delta_v S_v dv\right] - K\widetilde{E}_t\left[R_{t,T}\right] \\ &= S_t E^*_t\left[\exp\left(-\int_t^T \delta_v dv\right)\right] - K\widetilde{E}_t\left[R_{t,T}\right] \end{aligned}$$

for $t \in [0,T]$, where E^*_t is the expectation under the measure Q^* defined above. The forward price $f(t,S)$, representing the delivery price at which the contract value is zero, is given by

$$f(t,S_t) = S_t \frac{E^*_t\left[\exp\left(-\int_t^T \delta_v dv\right)\right]}{\widetilde{E}_t\left[R_{t,T}\right]}$$

for $t \in [0,T]$.

2.9 Markets with Deterministic Coefficients

When the interest rate is constant, the price of an option written on a nondividend-paying stock whose price follows a geometric Brownian motion process satisfies the Black and Scholes [1973] formula (see also Merton [1973a]).

Corollary 13 *(Black and Scholes [1973]) Suppose that the interest rate r is constant and that the stock price follows a geometric Brownian motion process* *without dividends ((μ, σ) constants, $\delta = 0$). Then the price of a European call option simplifies to*

$$c_t = S_t N(d) - e^{-r\tau} K N\left(d - \sigma\sqrt{\tau}\right) \tag{2.24}$$

where $\tau \equiv T - t$ is the time to maturity, $N(\cdot)$ is the cumulative standard normal distribution function and $d \equiv \left(\sigma\sqrt{\tau}\right)^{-1}\left[\log(S_t/K) + \left(r + \frac{1}{2}\sigma^2\right)\tau\right]$. The price of a European put option with same maturity and exercise price can be obtained from the put-call parity relationship: $p_t = c_t - S_t + e^{-r\tau}K$, or from the PCS property.

An explicit formula for the option can also be computed when the coefficients of the model change deterministically over time.

Corollary 14 *(Black-Scholes with deterministic coefficients) Consider the financial market model with deterministic interest rate, drift and volatility coefficients (r_t, μ_t, σ_t) and without dividends ($\delta = 0$). Then, the price of a European call option is given by*

$$c_t = S_t N(d) - R_{t,T} K N\left(d - \sqrt{\sigma_{t,T}^2}\right)$$

where $N(\cdot)$ is the cumulative standard normal distribution function and

$$\sigma_{t,T}^2 = \int_t^T \sigma_v^2 dv$$

$$d \equiv \left(\sqrt{\sigma_{t,T}^2}\right)^{-1}\left[\log(S_t/K) + \int_t^T \left(r_v + \frac{1}{2}\sigma_v^2\right) dv\right].$$

The next result provides the price of a European option on a dividend-paying stock in a financial market with deterministic coefficients.

Corollary 15 *(Black-Scholes with dividend adjustment) Consider the financial market model with deterministic interest rate, drift and volatility coefficients, and dividend rate $(r_t, \mu_t, \sigma_t, \delta_t)$. The price of a European call option is given by*

$$c_t = S_t D_{t,T} N(d) - R_{t,T} K\, N\left(d - \sqrt{\sigma_{t,T}^2}\right)$$

where $D_{t,T} \equiv \exp\left(-\int_t^T \delta_v dv\right)$, $N(\cdot)$ is the cumulative standard normal distribution function and

$$d \equiv \left(\sqrt{\sigma_{t,T}^2}\right)^{-1}\left[\log(S_t/K) + \int_t^T \left(r_v - \delta_v + \frac{1}{2}\sigma_v^2\right) dv\right]. \tag{2.25}$$

Figure 2.1 shows the behavior of the call price function when the coefficients r, δ, σ are constants. Convexity with respect to the underlying asset price is apparent. This property can be easily proved from the convexity of the payoff function. The call price converges to zero as the underlying price converges to zero, reflecting the vanishing probability of exercise. At the other extreme, the call price converges to the discounted value of the difference between the asset price and the strike, $S_t e^{-\delta(T-t)} - K e^{-r(T-t)} = \widetilde{E}_t \left[R_{t,T} \left(S_T - K \right) \right]$. For intuition note that the probability of exercise converges to one as S_t becomes large and therefore that the call value converges to the present value of $S_T - K$. As the call price is non-negative and the call payoff dominates $S_T - K$, we also have the lower bound $c_t \geq \max\left(0, S_t e^{-\delta(T-t)} - K e^{-r(T-t)}\right)$.

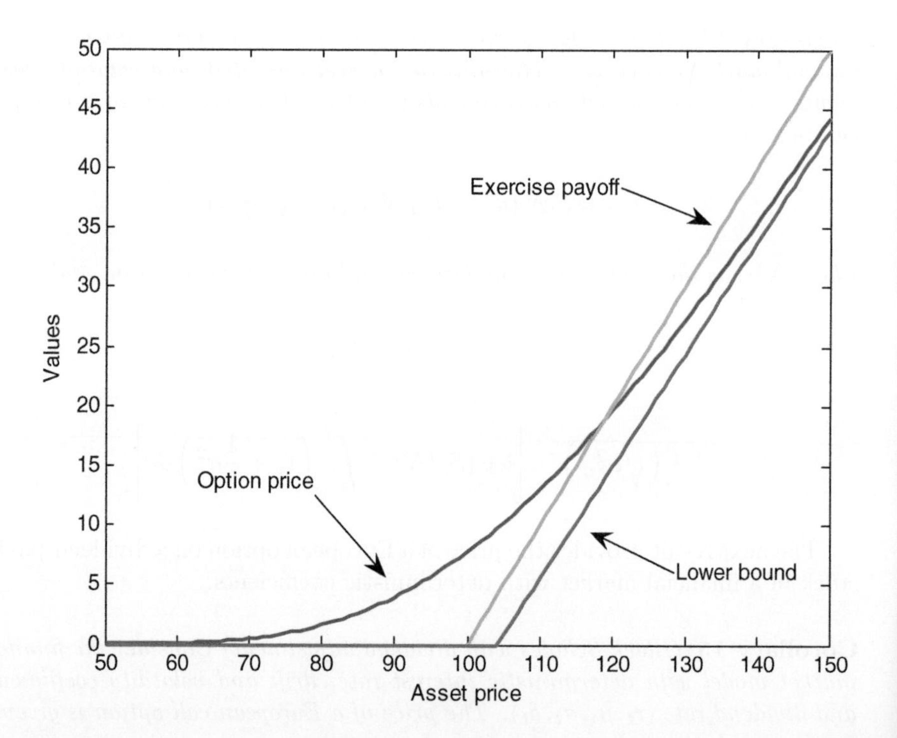

Figure 2.1: Black-Scholes model. This figure graphs the European call price c_t, the payoff function $(S_t - K)^+$ and the lower bound $\max\left(0, S_t e^{-\delta(T-t)} - K e^{-r(T-t)}\right)$ (y-axis) with respect to the underlying asset price (x-axis). Parameter values are $r = 2\%$, $\delta = 6\%$, $\sigma = 0.25$ and $T - t = 1$.

For risk management purposes it is important to identify the sensitivity of the option price with respect to the underlying asset price. The *delta hedge* of the option, given by the derivative $\partial c(S,t)/\partial S$, quantifies this sensitivity. In the context of the model with dividend adjustment one obtains,

Corollary 16 *Consider the model with dividend adjustment of Corollary 15 and let $c(S,t)$ denote the value of the European call option. The delta hedge ratio $h(S,t) = \partial c(S,t)/\partial S$ is*

$$h(S_t,t) = D_{t,T}N(d)$$

with d as in (2.25).

Figure 2.2 graphs the delta hedge ratio as a function of the underlying asset price. As expected the delta hedge converges to zero as S approaches zero and to $e^{-\delta(T-t)}$ as S becomes large. Convexity of the price function ensures that $h(S_t,t)$ is bounded above by $e^{-\delta(T-t)}$.

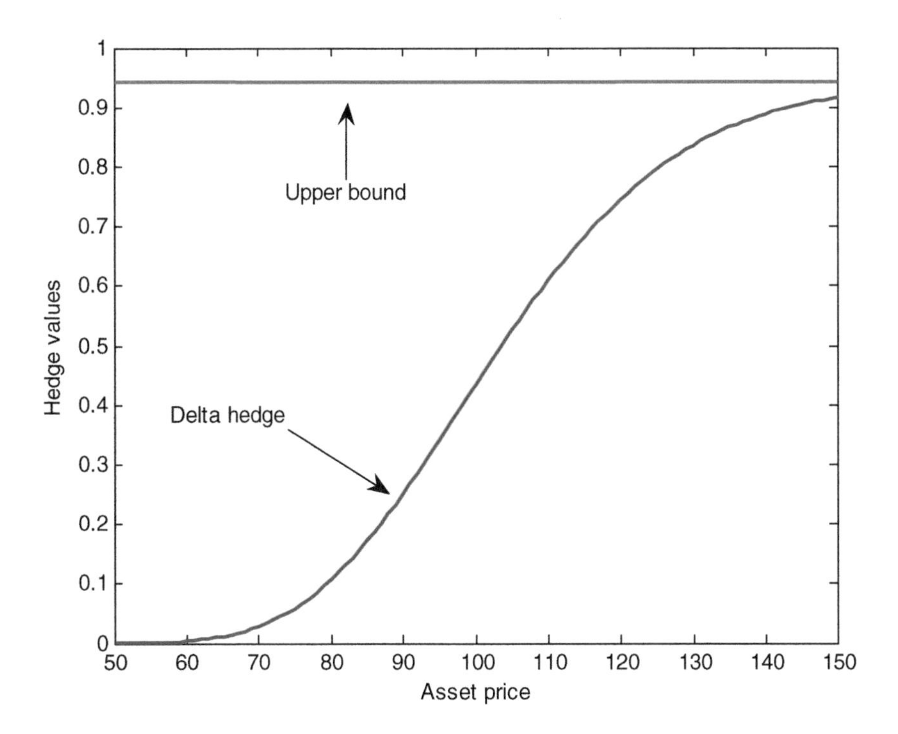

Figure 2.2: Hedging in the Black-Scholes model. This figure displays the behavior of the call hedge ratio h_t (y-axis) with respect to the underlying asset price (x-axis). Parameter values are $r = 2\%$, $\delta = 6\%$, $\sigma = 0.2$ and $T - t = 1$. The upper bound is $e^{-\delta(T-t)}$.

To complete this section we record the value of a forward contract in the simple setting of Corollary 15. This formula and the corresponding expression for the forward price follow immediately from Corollary 12 and are straightforward adaptations of standard results (see, for instance, Hull [2002]).

Corollary 17 *Consider the model with dividend adjustment of Corollary 15 and let $v^f(S,t)$ denote the value of a forward contract with delivery date T, delivery price K and written on the asset price S. Then*

$$v_t^f = S_t \exp\left(-\int_t^T \delta_v dv\right) - KR_{t,T}$$

for $t \in [0,T]$. The forward price is $f(t,S_t) = S_t \exp\left(\int_t^T (r_v - \delta_v)dv\right)$ for $t \in [0,T]$.

Lastly, we point out that all the results in this section are also valid when the coefficient μ of the asset price process is an adapted stochastic process. In the risk neutral environment only the properties of the interest rate and of the volatility of the asset return matter for pricing payoffs that are not explicitly tied to the drift of the asset return. As long as r and σ are constants, or deterministic functions of time, the formulas displayed above will apply.

2.10 Markets with Multiple Assets

We now outline an extension to a financial market with d risky securities. The underlying sources of uncertainty are represented by a d-dimensional vector of Brownian motion processes z. There is a riskless asset paying interest at the rate r where r is a positive, bounded and progressively measurable process. The vector of risky asset prices S satisfies

$$dS_t = I^S[(\mu_t - \delta_t)dt + \sigma_t dz_t], \quad S_0 \text{ given} \tag{2.26}$$

where I^S is a $d \times d$ diagonal matrix with vector of prices on its diagonal, δ is a $d \times 1$ vector of dividend rates, μ is a $d \times 1$ vector of drifts and σ a $d \times d$ matrix of volatility coefficients. All the coefficients are progressively measurable and bounded processes; the dividends are non-negative. We also assume that the volatility matrix σ is invertible and that the d-dimensional MPR process $\theta \equiv \sigma^{-1}(\mu - r)$ satisfies the Novikov condition

$$E\left[\exp\left(\frac{1}{2}\int_0^T \| \theta_s \|^2 ds\right)\right] < \infty$$

where $\| \theta \|^2 = \theta'\theta$. This last assumption ensures that the exponential process

$$\eta_t = \exp\left(-\int_0^t \theta_s' dz_s - \frac{1}{2}\int_0^t \| \theta_s \|^2 ds\right) \tag{2.27}$$

is a martingale and that the measure, $Q(A) \equiv E[\eta_T 1_A], A \in \mathcal{F}_T$, is a probability measure equivalent to P; Q is the equivalent martingale measure in the multiasset case.

All the definitions, properties and results demonstrated in the context of the single asset model generalize naturally to this multiasset financial market. In particular the value of a contingent claim can be written as the present value of its future discounted cash flows.

Proposition 18 *The rational price at time t of the European contingent claim* $(f, Y) \in \mathcal{I}_+$ *is uniquely given by*

$$V_t(f,Y) = \widetilde{E}_t\left[R_{t,T}Y + \int_t^T R_{t,s}df_s\right] = E_t\left[\xi_{t,T}Y + \int_t^T \xi_{t,s}df_s\right]$$

for $t \in [0,T]$ *where* $dQ = \eta_T dP$, η_T *is defined in (2.27) and where* $\xi_t = R_t\eta_t$.

The rational price of the contingent claim is calculated by discounting cash flows either at the riskfree rate (under the risk neutral measure) or at the risk-adjusted rate embedded in the SPD ξ (under the original measure). The structures of these valuation formulas are identical to those in the single asset case. Analogs of Proposition 6 and Theorem 7 also hold in this multiasset setting.

2.11 Appendix: Proofs

Proof of Theorem 1: Detailed proofs can be found in Karatzas and Shreve [1988], [1998]. Our demonstration below provides the key elements in the derivation of this central result.

Consider a consumption-portfolio policy (C, π) and let X be the wealth process generated by (C, π). An application of Ito's lemma gives

$$\xi_t X_t + \int_0^t \xi_s dC_s = x + \int_0^t \xi_s(\pi_s\sigma_s - X_s\theta_s)dz_s \tag{2.28}$$

for all $t \in [0,T]$.

(i) Necessity: Suppose now that the policy is admissible: $(C,\pi) \in \mathcal{A}(x)$. The right-hand side of (2.28) is a continuous local martingale. Admissibility of (C,π) implies that the left-hand side of (2.28) is non-negative. The combination of these two properties implies that the right-hand side is a non-negative supermartingale (Karatzas and Shreve [1988, Chapter 1, Problem 5.19]). Taking expectations on both sides of (2.28) and setting $t = T$ then yields

$$E\left[\xi_T X_T + \int_0^T \xi_s dC_s\right] \leq x.$$

Hence if (f, Y) is attainable ($X_T \geq Y$ and $dC_t \geq df_t$ for all $t \in [0,T]$) from initial wealth x then

$$E\left[\xi_T Y + \int_0^T \xi_s df_s\right] \le E\left[\xi_T X_T + \int_0^T \xi_s dC_s\right] \le x.$$

The static budget constraint (2.11) follows by passing to the Q-measure..

(ii) Sufficiency: conversely, suppose that $(f, Y) \in \mathcal{I}_+$ satisfies (2.11) and consider the P-martingale M defined by $M_t \equiv E_t[\xi_T Y] + E_t\left[\int_0^T \xi_s df_s\right]$. By the fundamental representation theorem for right continuous left limit (RCLL) Brownian martingales (Karatzas and Shreve [1988, Theorem 3.4.15, Problem 3.4.16]) M has the representation

$$M_t = M_0 + \int_0^t \phi_s dz_s \tag{2.29}$$

where ϕ is a one-dimensional, $\mathcal{F}_{(\cdot)}$-progressively measurable process such that $\int_0^T \phi_t^2 dt < \infty$, ($P-$a.s.). Selecting the portfolio process π_t such that $\xi_s(\pi_s \sigma_s - X_s \theta_s) = \phi_s$ and substituting in the wealth process X of (2.28) yields

$$\xi_t X_t + \int_0^t \xi_s dC_s = x + \int_0^t \phi_s dz_s$$

or, as $\int_0^t \phi_s dz_s = M_t - M_0$,

$$\xi_t X_t + \int_0^t \xi_s dC_s = x - E\left[\xi_T Y + \int_0^T \xi_s df_s\right] + E_t\left[\xi_T Y + \int_0^T \xi_s df_s\right] \tag{2.30}$$

for all $t \in [0, T]$. Evaluating this expression at $t = T$ shows

$$\xi_T X_T + \int_0^T \xi_s dC_s = x - E\left[\xi_T Y + \int_0^T \xi_s df_s\right] + \xi_T Y + \int_0^T \xi_s df_s$$

given that $\xi_T Y + \int_0^T \xi_s df_s$ is $\mathcal{F}_T$-measurable. The static budget constraint (2.11), that can also be written as $x - E\left[\xi_T Y + \int_0^T \xi_s df_s\right] \ge 0$, then implies

$$\xi_T X_T + \int_0^T \xi_s dC_s \ge \xi_T Y + \int_0^T \xi_s df_s.$$

Selecting $C = f$ yields $X_T \ge Y$. Moreover, (2.30) and the static budget constraint (2.11) also give $\xi_t X_t + \int_0^t \xi_s dC_s \ge E_t\left[\xi_T Y + \int_0^T \xi_s df_s\right]$ for all $t \in [0, T]$. Combining this inequality with $C = f$, where f is non-decreasing, and with $Y \ge 0$ then yields $\xi_t X_t \ge E_t\left[\xi_T Y + \int_t^T \xi_s df_s\right] \ge 0$, for all $t \in [0, T]$. This shows that $(C, \pi) \in \mathcal{A}(x)$. We conclude that (f, Y) is attainable from x by using the policy (C, π).

Finally, note that $X_T = Y$, (P-a.s.) if (2.11) holds with equality. It follows that (C, π) replicates (f, Y) from an initial investment x (i.e., at cost x). When

$\xi_T Y + \int_0^T \xi_s df_s \in \mathcal{L}^2$ the martingale M, defined above, is square-integrable. The process ϕ in the representation (2.29) is then unique. The uniqueness of the replicating consumption-portfolio policy follows. ■

Proof of Proposition 4: The contingent claim (f, Y) is attainable from all initial investments x satisfying the budget constraint (2.11). Minimizing over this set yields the (unique) minimum investment from which (f, Y) is attainable: $x^* = \widetilde{E}\left[R_{0,T}Y + \int_0^T R_{0,s}df_s\right]$. Let (C, π) be the replicating strategy and X the associated wealth process. Note that withdrawals associated with this replicating policy can be reinvested in the account. In this case the balance of the account evolves according to $dX + dC$. To prove the proposition we show that $V_0(f, Y) = x^*$ for $t = 0$. A straightforward adaptation of the argument will then establish the result for an arbitrary time $t \in [0, T]$.

So let us first suppose that $V_0(f, Y) > x^*$. Under this condition the strategy of selling (f, Y) at a price $V_0(f, Y)$, and undertaking the policy (C, π) that generates (f, Y) from initial wealth x^*, is an arbitrage opportunity. Indeed, this strategy produces an immediate inflow equal to $V_0(f, Y) - x^* > 0$ and subsequent cash flows equal to $d\widehat{C}_t \equiv dC_t - df_t = 0$ and $X_T - Y = 0$. The wealth process associated with the strategy is

$$\begin{aligned} d\widehat{X}_t &= \left(\widehat{X}_t - X_t + V_t\right) r_t dt + (dX_t + dC_t) - (dV_t + df_t) \\ &= \left(\widehat{X}_t - X_t + V_t\right) r_t dt + dX_t - dV_t \end{aligned}$$

with initial condition $\widehat{X}_0 = 0$, as no external cash infusions are needed to start the strategy. Solving this stochastic differential equation yields $\widehat{X}_T = R_{0,T}^{-1}\left(\widehat{X}_0 - X_0 + V_0\right) + X_T - V_T = R_{0,T}^{-1}(V_0 - x^*) > 0$. To preclude this arbitrage it must be the case that $V_0(f, Y) \leq x^*$.

Assume then that $V_0(f, Y) < x^*$. In this instance, buying the claim and investing in the replicating portfolio yields cash flows $x^* - V_0(f, Y) > 0$, $d\widehat{C}_t \equiv dC_t - df_t = 0$ and $X_T - Y = 0$. This is again an arbitrage strategy, by the same arguments. We conclude that $V_0(f, Y) = x^*$ to preclude the existence of arbitrage opportunities.

As the sum of discounted wealth plus cumulative discounted consumption is a Q-martingale, a similar reasoning establishes that the minimum amount of wealth needed at date t to generate (f, Y) is $X_t = \widetilde{E}_t\left[R_{t,T}Y + \int_t^T R_{t,s}df_s\right]$. The price of the claim at date t follows. ■

Proof of Theorem 5: The discounted wealth process generated by a policy (C, π) satisfies

$$R_{0,t}X_t + \int_0^t R_{0,s}dC_s = x + \int_0^t R_{0,s}\pi_s\sigma_s d\widetilde{z}_s \tag{2.31}$$

for all $t \in [0, T]$. If we use the state price density as a deflator we obtain the alternative expression (2.28).

(i) Necessity: Suppose now that the policy is K-admissible: $(C, \pi) \in \mathcal{A}(x; K)$. The right-hand side of (2.31) is a continuous Q-local martingale. K-admissibility of (C, π) and boundedness of the interest rate imply that the left-hand side is bounded below. We conclude that the right-hand side is a supermartingale. Taking the Q-expectation on each side and setting $t = T$ then yields

$$\widetilde{E}\left[R_{0,T}X_T + \int_0^T R_{0,s}dC_s\right] \leq x.$$

Hence if (f, Y) is K-attainable ($X_T \geq Y$ and $dC_t \geq df_t$ for all $t \in [0, T]$) from initial wealth x then

$$\widetilde{E}\left[R_{0,T}Y_T + \int_0^T R_{0,s}df_s\right] \leq \widetilde{E}\left[R_{0,T}X_T + \int_0^T R_{0,s}dC_s\right] \leq x,$$

which establishes the budget constraint (2.15).

(ii) Sufficiency: conversely, suppose that $(f, Y) \in \mathcal{I}_{ubb}$ satisfies (2.15) and consider the P-martingale M defined by $M_t \equiv E_t[\xi_T Y] + E_t\left[\int_0^T \xi_s df_s\right]$. Proceeding along the same lines as in the proof of Theorem 1 leads to

$$\xi_t X_t + \int_0^t \xi_s dC_s = x - E\left[\xi_T Y + \int_0^T \xi_s df_s\right] + E_t\left[\xi_T Y + \int_0^T \xi_s df_s\right] \quad (2.32)$$

for all $t \in [0, T]$. Evaluating this expression at $t = T$ establishes

$$\xi_T X_T + \int_0^T \xi_s dC_s = x - E\left[\xi_T Y + \int_0^T \xi_s df_s\right] + \xi_T Y + \int_0^T \xi_s df_s$$

given that $\xi_T Y + \int_0^T \xi_s df_s$ is $\mathcal{F}_T$-measurable. The budget constraint (2.15) then implies

$$\xi_T X_T + \int_0^T \xi_s dC_s \geq \xi_T Y + \int_0^T \xi_s df_s.$$

Selecting $C = f$ yields $X_T \geq Y$. Moreover, (2.32) and the static budget constraint (2.15) also give $\xi_t X_t + \int_0^t \xi_s dC_s \geq E_t\left[\xi_T Y + \int_0^T \xi_s df_s\right]$ for all $t \in [0, T]$. As $C = f$ we can also write

$$X_t \geq E_t\left[\xi_{t,T} Y + \int_t^T \xi_{t,s} df_s\right] = \widetilde{E}_t\left[R_{t,T}Y_T + \int_t^T R_{t,s}df_s\right]$$

for all $t \in [0, T]$. Given that r is bounded and (f, Y) is bounded from below, the random variable $R_{0,T}Y_T + \int_0^T R_{0,s}df_s$ is bounded from below as well. It then follows that $X_t \geq -K$ for all $t \in [0, T]$. We conclude that $(C, \pi) \in \mathcal{A}(x; K)$ and that the policy (C, π) K-superreplicates the claim (f, Y). Note also that $X_T = Y$, (P-a.s.) if (2.15) holds with equality. In this case (C, π) is seen to K-replicate (f, Y) at cost x.

The last statement is proved as in Theorem 1. ■

Proof of Proposition 6: The proof parallels the proof of Proposition 4. Set $x^* \equiv \widetilde{E}\left[R_{0,T}Y + \int_0^T R_{0,s}df_s\right]$ and note that Theorem 5 ensures the existence of a K-replicating consumption-portfolio policy starting from initial wealth x^*. The same no-arbitrage argument can then be applied to show that $V_0(f, Y) = x^*$. ■

Proof of Theorem 7: The no-arbitrage argument sketched in the text shows that

$$e_t \equiv -\sigma_t^{-1}\rho_t(\mu_t - r_t) + (\alpha_t - r_t) = 0.$$

It follows immediately that $\alpha_t - r_t = \rho_t\theta_t = \beta_t(\mu_t - r_t)$ where $\beta_t \equiv \rho_t/\sigma_t$.

If the underlying asset satisfies the CAPM then $\mu_t - r_t = (\sigma_t/\sigma_t^m)(\mu_t^m - r_t)$, where μ^m, σ^m are the coefficients of the process followed by the price of the market portfolio of risky assets. Simple algebra establishes that $\alpha_t - r_t = \beta_t^m(\mu_t^m - r_t)$ where $\beta^m = \rho_t/\sigma_t^m$. ■

Proof of Theorem 8: The proof is outlined in the text leading up to the proposition. ■

Proof of Corollary 9: Substitution of the call and put payoff functions in the formula of Proposition 4 yields the results stated. ■

Proof of Theorem 10: Let $p_t \equiv p(S, K, r, \delta; \mathcal{F}_t)$ be the price of the put option with characteristics (K, T), in the original financial market. This price has the (present value) representation

$$p_t = \widetilde{E}_t\left[\exp\left(-\int_t^T r_v dv\right)\left(K - S\exp\left(\int_t^T \alpha_v dv + \int_t^T \sigma_v d\widetilde{z}_v\right)\right)^+\right]$$

where $\alpha \equiv r - \delta - \frac{1}{2}\sigma^2$ and the expectation is taken relative to the equivalent martingale measure Q. Simple manipulations show that the right hand side of this equation equals

$$\widetilde{E}_t\left[\exp\left(-\int_t^T \delta_v dv\right)\eta_{t,T}^*\left(K\exp\left(-\int_t^T \alpha_v dv - \int_t^T \sigma_v d\widetilde{z}_v\right) - S\right)^+\right] \tag{2.33}$$

where $\eta_t^* \equiv \exp\left(-\frac{1}{2}\int_0^t \sigma_v^2 dv + \int_0^t \sigma_v d\widetilde{z}_v\right)$.

Consider the new measure $dQ^* = \eta_T^* dQ$, which is equivalent to Q. Girsanov's Theorem (Karatzas and Shreve [1988, Theorem 5.1]) implies that the process

$z_t^* = -\widetilde{z}_t + \int_0^t \sigma_v dv$ is a Q^*-Brownian motion. Substituting z^* in (2.33) and passing to the measure Q^* yields

$$E_t^* \left[\exp\left(-\int_t^T \delta_v dv \right) \left(K \exp\left(\int_t^T (\delta_v - r_v - \frac{1}{2}\sigma_v^2) dv + \int_t^T \sigma_v dz_v^* \right) - S \right)^+ \right] \tag{2.34}$$

where E_t^* denotes the conditional expectation relative to Q^*. But the right hand side is the value of a call option with strike S, maturity date T in an economy with interest rate δ and pricing measure Q^*. The underlying asset price is

$$S_T^* = K \exp\left(\int_t^T (\delta_v - r_v - \frac{1}{2}\sigma_v^2) dv + \int_t^T \sigma_v dz_v^* \right)$$

at date T. An application of Ito's lemma shows that S_v^* satisfies (2.20), for all $v \in [t, T]$, with initial condition $S_t^* = K$. Inspection of this equation shows that the asset pays dividends at the rate r in the economy with interest rate δ and risk neutral measure Q^*. ■

Proof of Corollary 11: When the coefficients (r, δ, σ) are adapted to the filtration $\mathcal{F}_{(\cdot)}^*$ the conditional expectation in (2.34) with respect to the sigma-algebra $\mathcal{F}_t$ is the same as the conditional expectation with respect to $\mathcal{F}_t^*$, for all $t \in [0, T]$. The result follows. ■

Proof of Corollary 12: An application of Proposition 6 shows that $V_t = \widetilde{E}_t [R_{t,T}(S_T - K)]$. Using the stock price formula (2.8) establishes that $V_t = S_t - \widetilde{E}_t \left[\int_t^T R_{t,v} \delta_v S_v dv \right] - K \widetilde{E}_t [R_{t,T}]$. Solving the linear SDE (2.2) and passing to the Q-measure gives

$$R_{t,T} S_T = S_t \exp\left(-\int_t^T \delta_v dv - \frac{1}{2} \int_t^T \sigma_v^2 dv + \int_t^T \sigma_v d\widetilde{z}_v \right)$$

and, with the definition of the measure Q^*,

$$R_{t,T} S_T = S_t \exp\left(-\int_t^T \delta_v dv \right) \frac{dQ^*}{dQ}.$$

Substituting in $V_t = \widetilde{E}_t [R_{t,T}(S_T - K)]$ and simplifying gives

$$\begin{aligned} V_t &= S_t \widetilde{E}_t \left[\exp\left(-\int_t^T \delta_v dv \right) \frac{dQ^*}{dQ} \right] - K \widetilde{E}_t [R_{t,T}] \\ &= S_t E_t^* \left[\exp\left(-\int_t^T \delta_v dv \right) \right] - K \widetilde{E}_t [R_{t,T}] . \end{aligned}$$

The forward price $f(t,S)$ is obtained by solving the equation

$$0 = S_t E_t^* \left[\exp\left(-\int_t^T \delta_v dv \right) \right] - f(t,S_t) \widetilde{E}_t \left[R_{t,T} \right]$$

for $f(t,S_t)$. ■

Proof of Corollary 13: Let $\tau = T - t$. The call option price in Proposition 4 can be written as

$$C_t = e^{-r\tau} \widetilde{E}_t \left[(S_T - K)^+ \right] = e^{-r\tau} \left(\widetilde{E}_t \left[1_{\{S_T \geq K\}} S_T \right] - K \widetilde{E}_t \left[1_{\{S_T \geq K\}} \right] \right).$$

Under the risk neutral measure Q the stock price is

$$S_T = S_t \exp\left(\left(r - \frac{1}{2}\sigma^2 \right) \tau + \sigma \left(\tilde{z}_T - \tilde{z}_t \right) \right)$$

where $\tilde{z}_T - \tilde{z}_t$ is distributed as $\tilde{z}\sqrt{T-t}$ and $\tilde{z}$ is a standard normal random variable. Thus,

$$\begin{aligned} \widetilde{E}_t \left[1_{\{S_T \geq K\}} \right] &= Q\left(\tilde{z} \geq \left(\sigma\sqrt{\tau} \right)^{-1} \left[\log\left(K/S_t \right) - \left(r - \frac{1}{2}\sigma^2 \right) \tau \right] \right) \\ &= N\left(d - \sigma\sqrt{\tau} \right) \end{aligned} \quad (2.35)$$

where $N(\cdot)$ is the cumulative standard normal distribution. The first expectation is

$$\begin{aligned} \widetilde{E}_t \left[1_{\{S_T \geq K\}} S_T \right] &= e^{r\tau} S_t \widetilde{E}_t \left[1_{\{S_T \geq K\}} e^{-\frac{1}{2}\sigma^2\tau + \sigma\tilde{z}\sqrt{\tau})} \right] \\ &= e^{r\tau} S_t E_t^* \left[1_{\{S_T \geq K\}} \right] \end{aligned} \quad (2.36)$$

where $E_t^*[\cdot]$ is the expectation under Q^* in (2.23), specialized to the case under consideration. Under this measure the asset price is distributed as

$$S_t \exp\left(\left(r - \frac{1}{2}\sigma^2 + \sigma^2 \right) \tau - \sigma z^* \sqrt{\tau} \right)$$

where z^* is a standard normal variate. An adaptation of the computation in (2.35), to account for the new measure Q^*, shows that (2.36) equals $e^{r\tau} S_t N(d)$. Combining (2.35) and (2.36) gives (2.24).

To prove the put-call parity relationship, note that $(K - S_T)^+ = (S_T - K)^+ - S_T + K$. No arbitrage implies that the value of the put must equal the value of the portfolio of the securities on the right-hand side of the equality. The parity relationship follows. ■

Proof of Corollary 14: Under the assumptions stated we have

$$S_T = S_t \exp\left(\int_t^T \left(r_v - \frac{1}{2}\sigma_v^2 \right) dv + \int_t^T \sigma_v d\tilde{z}_v \right).$$

Furthermore, the stochastic integral $\int_t^T \sigma_v d\tilde{z}_v$ has normal distribution with mean zero and variance $\int_t^T \sigma_v^2 dv$. Performing the same computations as in the proof of Corollary 13 yields the result. ■

Proof of Corollary 15: The stock price is now given by

$$S_T = S_t \exp\left(\int_t^T \left(r_v - \delta_v - \frac{1}{2}\sigma_v^2\right) dv + \int_t^T \sigma_v d\tilde{z}_v\right).$$

Proceeding as in the proof of Corollary 13 leads to the result. ■

Proof of Corollary 16: Taking the derivative of the option price in Corollary 15 with respect to S and using the relation $S_t D_{t,T} n(d) - R_{t,T} K\, n\left(d - \sqrt{\sigma_{t,T}^2}\right) = 0$ proves the result. ■

Proof of Corollary 17: Given that r, δ are deterministic the formulas stated follow immediately from those in Corollary 12. ■

Proof of Proposition 18: The case of complete markets with multiple risky assets parallels the single dimensional case (single asset, single Brownian motion case). Both Theorem 1 and Proposition 4 hold in this setting and are proved along the same lines. ■

Chapter 3

American Contingent Claims

We now turn our attention to the valuation of American contingent claims. By definition these are contracts that can be exercised during certain prespecified periods of time, at the option of the holder. Their valuation, naturally, requires the determination of the optimal exercise policy. The absence of arbitrage opportunities implies that the value of the contract is the value under the optimal exercise policy.

This chapter presents several approaches that can be used to price American-style contingent claims. The analysis is cast in the general financial market model with Ito price processes and adapted interest rate process. The results described will be used in subsequent chapters to study more specific contractual forms, such as vanilla options, barrier and capped options, multiasset options and occupation time derivatives, in markets with constant coefficients. They also serve as starting points for some of the numerical methods presented in chapter 8.

In a preliminary step we extend the valuation formula in Proposition 4 to securities with payoffs at random times (section 3.1). We then value American contingent claims (sections 3.2 and 3.3) and specialize the results to options (section 3.5). The case of financial markets with multiple assets is considered last (section 3.6).

3.1 Contingent Claims with Random Maturity

The economic setting is the one presented in section 2.2. In order to describe claims with random timing of payments we need to introduce a few additional concepts. A random time τ is a *stopping time* of the (Brownian) filtration $\mathcal{F}_{(\cdot)}$ if the event $\{\tau \leq t\}$ belongs to the σ-field $\mathcal{F}_t$ for every $t \in [0, T]$. In other words τ is a stopping time if, at any time t, an individual observing the underlying source of uncertainty (thus endowed with the σ-algebra $\mathcal{F}_t$) can tell whether τ

has occurred or not. Let $\mathcal{S}_{0,T}$ denote the set of stopping times taking values in $[0, T]$.

Consider a contingent claim (f, Y) and an exogenously specified stopping time $\tau \in \mathcal{S}_{0,T}$. Here f is a cumulative payment process up to τ: this finite variation process is non-decreasing, progressively measurable, right-continuous and null at zero. Also, Y is used to represent a terminal, non-negative and $\mathcal{F}_\tau$-measurable cash flow Y_τ paid at the random time τ. To prepare the ground for the case of American claims we can actually think of Y as a right-continuous, adapted stochastic process with value Y_τ at the random maturity date τ. By analogy with condition (2.10) in section 2.2 we consider claims (f, Y) satisfying the integrability condition[1]

$$E\left[\sup_{t\in[0,T]}\left(\xi_t Y_t + \int_0^t \xi_s df_s\right)\right] < \infty. \tag{3.1}$$

Let $\mathcal{I}_+^*$ be the class of claims that satisfy this condition and $df \geq 0, Y \geq 0$.

Theorem 19 *Let τ be a stopping time in $\mathcal{S}_{0,T}$ and suppose that $(f, Y) \in \mathcal{I}_+^*$. The rational price of this contingent claim, $V_t(f, Y, \tau)$, is uniquely given by*

$$V_t(f, Y, \tau) = \widetilde{E}_t\left[\int_t^\tau R_{t,s} df_s + R_{t,\tau} Y_\tau\right] = E_t\left[\int_t^\tau \xi_{t,s} df_s + \xi_{t,\tau} Y_\tau\right]$$

for all $t \in [0, \tau]$.

This result shows that contingent claims that expire at a stopping time of the filtration are valued in the same manner (using the same pricing operator) as contracts with fixed maturity date. The key ingredient is the fact that the random maturity is a function of the Brownian motion that generates the information structure. The completeness of the financial market (relative to this Brownian motion) ensures that maturity risk can be properly hedged. The absence of arbitrage opportunities validates the equality between the price of the claim and the value of the replicating portfolio.

Note that the class of claims covered by Theorem 19 includes contracts paying cash flows at multiple random dates. These discrete payments are, in fact, embedded in the cumulative cash flow process f, which may include discontinuities at random times. Letting Δf be the jumps in f we can write $f = f^c + f^d$ where f^c is the purely continuous part of the cumulative payment process and $f^d = \sum \Delta f$ its purely discontinuous part. The valuation formula can then be

[1] Note that the supremum inside the expectation is taken path-by-path. Condition (3.1), therefore, implies that $\sup_{\tau\in\mathcal{S}_{t,T}} E[D_t] < \infty$, where $D_t \equiv \xi_t Y_t + \int_0^t \xi_s df_s$ is the deflated payoff process.

restated as

$$\begin{aligned} V_t(f,Y,\tau) &= \widetilde{E}_t\left[\int_t^\tau R_{t,s}df_s^c + \sum_{t\le s\le \tau} R_{t,s}\Delta f_s + R_{t,\tau}Y_\tau\right] \\ &= \widetilde{E}_t\left[\int_t^\tau R_{t,s}df_s^c + \sum_j R_{t,\tau_j}\Delta f_{\tau_j} + R_{t,\tau}Y_\tau\right] \end{aligned}$$

where τ_j is the random time of the jth-jump in the interval $[t,\tau]$. This result is already implicit in the formulas provided in the previous chapter, which treat the special case $\tau = T$.

3.2 American Contingent Claims

If, instead of being exogenously specified, the stopping time τ can be chosen by the holder of the claim, then (f,Y) is an *American-style contingent claim.* A stopping time is also called an *exercise time* or *exercise policy* for the claim. In fact, as the choice of an exercise time can only be based on the information available (and given that information is assumed to be homogeneous among participants in the financial market), the class of exercise policies available to the claimholder naturally coincides with the class of stopping times of the filtration. The decision to exercise can then be thought of as the selection of the best stopping time τ with values in $[0,T]$, i.e., the one that maximizes the value of the claim for the owner. This is the optimal exercise policy.

In order to state the first pricing result we introduce the following notation. For $t \in [0,T]$ define the deflated payoff process

$$D_t \equiv \xi_{0,t}Y_t + \int_0^t \xi_{0,s}df_s \tag{3.2}$$

where $\xi_{0,t} = R_{0,t}\eta_t$ is the state price density introduced in chapter 2. To simplify matters, assume that D is a continuous process, for the remainder of this chapter. Consider the associated process Z such that

$$Z_t \equiv \sup_{\tau\in\mathcal{S}_{t,T}} E_t\left[D_\tau\right], \tag{3.3}$$

which is known as the *Snell envelope* of D. The process Z plays an important role in the characterization of the optimal exercise time and the valuation of the American claim. It has a number of useful properties (see El Karoui [1981] and Karatzas and Shreve [1998, Appendix D] for detailed analyses). Clearly Z is non-negative as D is non-negative. In fact, it can be shown that Z is a supermartingale with a right-continuous left-limit modification.[2] Below, we

[2] To see the supermartingale property let τ_t be the optimal stopping time for Z_t and note that $E_t[Z_s] = E_t[E_s[D_{\tau_s}]] = E_t[D_{\tau_s}] \le E_t[D_{\tau_t}] = Z_t$, for all $s \ge t$.

will assimilate it with its modification. It is also a majorant of D: as $\tau = t$ is a stopping time in $\mathcal{S}_{t,T}$ we obviously have $Z_t \geq D_t$, for all $t \in [0, T]$. In fact Z is the smallest supermartingale majorant of D (i.e., it is bounded above by any supermartingale that majorizes D).[3] Finally, for each $t \in [0, T]$, the optimization problem on the right hand side of (3.3) is solved by the stopping time

$$\tau_t \equiv \inf \{s \in [t, T] : Z_s = D_s\} \tag{3.4}$$

(as $Z_s = D_s$ means that immediate exercise is optimal at s). Given that D is continuous and $E\left[\sup_{t\in[0,T]} D_t\right] < \infty$ (because $(f, Y) \in \mathcal{I}_+^*$) a solution exists (see Karatzas and Shreve [1998, Appendix D]). With these definitions we are in a position to identify the value of an American contingent claim.

Theorem 20 *(Bensoussan [1984], Karatzas [1988]). Consider an American-style contingent claim $(f, Y) \in \mathcal{I}_+^*$. For all $t \in [0, \tau_0]$ the value of the claim, $V_t(f, Y)$, is uniquely given by*

$$V_t(f, Y) = \xi_{0,t}^{-1}\left(Z_t - \int_0^t \xi_{0,s} df_s\right) = \sup_{\tau\in\mathcal{S}_{t,T}} E_t\left[\int_t^\tau \xi_{t,s} df_s + \xi_{t,\tau} Y_\tau\right]. \tag{3.5}$$

The optimal exercise policy is given by

$$\tau_0 = \inf \{t \in [0, T] : Z_t = D_t\}. \tag{3.6}$$

Note that the right hand side of (3.5) can also be written relative to the Q-measure as

$$V_t(f, Y) = \sup_{\tau\in\mathcal{S}_{t,T}} \widetilde{E}_t\left[\int_t^\tau R_{t,s} df_s + R_{t,\tau} Y_\tau\right].$$

Thus, the theorem formalizes the intuitive notion that the value of an American contingent claim ought to equal the highest value that can be achieved by selecting exercise policies in the feasible set. In effect the proof demonstrates the existence of an admissible consumption-portfolio policy that replicates the right hand side of (3.5). The value of the claim is the value of this replicating policy. Formula (3.5) also demonstrates that the same pricing principles, as in the case of European claims or claims with exogenously specified random maturity, apply.

Theorem 20 suggests a separation of the valuation problem into two steps. The first step consists in solving the auxiliary problem embedded in the definition of the Snell envelope (3.3). This problem is well defined for all times $t \in [0, T]$. Its resolution produces the supermartingale Z along with the set of stopping times (3.4) indexed by t. In the second step the valuation problem

[3]Let S be an arbitrary supermartingale majorant of D. Then, for any $\tau \in \mathcal{S}_{t,T}$ we have $E_t D_\tau \leq E_t S_\tau \leq S_t$, where the first inequality follows from the majorizing property of S and the second one from the supermartingale property of S. Optimizing over the set of stopping times $\mathcal{S}_{t,T}$ establishes that $Z_t \leq S_t$.

is expressed in terms of the Snell envelope. The optimal exercise policy is the specific stopping time τ_0, i.e., the first time in the interval $[0, T]$ at which the Snell envelope equals the discounted payoff D. The value of the American-style contingent claim prior to this exercise date is the value under this exercise policy. In a rational financial market the American contingent claim issued at date 0 matures endogenously at this exercise time.

Remark 3 *The actualized cash flow-corrected Snell envelope*

$$\xi_{0,t}^{-1}\left(Z_t - \int_0^t \xi_{0,s} df_s\right) \tag{3.7}$$

prices newly issued claims at any point in time $t \in [0, T]$. *For example, an American contingent claim* (f, Y) *marketed (or remarketed) at time* $t \geq \tau_0$ *and whose cash flows are those in the original contract that pertain to the sub-period* $[t, T]$ *is worth* $V_t(f, Y) = \xi_{0,t}^{-1}\left(Z_t - \int_0^t \xi_{0,s} df_s\right)$. *The optimal exercise time is the first time post-introduction at which the Snell envelope equals the deflated payoff of the claim* $\tau_t \equiv \inf\{s \in [t, T] : Z_s = D_s\}$. *If* $\tau_t > t$ *the newly issued claim will survive until its endogenous expiration date. If* $\tau_t = t$ *immediate exercise is optimal: the claim has no material existence. A claim that is continuously marketed (or quoted) during the period* $[0, T]$ *carries a price given by (3.7). This provides further motivation for the study of the Snell envelope.*

Remark 4 *For claims* $(0, Y) \in \mathcal{I}_+^*$ *consisting solely of a terminal payoff, Theorem 20 states that* $V_t(f, Y) = \sup_{\tau \in \mathcal{S}_{t,T}} \widetilde{E}_t[R_{t,\tau} Y_\tau] = \widetilde{E}_t[R_{t,\tau_0} Y_{\tau_0}]$ *for* $t \leq \tau_0$. *One concludes immediately that the discounted price,* $R_{0,t} V_t$, *of an American contingent claim without intermediate payments is a* Q*-martingale prior to the optimal exercise time (where discounting is at the riskfree rate). For claims* $(f, Y) \in \mathcal{I}_+^*$ *with intermediate cash flows it is the discounted price augmented by the sum of discounted intermediate cash flows that has the* Q*-martingale property.*

Although intuitive, the representation of the price provided in (3.5)-(3.6) is incomplete, because the optimal stopping time defined via the Snell envelope is not expressed in an explicit form. A further characterization of this optimal policy is therefore needed to proceed with valuation. A decomposition, that builds upon the Snell envelope and emphasizes the gains from early exercise (prior to the maturity date T), proves useful in that regard. It often produces additional insights about the price structure of an American claim and leads to a characterization of the optimal exercise time that can be exploited in specialized contexts. This decomposition is examined next.

3.3 Exercise Premium Representations

The formula identifying the gains realized by optimally exercising before the maturity date of the contract is known as the *early exercise premium representation* (EEP) of the American contingent claim price. The EEP formula relates to a general result valid for a large class of supermartingales, known as the *Riesz decomposition* (see Karatzas and Shreve [1988, Chapter 1, Exercise 3.18]): it corresponds to the Riesz decomposition of the Snell envelope.

The Riesz decomposition states that any right-continuous, uniformly integrable supermartingale can be written as the sum of a right-continuous, uniformly integrable martingale and a right-continuous, non-negative supermartingale whose expected value converges to zero as time goes to infinity (i.e., a *potential*). This decomposition of the Snell envelope was initially established by El Karoui and Karatzas [1991] for stopping time problems defined in the context of models with Brownian filtration. An adaptation of their result, to the valuation of American puts in the Black and Scholes framework (constant interest rate and geometric Brownian motion for the underlying asset price), can be found in Myneni [1992]. As shown by Rutkowski [1994], the decomposition of the Snell envelope generalizes to settings where payoffs are continuous semimartingales adapted to a general filtration satisfying the "usual conditions". The results reported below are special cases of Rutkowski as the underlying uncertainty-information structure, in the economy under consideration, is given by the Brownian filtration introduced in section 2.2.

We consider a class of contingent claims $(f, Y) \in \mathcal{I}_+^*$ such that the payoff Y, under the Q-measure, satisfies

$$Y_t = Y_0 + A_t^Y + M_t^Y, \qquad t \in [0, T] \tag{3.8}$$

where M^Y is a Q-martingale and A^Y is a non-decreasing process null at 0; both M^Y and A^Y are progressively measurable processes of the Brownian filtration. For the example of a call option the exercise payoff is $Y = (S - K)^+$ and $f = 0$. This payoff can be written in the form (3.8) by an application of the Tanaka-Meyer formula (Karatzas and Shreve [1988, Chapter 3, Proposition 6.8]).

Theorem 21 *Consider an American-style contingent claim $(f, Y) \in \mathcal{I}_+^*$ and such that the payoff Y has the decomposition (3.8). The value of the claim at $t \in [0, \tau_0]$ has the early exercise premium representation*

$$\begin{aligned} V_t(f, Y) &= \widetilde{E}_t \left[R_{t,T} Y_T + \int_t^T R_{t,s} df_s \right] \\ &\quad + \widetilde{E}_t \left[\int_{\tau_0}^T R_{t,s} 1_{\{\tau_s = s\}} \left(r_s Y_s ds - dA_s^Y - df_s \right) \right] \end{aligned} \tag{3.9}$$

where $\tau_t = \inf\{v \in [t, T] : Z_v = D_v\}$. The first component in (3.9) is the value of a European-style claim with the same characteristics (f, Y); the second component is the Early Exercise Premium (EEP). If we reinterpret the claim as being continuously marketed throughout the interval $[0, T]$ then (3.9) holds for all $t \in [0, T]$, with the substitution of τ_t in place of τ_0 in the lower bound of the integral in the EEP.

The EEP representation formula (3.9) provides an intuitive decomposition of the price of the American-style contingent claim. It indicates that the price of the contract is the value of a European contingent claim with matching characteristics augmented by the present value of the gains from early exercise (the early exercise premium). The (local) gains from early exercise are given by the cash flow $r_s Y_s ds - dA_s^Y - df_s$. The first component, $r_s Y_s ds$, represents the interest collected over the next increment of time, by exercising immediately and placing the proceeds in the riskless account (this is also the opportunity cost saved by exercising immediately instead of waiting, i.e., the time value of money embedded in the discount rate r). The second component, $-dA_s^Y$, is the loss incurred, upon exercise, due to the natural appreciation of the payoff Y. Indeed, by exercising, the claimholder forgoes further appreciation in the payoff. The last component, $-df_s$, is the loss incurred, upon exercise, by forgoing the cash flow paid by the claim. As we shall see in the next section the early exercise premium and the gains from early exercise have a more specialized interpretation in the case of an American option.

The EEP representation provides useful insights into the determinants of the exercise decision. The following result is a direct consequence of the formula.

Corollary 22 *Contingent claims such that $r_v Y_v dv - dA_v^Y - df_v \leq 0$ for all $v \in [0, T]$ will never be exercised prior to the maturity date.*

Under the condition of the Corollary the net benefit of exercising is nonpositive at all times. As a result, incentives to exercise are insufficient and the early exercise premium will be null. A special case of this result is obtained when the claim is a call option on a non-dividend-paying stock. It is well known that it is suboptimal to exercise an American call prior to maturity under these conditions (Merton [1973a]). For this contract $Y = (S - K)^+$and $f = 0$, and, in the exercise region where $S > K$, we have $r_v Y_v dv - dA_v^Y = r_v (S_v - K) dv - S_v r_v dv = -r_v K dv < 0$. Corollary 22 then applies and shows that early exercise is a suboptimal policy as it can only reduce the value of the contract.

A complement to the EEP representation of an American contingent claim is a decomposition emphasizing the gains from delaying exercise. This formula is known as the *delayed exercise premium representation* (DEP). The DEP of an American put option on a non-dividend-paying asset and in a financial market with constant coefficients (constant interest rate and geometric Brownian motion for the stock price) was derived by Carr, Jarrow and Myneni [1992]. The next theorem extends their analysis to the more general class of American claims discussed in this section.

Theorem 23 *Consider an American-style contingent claim $(f, Y) \in \mathcal{I}_+^*$ and such that the payoff Y has the decomposition (3.8). The value of the claim at $t \in [0, \tau_0]$ has the delayed exercise premium representation*

$$V_t(f, Y) = Y_t + \widetilde{E}_t \left[\int_t^T R_{t,s} 1_{\{\tau_0 > s\}} \left(dA_s^Y + df_s - r_s Y_s ds \right) \right] \qquad (3.10)$$

where $\tau_t = \inf\{v \in [t, T] : Z_v = D_v\}$. The first component in (3.10) is the immediate exercise value; the second one is the Delayed Exercise Premium (DEP).

If we reinterpret the claim as being continuously marketed throughout the interval $[0,T]$ *then (3.10) holds for all* $t \in [0,T]$*, with the substitution of* τ_t *in place of* τ_0 *in the indicator function appearing in the DEP.*

The delayed exercise premium representation can be viewed as the counterpart of the EEP representation to the extent that it emphasizes the gains from waiting to exercise. The net gains consist of three parts: the appreciation of the payoff dA_s^Y, the cash inflow collected df_s and the interest foregone by postponing exercise $-r_s Y_s ds$ (an opportunity cost). The present value of these gains is captured by the second component on the right hand side of (3.10). This value component is also known as the *time value* of the contingent claim.

3.4 A Duality Formula: Upper Price Bounds

The Snell envelope (3.3) and the valuation formula (3.5) immediately suggest that the price of an American-style contingent claim is bounded below by

$$V_0(f,Y) \geq E[D_\tau]$$

where τ is any arbitrary stopping time in $\mathcal{S}_{0,T}$. This simple insight will, in fact, be exploited in chapter 8, in order to manufacture interesting lower bound approximations for option prices. The deduction of an upper bound for the claim's price is much less evident, but nevertheless feasible. It relies on duality results established by Davis and Karatzas [1994] and Rogers [2002].

Theorem 24 *(Davis and Karatzas [1994], Rogers [2002]) Consider an American-style contingent claim* $(f,Y) \in \mathcal{I}_+^*$*. The value of the claim,* $V_0(f,Y)$*, is given by*

$$V_0(f,Y) = Z_0 \equiv \sup_{\tau \in \mathcal{S}_{0,T}} E[D_\tau] = \inf_{M \in \mathcal{H}_0^1(P)} E\left[\sup_{t\in[0,T]} (D_t - M_t)\right] \tag{3.11}$$

where $\mathcal{H}_0^1(P)$ *is the space of* P*-martingales that are null at* $t = 0$ *and such that* $\sup_{t\in[0,T]} |M_t| \in \mathcal{L}^1(P)$.

Theorem 24 shows that the claim's price is bounded above by

$$V_0(f,Y) \leq E\left[\sup_{t\in[0,T]} (D_t - M_t)\right]$$

where M is any arbitrary martingale that is integrable in the sense of $\mathcal{H}_0^1(P)$. Taking the minimum upper bound over this class, in fact, produces the price of the claim. Thus, instead of maximizing over a class of stopping times one can, equivalently, minimize over a set of martingales, in order to price the claim.

The reasons underlying this duality formula are deep. It is fairly clear that the quantity $E\left[\sup_{t\in[0,T]} D_t\right]$ is an upper bound for the price, as it maximizes the payoff pathwise (using all the information up to maturity) and then averages the resulting random variable. Subtracting a martingale M, that is null at zero, from D is neutral from a pricing point of view given that the expectation of M is zero. The upper bound $E\left[\sup_{t\in[0,T]}(D_t - M_t)\right]$ then follows for the same reasons. It is much less apparent why the smallest upper bound over the class of martingales described will actually equal the claim's price.

Some insights emerge when we think of the structure of the Snell envelope. Under the conditions of the theorem the Doob-Meyer decomposition holds and establishes that Z is the difference of a martingale and of a non-decreasing process, both null at zero (i.e., $Z_t = Z_0 + M_t^Z - A_t^Z$). Using the martingale M_t^Z produces the particular upper price bound $E\left[\sup_{t\in[0,T]}\left(D_t - M_t^Z\right)\right]$. But, as immediate exercise is a feasible policy, it must be that $D_t \leq Z_t$. It follows that $D_t - M_t^Z \leq Z_t - M_t^Z = Z_0 - A_t^Z$, and given that A^Z is non-decreasing and null at zero we conclude that $D_t - M_t^Z \leq Z_0$. That is, $E\left[\sup_{t\in[0,T]}\left(D_t - M_t^Z\right)\right] \leq Z_0 = V_0(f, Y)$: our particular upper bound is, in fact, bounded above by the claim's price!

The bounds suggested by the theorem are of practical relevance because they are easy to compute (for instance by Monte Carlo simulation) and can be used to derive upper bound approximations to a claim's value. Such a construction may prove particularly useful when the computation of the exact solution of the stopping time problem is costly. Claims with path-dependent payoffs or multiple underlying assets are natural candidates for applications of this approach.

Finally, before turning to applications to specific contracts, we note that analogs of Theorems 20, 21, 23 and 24 also hold for claims $(f, Y) \in \mathcal{I}_{ubb}^*$. The proofs of these results rely on the replicability of these claims, a property that can be established along the lines of Theorem 5. Similarly, the analog of Theorem 8 holds for American claims. Our next section elaborates on that point in the context of option pricing.

3.5 American Options and Forward Contracts

In the case of an American call option the exercise payoff is $Y = S - K$ and there are no intermediate cash flows ($f = 0$). For this contract the EEP representation formula of Theorem 21 simplifies as follows,

Proposition 25 *Consider an American-style call option with exercise payoff $Y = (S-K)^+$ in the financial market (2.1)-(2.2). The option value at $t \in [0, \tau_0]$ has the early exercise premium representation*

$$C_t = \widetilde{E}_t\left[R_{t,T}(S_T - K)^+\right] + \widetilde{E}_t\left[\int_{\tau_0}^T R_{t,s} 1_{\{\tau_s = s\}}(\delta_s S_s - r_s K)\, ds\right] \tag{3.12}$$

where $\tau_t = \inf\left\{v \in [t, T] : Z_v = \xi_{0,v}(S_v - K)^+\right\}$ *and* $Z_t \equiv \sup_{\tau\in\mathcal{S}_{t,T}} E_t[D_\tau]$ *with* $D_t \equiv \xi_{0,t}(S_t - K)^+$. *If the call option is being continuously marketed throughout the interval* $[0,T]$ *then (3.12) holds for all* $t \in [0,T]$, *with the substitution of* τ_t *in place of* τ_0 *in the EEP.*

Under the risk neutral measure the drift of the underlying asset price (2.2) is given by $S_s(r_s - \delta_s)ds$. In the exercise region we must have $S > K$. It follows that the local gains from exercise of a call option are given by

$$r_s Y_s ds - dA_s^Y = r_s(S_s - K)ds - S_s(r_s - \delta_s)ds = (\delta_s S_s - r_s K)ds$$

in the event $\{\tau_s = s\}$. These gains are equal to the dividend benefits collected upon receipt of the underlying asset net of the interest forgone on the cash paid to exercise the option. The optimal exercise time is determined by the supermartingale Z corresponding to the deflated option payoff. Clearly, the payment of dividends on the underlying asset provides incentives for early exercise. Intuition suggests that immediate exercise will take place when these incentives are sufficiently large.

For American options the simple put-call parity relation (2.19) for contracts with identical characteristics (T,K) fails. Indeed, the optimal exercise policies associated with the put and the call differ and, as a result, the simple replicating argument underlying (2.19) cannot be invoked. Yet one can still use no-arbitrage arguments to establish bounds on the differential between the put and call prices (see, for instance, Hull [2002]).

In contrast, put-call symmetry holds for American options. To see this, note that at any time prior to the optimal exercise date the put value $P_t \equiv P(S,K,r,\delta;\mathcal{F}_t)$ is

$$\begin{aligned} P_t &= \sup_{\tau \in \mathcal{S}_{t,T}} \widetilde{E}_t \left[\exp\left(-\int_t^\tau r_v dv \right) \right. \\ &\quad \left. \times \left(K - S \exp\left(\int_t^\tau \left(r_v - \delta_v - \frac{1}{2}\sigma_v^2 \right) dv + \int_t^\tau \sigma_v d\widetilde{z}_v \right) \right)^+ \right]. \end{aligned}$$

Using the same arguments as in the proof of Theorem 10 enables us to rewrite this expression as

$$\begin{aligned} P_t &= \sup_{\tau \in \mathcal{S}_{t,T}} E_t^* \left[\exp\left(-\int_t^\tau \delta_v dv \right) \right. \\ &\quad \left. \times \left(K \exp\left(\int_t^\tau \left(\delta_v - r_v - \frac{1}{2}\sigma_v^2 \right) dv + \int_t^\tau \sigma_v dz_v^* \right) - S \right)^+ \right] \end{aligned}$$

where the expectation is relative to the equivalent measure Q^*, defined in (2.23), and conditional on the information $\mathcal{F}_t$. As the change of measure performed does not affect the set of stopping times over which the claimholder optimizes the following result holds (see Schroder [1999] and Detemple [2001]).

Theorem 26 *(American PCS). Consider an American-style put with characteristics* K *and* T *written on an asset with price* S *given by (2.2) in the market* *with stochastic interest rate r. Let $P(S, K, r, \delta; \mathcal{F}_t)$ denote the American put price and $\tau^p(K, r, \delta)$ the optimal exercise time. Then, prior to exercise, the put price is*

$$P(S, K, r, \delta; \mathcal{F}_t) = C(K, S, \delta, r; \mathcal{F}_t) \tag{3.13}$$

where $C(K, S, \delta, r; \mathcal{F}_t)$ is the value of an American call with strike price S and maturity date T in a financial market with interest rate δ and in which the underlying asset price follows the Ito process (2.20) with initial value K and with z^* defined by (2.22). The optimal exercise time for the put option is

$$\tau^p(S, K, r, \delta) = \tau^c(K, S, \delta, r) \tag{3.14}$$

where $\tau^c(K, S, \delta, r)$ denotes the optimal exercise time for the (symmetric) call option.

Consider now a forward contract that has an American-style exercise provision before the delivery date T, but must be executed in all circumstances at the delivery date T.

Proposition 27 *Consider an American-style forward contract with payoff $Y = S - K$ and delivery date at or before T. This contract can be exercised at the option of the holder before T. In the event that exercise does not take place before T the contract must be executed at the prescribed delivery date. The value of this contract in the financial market (2.1)-(2.2) has, for $t \in [0, \tau_0]$, the early exercise premium representation*

$$V_t^f(S, K) = \widetilde{E}_t\left[R_{t,T}(S_T - K)\right] + \widetilde{E}_t\left[\int_{\tau_0}^{T} R_{t,s} 1_{\{\tau_s = s\}} (\delta_s S_s - r_s K)\, ds\right] \tag{3.15}$$

where $\tau_t = \inf\left\{v \in [t, T] : Z_v = \xi_{0,v}(S_v - K)\right\}$ and $Z_t \equiv \sup_{\tau \in \mathcal{S}_{t,T}} E_t[D_\tau]$ with $D_t \equiv \xi_{0,t}(S_t - K)$. If the contract is being continuously marketed throughout the interval $[0, T]$ then (3.15) holds for all $t \in [0, T]$, with the substitution of τ_t in place of τ_0 in the EEP.

A forward contract with an early exercise provision can be optimally exercised before the maturity date. This is easy to see as the value of a standard (European) forward contract $\widetilde{E}_t\left[R_{t,T}(S_T - K)\right]$ will fall below the immediate exercise value $S_t - K$ if the underlying asset pays dividends at a high rate (see the formulas in Corollary 12). This simple argument shows that the EEP component of the contract value (3.15) will be positive in some circumstances. Further analysis of the claim will be provided in our next chapter for the model with constant coefficients.

3.6 Multiple Underlying Assets

Consider again the model of section 2.10 with d risky securities and a locally riskless asset. Completeness of this financial market suggests that the results above ought to generalize to this setting as well. Indeed, consider an American contingent claim (f, Y) where f is a cash flow process and Y an exercise payoff, both of which are progressively measurable with respect to the d-dimensional Brownian filtration. As before assume that the process D is continuous. The value of the claim is given by,

Theorem 28 *The rational price of the American contingent claim* $(f, Y) \in \mathcal{I}_+^*$ *is uniquely given by*

$$V_t(f, Y) = \widetilde{E}_t\left[R_{t,\tau_0} Y_{\tau_0} + \int_t^{\tau_0} R_{t,s} df_s\right] = E_t\left[\xi_{t,\tau_0} Y_{\tau_0} + \int_t^{\tau_0} \xi_{t,s} df_s\right]$$

for $t \in [0, \tau_0]$. *The optimal exercise time is* $\tau_0 = \inf\{v \in [0,T] : Z_v = D_v\}$ *where* $Z_t = \sup_{\tau \in S_{t,T}} E_t[D_\tau]$.

The valuation formulas in Theorem 28 are identical to those in the single asset (single Brownian motion) case. As before the price of the claim is obtained by discounting the cash flows collected up to the optimal exercise date. Discounting is at the riskfree rate if the expectation is calculated under the risk neutral measure. If computations are performed under the original probability measure a risk-adjusted rate is employed. The results in the theorem also apply to claims $(f, Y) \in \mathcal{I}_{ubb}^*$, with bounded negative cash flows. Representation formulas such as the EEP, the DEP and the duality formula characterize prices in the multidimensional case as well.

3.7 Appendix: Proofs

Proof of Theorem 19: The proof is a special case of the proof of Theorem 20. ∎

Proof of Theorem 20: We prove the theorem for $f \neq 0$. The proof parallels Karatzas [1988], who deals with the case $f = 0$. For $t \in [0, T]$ consider the deflated payoff process $D_t \equiv \xi_{0,t} Y_t + \int_0^t \xi_{0,s} df_s$ and its expectation $E_t[D_\tau]$. Also consider the family (indexed by t) of stopping time problems

$$Z_t = \sup_{\tau \in S_{t,T}} E_t[D_\tau], \text{ for } t \in [0, T]. \tag{3.16}$$

Given the results of chapter 2 and section 3.1 it is natural to conjecture that the value of the American claim equals $\xi_{0,t}^{-1}\left(Z_t - \int_0^t \xi_{0,s} df_s\right)$ at any time t prior to the optimal exercise time. The proof below consists in demonstrating the validity of this intuition.

The process Z is the Snell envelope of D. This process exists and is a nonnegative, right-continuous with left-hand limits supermartingale adapted to the filtration (El Karoui [1981]). The optimal stopping times τ_t that solve this family of optimization problems are given by

$$\tau_t \equiv \inf\{s \in [t, T] : Z_s = D_s\}.$$

Because D is continuous and $E\left[\sup_{t\in[0,T]} D_t\right] < \infty$, the stopping times $\tau_t, t \in [0,T]$, exist, i.e., the maximum in (3.16) is attained (Karatzas and Shreve [1998, Appendix D, Theorem D.12]). In order to show that (3.5) correctly values the American contingent claim it must be shown that $\xi_{0,t}^{-1}\left(Z_t - \int_0^t \xi_{0,s} df_s\right)$ is the wealth process corresponding to an admissible consumption-portfolio policy (C,π) that replicates (f,Y).

Under condition (3.1) the Snell envelope Z is a process of class $\mathbb{D}$ and is regular (Karatzas and Shreve [1988, Chapter 1, Definitions 4.8 and 4.12]). Hence the Doob-Meyer decomposition

$$Z_t = Z_0 + M_t^Z - A_t^Z$$

applies, where M^Z is a uniformly integrable, RCLL, P-martingale and A^Z is a right continuous, non-decreasing, adapted process with $M_0^Z = A_0^Z = 0$. Moreover, as D is continuous, A is continuous as well (Karatzas and Shreve [1998, Appendix D, Theorem D. 13]). The Martingale Representation Theorem gives

$$M_t^Z = \int_0^t \phi_s dz_s, \ t \in [0,T]$$

where ϕ is a one-dimensional, $\mathcal{F}_{(\cdot)}$-progressively measurable process such that $\int_0^t \phi_s^2 ds < \infty$.

Define now the process

$$X_t \equiv \xi_{0,t}^{-1}\left(Z_t - \int_0^t \xi_{0,s} df_s\right).$$

Substituting $Z_t = \sup_{\tau\in S_{t,T}} E_t\left[D_\tau\right]$, using the definition of D, and simplifying shows that

$$\begin{aligned} X_t &= \xi_{0,t}^{-1}\left(\sup_{\tau\in S_{t,T}} E_t\left[\xi_{0,\tau}Y_\tau + \int_0^\tau \xi_{0,s} df_s\right] - \int_0^t \xi_{0,s} df_s\right) \\ &= \sup_{\tau\in S_{t,T}} E_t\left[\xi_{t,\tau}Y_\tau + \int_t^\tau \xi_{t,s} df_s\right] \end{aligned}$$

is a non-negative process.

An application of Ito's lemma now yields

$$\begin{aligned} dX_t &= \left(Z_t - \int_0^t \xi_{0,s} df_s\right) d\xi_{0,t}^{-1} + \xi_{0,t}^{-1}\left(dM_t^Z - dA_t^Z - \xi_{0,t} df_t\right) + d\left[\xi^{-1}, M^Z\right]_t \\ &= X_t\left(r_t + \theta_t^2\right) dt + X_t\theta_t dz_t + \xi_{0,t}^{-1}\left(\phi_t dz_t - dA_t^Z - \xi_{0,t} df_t\right) + \xi_{0,t}^{-1}\theta_t\phi_t dt \\ &= r_t X_t dt + \left(\xi_{0,t}^{-1}\phi_t + X_t\theta_t\right) d\tilde{z}_t - \xi_{0,t}^{-1}\left(dA_t^Z + \xi_{0,t} df_t\right) \end{aligned}$$

for all $t \in [0,T]$. Selecting the portfolio and consumption processes

$$\pi_{1t} \equiv \sigma_t^{-1}\left(\xi_{0,t}^{-1}\phi_t + X_t\theta_t\right)$$

$$C_t \equiv \int_0^t \xi_{0,s}^{-1} \left(dA_s^Z + \xi_{0,s} df_s \right),$$

then establishes that

$$dX_t = r_t X_t dt + \pi_{1t} \sigma_t d\tilde{z}_t - dC_t.$$

This shows that X is a well-defined wealth process, corresponding to the admissible policy (C, π_1). Note also that

$$X_{\tau_t} = \xi_{\tau_t}^{-1} \left(Z_{\tau_t} - \int_0^{\tau_t} \xi_{0,s} df_s \right) = \xi_{\tau_t}^{-1} \left(D_{\tau_t} - \int_0^{\tau_t} \xi_{0,s} df_s \right) = Y_{\tau_t}$$

and that $dC = df$ on the event $\{t < \tau_t\}$ because $dA^Z = 0$ on that event (the Snell envelope is a martingale on $\{t < \tau_t\}$).

We conclude that (C, π) is an admissible policy that replicates the claim (f, Y) from initial wealth $X_0 = \xi_0^{-1} Z_0 = Z_0$, and that X represents the corresponding wealth process. An application of the no-arbitrage principle then establishes the valuation formula (3.5) in the theorem. ■

Proof of Theorem 21: Recall that the payoff Y has the representation $Y_t = Y_0 + A_t^Y + M_t^Y$, $t \in [0, T]$ under the risk neutral measure Q. Passing to the P-measure gives, for $t \in [0, T]$,

$$Y_t = Y_0 + A_t^Y + \int_0^t \phi_v^Y \theta_v dv + M_t^{Y,P} \equiv Y_0 + A_t^{Y,P} + M_t^{Y,P}$$

where ϕ_v^Y is a progressively measurable process, $M_t^{Y,P}$ is a P-martingale and the expression on the right hand side defines the process $A^{Y,P}$. Ito's lemma produces the following dynamics for the deflated exercise and total payoffs

$$d\left(\xi_{0,v} Y_v\right) = \xi_{0,v} \left(dA_v^{Y,P} - \left(\phi_v^Y \theta_v + Y_v r_v \right) dv + dM_v^{Y,P} - Y_v \theta_v dz_v \right)$$

$$\begin{aligned} dD_v &= d\left(\xi_{0,v} Y_v\right) + \xi_{0,v} df_v \\ &= \xi_{0,v} \left(dA_v^{Y,P} - \phi_v^Y \theta_v dv - Y_v r_v dv + df_v + dM_v^{Y,P} - Y_v \theta_v dz_v \right) \end{aligned}$$

under the P-measure.

Recall that the Snell envelope Z is a supermartingale. In fact, as shown in Lemma 30 below, it is a martingale in the event that waiting is optimal (prior to exercise). The supermartingale property is therefore associated with the behavior of the process in the event of exercise. This suggests that one can construct a martingale, on that event, by compensating Z in an appropriate manner. Because $Z = D$, when immediate exercise is optimal, the required compensation is the negative of the finite variation part of the deflated payoff process D (see the expression for the evolution of D above). The compensated process, M, is therefore defined by

$$M_t \equiv Z_t + \int_0^t 1_{\{\tau_v = v\}} \xi_{0,v} \left[r_v Y_v dv - dA_v^{Y,P} + \phi_v^Y \theta_v dv - df_v \right], \quad t \in [0, T] \quad (3.17)$$

and is a P-martingale as verified in Lemma 30. The proof of Theorem 21 now follows from our next result in Lemma 29. ■

Lemma 29 *Let $Z_t \equiv \sup_{\tau \in S_{t,T}} E_t[D_\tau], t \in [0,T]$ and suppose that the process M defined in (3.17) is a P-martingale. Then the representation (3.9) holds.*

Proof of Lemma 29: Given that M is a P-martingale one can write

$$E\left[Z_T + \int_0^T 1_{\{\tau_v = v\}} \xi_{0,v} \left(r_v Y_v dv - dA_v^{Y,P} + \phi_v^Y \theta_v dv - df_v\right)\right] = E[Z_0]. \tag{3.18}$$

By definition

$$Z_T = \sup_{\tau \in S_{T,T}} E_T[D_\tau] = E_T[D_T] = D_T \tag{3.19}$$

$$Z_0 = \sup_{\tau \in S_{0,T}} E_0[D_\tau] = E_0[D_{\tau_0}]. \tag{3.20}$$

Substituting (3.19) and (3.20) in (3.18) and using the definition $dA_v^{Y,P} = dA_v^Y + \phi_v^Y \theta_v dv$ yields

$$E[D_T] + E\left[\int_0^T 1_{\{\tau_v = v\}} \xi_{0,v} \left(r_v Y_v dv - dA_v^Y - df_v\right)\right] = E[D_{\tau_0}]. \tag{3.21}$$

By Theorem 20 the right-hand side of (3.21) equals $V_0(f, Y)$. As $1_{\{\tau_v = v\}} = 0$ in the random interval $[0, \tau_0]$ we conclude that the assertion of the lemma holds. ■

Lemma 30 *Let $Z_t \equiv \sup_{\tau \in S_{t,T}} E_t[D_\tau], t \in [0,T]$. Then Z_s is a martingale on the event $\{t \leq s < \tau_t\}$ and the process M in (3.17) is a P-martingale.*

Proof of Lemma 30: By definition of the optimal stopping time we have $D_{\tau_t} = Z_{\tau_t}$. It follows that $Z_t = E_t[D_{\tau_t}] = E_t[Z_{\tau_t}]$. Moreover, on the event $\{t \leq s < \tau_t\}$ it must be that $\tau_s = \tau_t$. Indeed, if not then τ_t is dominated by some other policy τ_s on some event $A \in \mathcal{F}_s$, i.e.,

$$E_s[Z_{\tau_t}] < E_s[Z_{\tau_s}], \qquad \text{for some } A \in \mathcal{F}_s$$

and therefore, taking expectations conditional on $\mathcal{F}_t$ and letting 1_A denote the indicator of the set A,

$$E_t[Z_{\tau_t}] < E_t[Z_{\tau_s} 1_A + Z_{\tau_t}(1 - 1_A)]$$

in contradiction with the optimality of τ_t. Thus $Z_s = E_s[D_{\tau_s}] = E_s[Z_{\tau_t}]$ for all $s \in [t, \tau_t)$, as claimed. From the Doob-Meyer decomposition we conclude that $dZ_s = dM_s^Z$ and $dA_s^Z = 0$ for all $s \in [t, \tau_t)$.

To see that the process M in (3.17) is a P-martingale note that $D_{\tau_t} = Z_{\tau_t}$ on the event $\{t = \tau_t\}$ and therefore

$$\int_0^T 1_{\{t=\tau_t\}} dZ_t = \int_0^T 1_{\{t=\tau_t\}} dD_t \tag{3.22}$$

where

$$dD_t = \xi_t \left(dA_t^{Y,P} - \phi_t^Y \theta_t dt - Y_t r_t dt + df_t \right) + \xi_t \left(dM_t^{Y,P} - Y_t \theta_t dz_t \right)$$

(for some of the subtleties underlying (3.22) see Rutkowski [1994, Lemmas A.2, A.3, and A.4]). It follows that

$$\begin{aligned} & \int_0^s 1_{\{t=\tau_t\}} dZ_t - \int_0^s 1_{\{t=\tau_t\}} \xi_t \left(dA_t^{Y,P} - \phi_t^Y \theta_t dt - Y_t r_t dt + df_t \right) \\ = & \int_0^s 1_{\{t=\tau_t\}} \xi_t \left(dM_t^{Y,P} - Y_t \theta_t dz_t \right) \end{aligned}$$

for all $s \in [0, T]$. Using Ito's lemma enables us to conclude that

$$\begin{aligned} M_s & \equiv Z_s - \int_0^s 1_{\{t=\tau_t\}} \xi_t \left(dA_t^{Y,P} - \phi_t^Y \theta_t dt - Y_t r_t dt + df_t \right) \\ & = Z_0 + \int_0^s 1_{\{t=\tau_t\}} dZ_t + \int_0^s 1_{\{t<\tau_t\}} dZ_t \\ & \quad - \int_0^s 1_{\{t=\tau_t\}} \xi_t \left(dA_t^{Y,P} - \phi_t^Y \theta_t dt - Y_t r_t dt + df_t \right) \\ & = Z_0 + \int_0^s 1_{\{t=\tau_t\}} \xi_t \left(dM_t^{Y,P} - Y_t \theta_t dz_t \right) + \int_0^s 1_{\{t<\tau_t\}} dM_t^Z \end{aligned}$$

and $E[M_s] = 0$. Rearranging yields the result announced. ■

Proof of Corollary 22: Under the condition stated exercise prior to T can only lead to a reduction in the value of the contract. It follows that exercise prior to the terminal date is never optimal. ■

Proof of Theorem 23: The value of the contingent claim can always be written as

$$\begin{aligned} V_t(f, Y) & = \widetilde{E}_t \left[\int_t^{\tau_t} R_{t,s} df_s + R_{t,\tau_t} Y_{\tau_t} \right] \\ & = Y_t + \widetilde{E}_t \left[\int_t^{\tau_t} R_{t,s} df_s + (R_{t,\tau_t} Y_{\tau_t} - Y_t) \right] \end{aligned}$$

where, by Ito's lemma,

$$\begin{aligned} R_{t,\tau_t} Y_{\tau_t} - Y_t & = \int_t^{\tau_t} (R_{t,s} dY_s + Y_s dR_{t,s}) \\ & = \int_t^{\tau_t} R_{t,s} \left(dA_s^Y + dM_s^Y - r_s Y_s ds \right). \end{aligned}$$

Substituting in the prior expression and eliminating the Q-martingale component gives

$$V_t(f,Y) = Y_t + \widetilde{E}_t\left[\int_t^{\tau_t} R_{t,s}\left(dA_s^Y - r_s Y_s ds + df_s\right)\right]$$

for all $t \in [0,T]$. The representation (3.10) follows ■

Proof of Theorem 24: Using the definition of the Snell envelope we can write, for any $M \in \mathcal{H}_0^1(P)$, that

$$Z_0 \equiv \sup_{\tau \in \mathcal{S}_{0,T}} E[D_\tau] = \sup_{\tau \in \mathcal{S}_{0,T}} E[D_\tau - M_\tau] \leq E\left[\sup_{\tau \in [0,T]} (D_\tau - M_\tau)\right].$$

Taking the infimum, on the right hand side, over martingales in $\mathcal{H}_0^1(P)$ shows that the expression on the right hand side of (3.11) is an upper bound for the claim's price.

To establish the result in the theorem it remains to show that the right hand side of (3.11) is also a lower bound for the claim's price. Recall the definition of the Snell envelope and that the assumptions on the discounted payoff imply that Z is a supermartingale of class $\mathbb{D}$. As indicated before, the Doob-Meyer decomposition shows that $Z_t = Z_0 + M_t^Z - A_t^Z$ where M^Z is a martingale that is null at zero and A^Z is a continuous, non-decreasing, adapted process with $EA_T^Z < \infty$, also null at zero. Given our assumptions it is easy to show that $M^Z \in \mathcal{H}_0^1(P)$. Indeed $M_t^Z = Z_t - Z_0 + A_t^Z$, combined with the properties of A_t^Z and the facts that $Z_0 \geq 0$ and $D \geq 0$ implies

$$\begin{aligned}
\sup_{t \in [0,T]} \left|M_t^Z\right| &\leq \sup_{t \in [0,T]} |Z_t| + \sup_{t \in [0,T]} \left|A_t^Z\right| \\
&\leq \sup_{t \in [0,T]} \left|E_t\left[\sup_{v \in [0,T]} |D_v|\right]\right| + \left|A_T^Z\right| \\
&= \sup_{t \in [0,T]} E_t\left[\sup_{v \in [0,T]} D_v\right] + A_T^Z \equiv H
\end{aligned}$$

where $H \in \mathcal{L}^1(P)$ because $\sup_{v \in [0,T]} D_v \in \mathcal{L}^1(P)$ and $A_T^Z \in \mathcal{L}^1(P)$.

Now, as $M^Z \in \mathcal{H}_0^1(P)$, we can write

$$\begin{aligned}
\inf_{M \in \mathcal{H}_0^1(P)} E\left[\sup_{t \in [0,T]} (D_t - M_t)\right] &\leq E\left[\sup_{t \in [0,T]} \left(D_t - M_t^Z\right)\right] \\
&\leq E\left[\sup_{t \in [0,T]} \left(Z_0 - A_t^Z\right)\right] \\
&= Z_0 - A_0 = Z_0
\end{aligned}$$

where the second inequality follows from the fact that $D_t \leq Z_t$ (immediate exercise is a feasible policy at t) and therefore $D_t - M_t^Z \leq Z_t - M_t^Z = Z_0 - A_t^Z$. This completes the proof of the theorem. ■

Proof of Proposition 25: The proof follows immediately from the EEP representation (Theorem 21) and from the option payoff function $f = 0, Y = (S - K)^+$. In the exercise region it must be that $S > K$. Therefore $r_s Y_s ds - dA_s^Y = r_s (S_s - K) ds - S_s (r_s - \delta_s) ds = (\delta_s S_s - r_s K) ds$. ■

Proof of Theorem 26: The proof parallels the proof of Theorem 10. Let $S_t = S$. The put option value is

$$\begin{aligned}
P_t &= \sup_{\tau \in \mathcal{S}_{t,T}} \widetilde{E}_t \left[\exp\left(-\int_t^\tau r_v dv \right) (K - S_\tau)^+ \right] \\
&= \sup_{\tau \in \mathcal{S}_{t,T}} \widetilde{E}_t \left[\exp\left(-\int_t^\tau r_v dv \right) \frac{S_\tau}{S} \left(\frac{K}{(S_\tau / S)} - S \right)^+ \right] \\
&= \sup_{\tau \in \mathcal{S}_{t,T}} E_t^* \left[\exp\left(-\int_t^\tau \delta_v dv \right) \left(\frac{K}{(S_\tau / S)} - S \right)^+ \right]
\end{aligned}$$

where $E_t^* [\cdot]$ is the expectation relative to the measure Q^* (see (2.23)). Note that in the third line the set of stopping times $\mathcal{S}_{t,T}$ is relative to the filtration generated by z. The remainder of the proof follows the proof for the European option case. ■

Proof of Proposition 27: The proof follows from the EEP representation in Theorem 21, specialized to the contract under consideration. ■

Proof of Theorem 28: The arguments in the proof of Theorem 20 apply in the multidimensional case. ■

Chapter 4

Standard American Options

We now focus on American-style plain vanilla options in a market where the underlying asset price follows a geometric Brownian motion (GBM) process. The structures of the immediate exercise region (section 4.1) and of the price function (section 4.2) will be examined first. These results will help us to specialize and simplify the EEP representation formula and to derive an integral equation for the exercise boundary (section 4.3). We then proceed to discuss a reduction in the dimensionality of the integral equation (section 4.4) and to analyze the delta hedge ratio (section 4.5). Extensions to general diffusions are sketched next (section 4.6). An application of the results to a path-dependent contract, the floating strike Asian option, that can be reduced to a standard option by a change of measure, is also provided (section 4.7). The optimal exercise decision and the pricing of American forward contracts are examined last (section 4.8).

4.1 The Immediate Exercise Region

Consider an American-style call option with exercise price $K > 0$ and maturity date T, written on an underlying asset whose price S satisfies the stochastic differential equation (under the risk neutral measure)

$$dS_t = S_t \left[(r - \delta) \, dt + \sigma d\tilde{z}_t \right], \; t \in [0, T] ; \; S_0 \text{ given.} \tag{4.1}$$

Here r, δ and σ are constant parameters: the price is a geometric Brownian motion (GBM) process. As exercise cannot be optimal when $S < K$ it has become customary to write the option payoff in the form $Y = (S - K)^+$. Our first objective is to characterize the structure of the exercise region and its boundary. Given that the environment is Markovian, the state of nature is completely described by the asset price S and the pair (S, t) contains all the information required for pricing and decision-making purposes. Let $C(S, t)$ be the option price at the point (S, t). The *immediate exercise region*, denoted by $\mathcal{E}$, is the set of pairs (S, t) at which immediate exercise is an optimal policy.

That is[1]

$$\mathcal{E} \equiv \left\{ (S,t) \in \mathbb{R}_+ \times [0,T] : C(S,t) = (S-K)^+ \right\}.$$

Its complement, $\mathcal{C} \equiv \left\{ (S,t) \in \mathbb{R}_+ \times [0,T] : C(S,t) > (S-K)^+ \right\}$, is the *continuation region*, i.e., the set of prices-dates at which immediate exercise is suboptimal. Our first proposition describes elementary properties of the exercise region.

Proposition 31 *The immediate exercise region has the following properties*
(i) $(S,t) \in \mathcal{E}$ *implies* $(S,s) \in \mathcal{E}$ *for all* $t \in [0,T]$ *and* $s \in [t,T]$.
(ii) $(S,t) \in \mathcal{E}$ *implies* $(\lambda S,t) \in \mathcal{E}$ *for all* $\lambda \geq 1$, *and for all* $t \in [0,T]$.
(iii) Suppose that $S < \max\{K, (r/\delta)K\}$. *Then* $(S,t) \notin \mathcal{E}$, *for all* $t \in [0,T)$.
(iv) $(S,T) \in \mathcal{E}$ *if and only if* $S \geq K$.

The first property establishes that immediate exercise, at a given price S, remains optimal when time to maturity decreases, if it is optimal at (S,t). The intuition for this result is clear. Indeed, as an American option with shorter maturity has a smaller set of feasible exercise opportunities, its value cannot exceed the value of a longer maturity option. As a result, immediate exercise, if optimal for the long maturity option, will also be optimal for the shorter maturity option. From a geometric point of view the property implies that the exercise region is connected as time moves forward, or equivalently, as time to maturity decreases. Its graph (see Figure 4.1) is left-connected in the time-to-maturity dimension.

Property (ii) shows that the exercise region is also connected when the underlying asset price S increases (up-connectedness). This follows because the exercise payoff at any exercise time $\tau \in [t,T]$, and for any constant $\lambda \geq 1$, satisfies the bound

$$(\lambda S_\tau - K)^+ \leq (S_\tau - K)^+ + (\lambda - 1) S_\tau$$

(see Appendix). Taking the present value on each side of this inequality and using the fact that the underlying asset pays dividends (i.e., the discounted price is a supermartingale under the risk neutral measure) shows that the option value at $(\lambda S,t)$ is bounded above by the option value at (S,t) plus the difference in asset prices $\lambda S - S$. That is, $C(\lambda S,t) \leq C(S,t) + (\lambda - 1) S$. The optimality of immediate exercise at (S,t) (i.e., $C(S,t) = S - K$) then implies that $\lambda S - K$ is an upper bound for the option value at $(\lambda S,t)$,

$$C(\lambda S,t) \leq C(S,t) + (\lambda - 1) S = S - K + (\lambda - 1) S = \lambda S - K.$$

As the upper bound is attained by exercising the option immediately, it must be the optimal exercise policy at the point $(\lambda S,t)$.

Property (iii) identifies regions in which immediate exercise is suboptimal. This is clearly the case when the underlying asset price is below the strike price

[1] It is understood that the point $(0,t) \notin \mathcal{E}$ because immediate exercise cannot be optimal when $S < K$.

of the call option. It is also true if the asset price is below the ratio $(r/\delta)K$. In this instance the instantaneous net benefit of exercising, equal to $\delta S - rK$, is negative: delaying exercise by an increment of time saves value and, therefore, dominates immediate exercise (recall the EEP formula (3.12)). Property (iv) is standard. It identifies the optimal exercise region at maturity with the set where the asset price is no less than the strike. Taken together (iii) and (iv) reveal an interesting phenomenon. If the interest rate exceeds the dividend rate, $r > \delta$, the continuation region "suddenly" shrinks as the maturity date of the contract approaches. At any moment prior to maturity it remains optimal to wait if the asset price lies below $(r/\delta)K$. At the maturity date, however, immediate exercise becomes optimal for all asset values above the strike, in particular those sandwiched between K and $(r/\delta)K$.

The exercise region is "separated" from the continuation region by a boundary, which is called the *immediate exercise boundary*. A formal definition of this boundary will be given below after we examine the structure of the option price function. An illustration of the exercise region for a standard American call option, and the corresponding boundary, is given in Figure 4.1 for times prior to maturity.

4.2 The Call Price Function

Additional insights about the immediate exercise region are obtained upon inspection of the call price function. The next proposition states some of its basic properties. Properties of the call and put price functions, in this and more general market models, can be found in Bergman, Grundy and Wiener [1996]. Properties of option prices in multiasset models are established in Jaillet, Lamberton and Lapeyre [1990]. The results presented below are special cases of theirs.

Proposition 32 *Let $C(S,t)$ be the value of the American call option. We have*
(i) $C(S,t)$ is continuous on $\mathbb{R}_+ \times [0,T]$.
(ii) $C(\cdot,t)$ is non-decreasing and convex on $\mathbb{R}_+$ for all $t \in [0,T]$.
(iii) $C(S,\cdot)$ is non-increasing on $[0,T]$ for all $S \in \mathbb{R}_+$.
(iv) $0 \leq \partial C(S,t)/\partial S \leq 1$ on $\mathbb{R}_+ \times [0,T]$; $\partial C(S,t)/\partial S = 1$ for $(S,t) \in \overset{o}{\mathcal{E}}$.[2]

The continuity of the call pricing function in (i) is a consequence of the continuity of the payoff function and the continuity of the solution of (4.1) with respect to the initial condition (i.e., the continuity of the flow of the stochastic differential equation (4.1)). Similarly, property (ii) is implied by the non-decreasing and convex structure of the payoff function and the fact that the solution of (4.1) is non-decreasing with respect to the initial condition. The

[2] $\overset{o}{\mathcal{E}}$ is the interior of the exercise region.

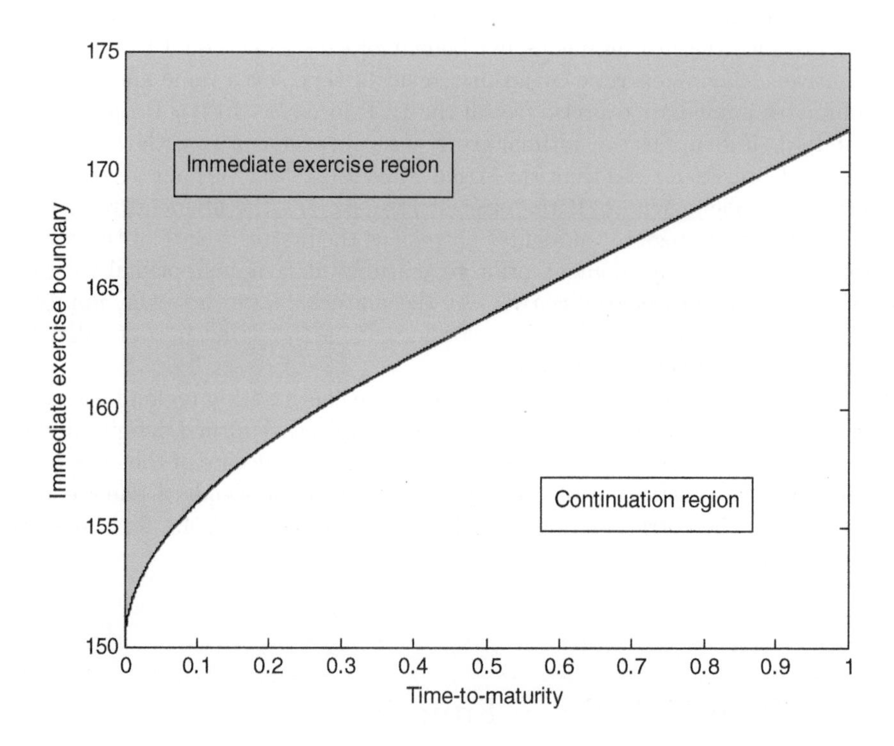

Figure 4.1: GBM model. Immediate exercise boundary for a call option (y-axis) versus time to maturity (x-axis). The strike price is 100, the interest rate 6%, the dividend rate 4% and the return volatility 0.2. Computations use the integral equation method with 2,000 time steps.

price behavior relative to time (iii) is the counterpart of property (i) in Proposition 31: as time increases the set of exercise opportunities shrinks and the value of the option falls. Property (iv) bounds the local change in the price and induces bounds on the delta hedge ratio, i.e., the derivative of the price with respect to S. The intuition for these bounds becomes apparent once we think of the option price as a weighted average taken over a convex payoff function whose slope is non-negative and bounded above by one.

The continuity of the call price function is a useful property as it implies that the immediate exercise region is a closed set (the continuation region is an open set). One concludes that the boundary of the immediate exercise region $B \equiv \{B_t : t \in [0, T]\}$ belongs to $\mathcal{E}$. Furthermore, up-connectedness of the exercise region (property (ii) in Proposition 31) implies that $B_t \equiv \inf \{S : (S, t) \in \mathcal{E}\}$

and that[3]

$$\mathcal{E} \equiv \{(S,t) \in \mathbb{R}_+ \times [0,T] : S \geq B_t\}.$$

It follows that the exercise boundary is unique and depends on time alone. It displays the following structure.

Proposition 33 *The boundary B of the immediate exercise region is continuous on $[0,T)$, non-increasing with respect to time (non-decreasing in time-to-maturity) and has the limiting values $\lim_{t \uparrow T} B_t = \max\{K, (r/\delta) K\}$ and $\lim_{T-t \uparrow \infty} B_t = B_{-\infty} \equiv K(b+f)/\left(b+f-\sigma^2\right)$ where $b \equiv \delta - r + \frac{1}{2}\sigma^2$ and $f \equiv \left(b^2 + 2r\sigma^2\right)^{1/2}$. At maturity $B_T = K$.*

The continuity and monotonicity properties of the boundary B follow from properties (i) and (ii) in Proposition 32 and from the fact that $\mathcal{E}$ is a closed set. Continuity, in particular, shows that the curve B coincides, at all times prior to maturity, with the boundary of the set $\mathcal{E}$ (see the discussion in footnote 3).

The limiting values are obtained from the recursive equation (4.6) for the exercise boundary displayed in Theorem 34 below. Note that the optimal exercise boundary for the deterministic problem with $\sigma = 0$ is a constant equal to $\max\{K, (r/\delta)K\}$: in the absence of uncertainty it pays to exercise if the payoff is non-negative ($S \geq K$) and the local gains are non-negative ($\delta S - rK \geq 0$). For the stochastic problem the uncertainty faced by the investor over the remaining life of the option, $\sigma^2(T-t)$, converges to zero as $t \uparrow T$ and one would naturally expect the optimal exercise boundary to converge to the boundary for the deterministic problem. This is the intuition underlying this limiting result stated in Proposition 33. At the other extreme, when $T - t \to \infty$, the boundary becomes a constant, independent of time. The optimal policy is to exercise at the first hitting time of this constant barrier level.

Additional properties of the American option exercise boundary are established in van Moerbeke [1976], Ait-Sahlia [1995] and Barles et al. [1995]. Explicit expressions for the boundary, valid near maturity, are derived by Evans, Kuske and Keller [2002]. These formulas can be used, in conjunction with a numerical scheme, to obtain faster and more accurate estimates of the exercise boundary and the option price.

4.3 Early Exercise Premium Representation

In the GBM case one can take advantage of the explicit nature of the underlying price distribution to compute the moments appearing in the EEP formula. Theorem 25 specializes as follows.

[3]Note that the curve defined by $B_t \equiv \inf\{S : (S,t) \in \mathcal{E}\}$ may not coincide exactly with the boundary of the set $\mathcal{E}$ (i.e., the set of limit points of $\mathcal{E}$). A discrepancy exists if $\mathcal{E}$ has a vertical boundary at some dates. If this situation arises the boundary of $\mathcal{E}$ is a correspondence with a thick component at those dates. It corresponds to the curve B at all other dates.

Theorem 34 *(Kim [1990], Jacka [1991], Carr, Jarrow and Myneni [1992]). Suppose that the underlying asset price follows the geometric Brownian motion process (4.1) and that the interest rate is constant. The value of an American call option has the early exercise premium representation*

$$C(S_t,t) = c(S_t,t) + \pi(S_t,t,B(\cdot)) \tag{4.2}$$

for $t \in [0,T]$*, where* $c(S,t)$ *represents the Black-Scholes value of a European call option (see (2.24) and (2.25)) and* $\pi(S_t,t,B(\cdot))$ *is the early exercise premium given by*

$$\begin{aligned}\pi(S_t,t,B(\cdot)) &= \int_t^T \Big(\delta S_t e^{-\delta(s-t)} N(d(S_t,B_s,s-t)) \\ &\quad -rKe^{-r(s-t)} N(d_1(S_t,B_s,s-t))\Big)\,ds \end{aligned} \tag{4.3}$$

with

$$d(S_t,B_s,s-t) = \frac{\log(S_t/B_s) + \left(r-\delta+\frac{1}{2}\sigma^2\right)(s-t)}{\sigma\sqrt{s-t}} \tag{4.4}$$

$$d_1(S_t,B_s,s-t) = d(S_t,B_s,s-t) - \sigma\sqrt{s-t}. \tag{4.5}$$

The immediate exercise boundary B *solves the recursive non-linear integral equation*

$$B_t - K = c(B_t,t) + \pi(B_t,t,B(\cdot)) \tag{4.6}$$

for $t \in [0,T)$*, subject to the boundary condition* $B_{T_-} = \max\{K,(r/\delta)K\}$*. At maturity* $B_T = K \le B_{T_-}$*.*

The EEP formula in (4.3)-(4.5) expresses the option price at (S_t,t) in terms of the (unknown) future values of the exercise boundary, $(B_s, s \in [t,T))$. As the formula holds at any point in time and for any value of the underlying asset price (recall the last statement in Proposition 25), it holds in particular on the boundary of the exercise region prior to maturity, where $S_t = B_t$ and $t < T$. This insight leads to the recursive integral equation (4.6). This equation, along with its associated boundary condition for B_{T_-}, provides a complete characterization of the exercise boundary prior to maturity. It lies at the core of one of the numerical procedures commonly used to price American options (see chapter 8).

It is also worth emphasizing that the integral equation (4.6), combined with the terminal point B_T, describes the immediate exercise boundary over the life of the option, the period $[0,T]$. The terminal point B_T must be added given our definition of the exercise boundary as the set of points $B \equiv \{B_t : t \in [0,T]\}$ such that $B_t \equiv \inf\{S : (S,t) \in \mathcal{E}\}$. The results in Theorem 34 show that the exercise boundary has a discontinuity at the maturity date if the interest rate exceeds the dividend rate: in that instance $B_T = K < B_{T_-} = (r/\delta)K$.

When the option maturity becomes infinite the option price formula (4.2) simplifies as follows (Samuelson [1965] and Merton [1973b]).

Corollary 35 *(American call with infinite maturity) Consider an American call option with infinite maturity. Its value is* $C(S,t) = (B_\infty - K)(S/B_\infty)^{2\alpha/\sigma^2}$, *where* $B_\infty = K(b+f)/(b+f-\sigma^2)$, $\alpha = \frac{1}{2}(b+f)$, $b = \delta - r + \frac{1}{2}\sigma^2$, *and* $f = \sqrt{b^2 + 2r\sigma^2}$.

Put-call symmetry enables us to infer the value of a put on a dividend-paying asset by a simple reparametrization of the American call pricing function. The general result of Theorem 26 naturally applies to the model under consideration. The symmetry result, in this particular GBM model, is a variation of the international put-call equivalence (Grabbe [1983]) and was originally proved by McDonald and Schroder [1990].[4]

Proposition 36 *(American put-call symmetry) Consider American put and call options written on the same underlying asset whose price satisfies (1.4.1). Suppose that these options have the same maturity and the same exercise price. Let* $P(S,K,r,\delta,T)$ *and* $C(S,K,r,\delta,T)$ *denote the respective price functions. Then* $P(S,K,r,\delta,T) = C(K,S,\delta,r,T)$. *The relationship between the respective exercise boundaries is* $B^p(K,r,\delta,t) = SK/B^c(S,\delta,r,t)$.

Proposition 36 implies that a put with exercise price K and maturity T, written on a stock with dividend rate δ and price S in a market with interest rate r, has the same value as a call with strike price S and maturity T written on a stock with dividend rate r and price K when the interest rate is δ. Moreover as the optimal exercise times of the two contracts must be the same ($\tau^p(K,r,\delta) = \tau^c(S,\delta,r)$), to preserve the equality of prices at exercise, it must be the case that

$$\inf\{v \in [0,T] : SN_v = B^p(K,r,\delta,v)\} = \inf\{v \in [0,T] : KN_v^{-1} = B^c(S,\delta,r,v)\}$$

where $N_v \equiv \exp\left(\left(r-\delta-\frac{1}{2}\sigma^2\right)v + \sigma\widetilde{z}_v\right)$. Simple algebra then establishes the relation between the two boundaries.

The model for the underlying asset price in (4.1) allows for dividends that are paid at a continuous rate. This type of model is commonly used to value foreign currency options, futures options and index options. See, for instance, Hull [1993] for a description of these contracts. Analytical solutions for American options in the case of discrete dividends are given in Roll [1977] and Geske [1979].

4.4 A One-Dimensional Integral Equation

The EEP formula (4.2) expresses the option price in terms of integrals taken over cumulative normal distribution functions. The integral equation (4.6), characterizing the optimal exercise boundary, inherits this structure: it depends on

[4] The proof provided by McDonald and Schroder [1990], [1998] relies on a binomial approximation of the lognormal model. Bjerksund and Stensland [1993] use a PDE approach to establish the same result.

double integrals. It can therefore be viewed as a two-dimensional characterization of the exercise boundary. A reduction to a one-dimensional integral equation can be achieved by exploiting the fact that the pricing formula (4.2) holds over the whole state space and, in particular, in the exercise region. This result, established by Little, Pant and Hou [2000], produces the following equation for the boundary,

Theorem 37 *The immediate exercise boundary B solves the recursive one-dimensional integral equation*

$$\log(B_t) = \frac{I(B_t, B(\cdot))}{J(B_t, B(\cdot))}$$

where the functions $J(B_t, B(\cdot))$ and $I(B_t, B(\cdot))$ are respectively given by

$$J(B_t, B(\cdot)) = \int_t^T e^{-\delta(s-t)} n(d(B_t, B_s, s-t)) \left(\frac{\delta B_s - rK}{B_s}\right) \frac{1}{\sigma^3 (s-t)^{3/2}} ds$$

$$\begin{aligned} I(B_t, B(\cdot)) &= e^{-\delta(T-t)} n(d(B_t, K, T-t)) \frac{1}{\sigma\sqrt{T-t}} \\ &\quad + \int_t^T \delta e^{-\delta(s-t)} n(d(B_t, B_s, s-t)) \frac{1}{\sigma(s-t)^{\frac{1}{2}}} ds \\ &\quad - \int_t^T e^{-\delta(s-t)} n(d(B_t, B_s, s-t)) \\ &\quad \times \frac{-\log(B_s) + \left(r - \delta + \frac{1}{2}\sigma^2\right)(s-t)}{\sigma^3 (s-t)^{3/2}} \left(\frac{\delta B_s - rK}{B_s}\right) ds \end{aligned}$$

and the equation is subject to the boundary condition $B_{T_-} = K \vee (r/\delta) K$. At maturity $B_T = K \leq B_{T_-}$.

The reduction to a one-dimensional integral equation is achieved by taking the second derivative of the American option price and setting it equal to zero when evaluated at the boundary. The motivation for this operation follows directly from the EEP representation stating that $S - K = C(S, t)$ for all prices in the exercise region.

The one-dimensional integral equation stated in the theorem is not the unique equation of this type that can be derived. Given that the functions $J(B_t, B(\cdot))$ and $I(B_t, B(\cdot))$ are infinitely differentiable with respect to B_t, differentiation can be performed over and over again. This leads to a whole family of one-dimensional equations that have the potential to be used for the computation of the exercise boundary.

4.5 Hedging

The EEP representation of the option price provides an explicit valuation formula, conditional on the optimal exercise boundary B. As B is independent of the stock price the formula can be used, in a straightforward manner, to compute the delta hedge associated with the option.

Proposition 38 *Let $C(S,t)$ denote the value of the American call option. The delta hedge ratio $H(S,t) = \partial C(S,t)/\partial S$ is continuous on $\mathbb{R}_+ \times [0,T)$. It is given by*

$$\begin{aligned} H(S_t,t) &= h(S_t,t) + \int_t^T \delta e^{-\delta(s-t)} N(d(S_t,B_s,s-t))\,ds \\ &\quad + \int_t^T e^{-\delta(s-t)} n(d(S_t,B_s,s-t)) \left(\frac{\delta B_s - rK}{B_s}\right) \frac{1}{\sigma(s-t)^{\frac{1}{2}}}\,ds \end{aligned} \tag{4.7}$$

where h is the hedge ratio for the European option (see Corollary 16 specialized to constant r and δ). In particular, along the exercise boundary, $H(B_t,t) = 1$.

As for the call price, the expression for the hedge ratio (4.7) is parametrized by the boundary B. Once the exercise boundary has been calculated all the components of (4.7) are identified and the implementation of the hedging formula becomes straightforward.

One property that bears emphasizing is the continuity of the hedge ratio along the boundary (this is the so-called *smooth pasting* condition that has been extensively discussed in the literature). Note that this feature of the option price is obtained as an outcome of the EEP representation that was derived using probabilistic arguments. This stands in contrast with derivations based on variational inequalities, that require a proof of smooth pasting as a step in the demonstration of the EEP formula.

Note that hedge ratios could also be derived in the general model with Ito processes of chapters 2 and 3. In this context hedging involves neutralization of the fluctuations in the value of the contract due to changes in the underlying Brownian motion. The relevant hedges can be expressed in terms of the Malliavin derivatives of the discounted payoff process. Ocone and Karatzas [1991] provide background information for this derivation: their paper contains an introduction to Malliavin calculus and an application to consumption-portfolio choice problems with Ito price processes. Implementations of the method, in the context of diffusion models, can be found in Detemple, Garcia and Rindisbacher [2003]. Extensions of the approach to hedging problems with discontinuous payoff functions and diffusion prices appear in Lasry, Lebuchoux, Lions and Touzi [1999].

4.6 Diffusion Processes

Suppose now that the underlying asset price S satisfies the stochastic differential equation (under the risk neutral measure)

$$\frac{dS_t}{S_t} = (r(S_t,t) - \delta(S_t,t))\,dt + \sigma(S_t,t)\,d\tilde{z}_t \tag{4.8}$$

where $r(S,t)$ represent the interest rate, $\delta(S,t)$ the dividend yield of the asset, $\sigma(S,t)$ the return volatility and $\widetilde{z}$ is a Q-Brownian motion. We assume that the coefficients of (4.8) are continuously differentiable and that (4.8) has a unique strong solution.[5] The interest rate $r(S,t)$ is positive, decreasing in S and increasing in t. The dividend $\delta(S,t)S$ is increasing in S. These conditions are natural and capture the well known inverse relation between the interest rate and the stock market and the positive association between dividends and stock prices.

We also assume that the solution to (4.8) satisfies a time-monotonicity condition. Specifically, for a fixed $s \in [0,T]$ and a given constant $h \geq 0$, let $S^{h,s}$ be the solution of (4.8) with initial condition $S_s = S$, at time s, and time-translated parameters $r(S_v, v+h), \delta(S_v, v+h)$ and $\sigma(S_v, v+h)$. We require that $S_v^{0,s} \geq S_v^{h,s}$ for all $h \geq 0$ and $s, v \in [0,T]$ with $v \geq s$. Given the direction of the inequality we say that the price process satisfies *decreasing time-monotonicity*. The reverse property is labelled *increasing time-monotonicity.* Decreasing time-monotonicity implies that a price trajectory issued from a given point S and corresponding to a given path of the underlying Brownian motion decreases if the initial date s is shifted forward by h. Decreasing time-monotonicity along with the assumptions on the interest rate function ensure that the prospects of the call option holder do not improve as maturity approaches. The condition is automatically satisfied when the coefficients of the model are independent of time.

The model described by (4.8) is fairly general and applies to currencies and indices. It also applies to individual stocks if we specialize it to a time-dependent (or constant) interest rate $r(t)$. The specification is flexible to the extent that it accommodates non-linear dependencies in dividend yields and in return volatilities.

The Markovian nature of this environment implies that the state space is completely described by the price S. The immediate exercise region of an American call is the set

$$\mathcal{E} = \{(S,t) \in \mathbb{R}_+ \times [0,T] : C(S,t) = \max(S-K, 0)\}.$$

In spite of the non-linear structure of the coefficients the exercise region has a simple structure. The following characterization appears in Detemple and Tian [2002],

Proposition 39 *The exercise region of the American call option with maturity T and strike K is $\mathcal{E} = \{(S,t) \in \mathbb{R}_+ \times [0,T] : S \geq B(t)\}$ for a non-increasing right-continuous function $t \to B(t)$.*[6]

---

[5]The existence of a unique strong solution of (4.8) can be ensured by imposing Growth and Lipshitz conditions on the coefficients of the equation (see Karatzas and Shreve [1988]).

[6]If the price function is sufficiently smooth (i.e., $C(S,t) \in \mathcal{C}^{2,1}(\mathbb{R}_+ \times [0,T])$ and the function $(\delta(x,t)x - rK) / \left(x^2\sigma(x,t)^2\right) \geq \varepsilon > 0$ for all $(x,t) \in \mathcal{E}$, then the arguments in the proof of Proposition 33 apply and show that the boundary is also left continuous.

As in the case of constant coefficients immediate exercise is optimal when the underlying asset price exceeds a boundary B that depends on time. Several aspects of this property may be surprising at first glance. The first one is the non-random nature of the exercise boundary that seems at odds with the price dependence of the coefficients of (4.8). This feature is a consequence of the Markovian structure of the environment in which all the relevant information is summarized in a single sufficient statistic, the current price. In this context immediate exercise will naturally involve thresholds that are functions of time alone. These critical prices will depend on the parameters of the model, including the structure of the asset price coefficients and the interest rate.

The second surprising aspect is the up-connectedness of the exercise region, implying a unique critical price (i.e., a unique boundary). This property is not obvious when the coefficients, in particular the interest rate, depend on the underlying asset price. In models where the call price is a convex function of the underlying asset price the exercise region is connected.[7] The time t-section of the exercise set is then either a finite or an infinite interval. In the first case the exercise region is defined by two boundaries (critical prices); in the second case there is a unique boundary (critical price). For contingent claims with non-convex prices, or claims with convex but non-monotone prices (decreasing-increasing price functions), the exercise region can even be the union of disjoint sets. In the model under consideration, convexity is not easily established, even for a call option, given that the underlying variable satisfies the diffusion (4.8) and the interest rate depends on it. The explanation for up-connectedness has to do with the assumptions embedded in the model, namely the negative relation between the interest rate and the asset price and the positive relation between the dividend and the asset price. As shown in the proof, these assumptions imply that the slope of the call price is bounded above by one: this prevents the price from increasing above the exercise payoff, in the exercise region, when the underlying asset price increases.[8]

Another feature of interest is the non-increasing behavior of the exercise boundary (right-connectedness in time of the exercise set). This property is implied by the decreasing time-monotonicity condition imposed on the solution of (4.8). Under this condition the set of possible price paths issued from a given point S does not improve as time passes. This implies a deterioration in the set of possible call payoff trajectories. When the interest rate is positively related to time and negatively related to the asset price interest rate trajectories are increased and the discount factor decreases (i.e., the time value of money

[7] See Bergman, Grundy and Wiener [1996], El Karoui, Jeanblanc-Picque and Shreve [1998] and Hobson [1998] for results on the convextity of European- and American-style contingent claims prices.

[8] Jacka and Lynn [1992] also provide results on the up-connectedness of the exercise region for general contingent claims written on diffusions. For the one dimensional case they show that up-connectedness holds for smooth (discounted) payoffs $g(S,t)$ if the Black-Scholes operator $\mathcal{L}g(S,t)$ decreases in S for all $(S,t) \in \mathbb{R}_+ \times [0,T]$. The result in Proposition 39 does not require smoothness of the discounted payoff over its whole domain and allows some form of history-dependence, i.e., $g\left(S_t, S_{(\cdot)}, t\right) = \exp\left(-\int_0^t r(S_v, v)\, dv\right)(S_t - K)^+$.

increases). The combination of all these effects implies a decrease in the value of any given call exercise policy as time passes.[9] It follows that immediate exercise will be optimal at future dates if it is optimal today (i.e., $(S,t) \in \mathcal{E}$ implies $(S,t') \in \mathcal{E}$ for all $t' > t$).

Given the structure of the exercise set we can now specialize the results in Chapter 2. The EEP formula for the option price becomes

$$C(S,t;B(\cdot)) = c(S,t) + \pi(S,t,B(\cdot)) \tag{4.9}$$

where $c(S,t)$ is the European option value and $\pi(S,t,B(\cdot))$ is the early exercise premium given by

$$\pi(S,t,B(\cdot)) = \widetilde{E}_t\left[\int_t^T e^{-\int_t^v r(S_v,v)dv}\left(\delta(S_v,v)S_v - r(S_v,v)K\right)1_{\{S_v \geq B(v)\}}dv\right] \tag{4.10}$$

with 1_A as the indicator of the set A. And given that immediate exercise is optimal when $S_t = B(t)$ we obtain the recursive integral equation for the exercise boundary

$$B(t) - K = c(B(t),t) + \pi(B(t),t,B(\cdot)) \tag{4.11}$$

subject to the boundary condition $B(T_-) = \max\left\{K, \frac{r(T,B(T_-))}{\delta(T,B(T_-))}K\right\}$, where $B(T_-) \equiv \lim_{t\to T} B(t)$. As the interest rate and the dividend yield depend on the underlying price the limiting boundary value, as maturity approaches, solves a non-linear equation.

A price process satisfying the assumptions of this section is the Constant Elasticity of Variance (CEV) process

$$dS_t = S_t(r-\delta)dt + \sigma S_t^{\theta/2}d\widetilde{z}_t$$

where r, δ, θ are positive constants and $\theta \in [0,2]$. This model was initially introduced by Cox and Ross (1976). It has proved popular because it captures a prominent feature of the data, namely the negative correlation between returns and return volatility innovations. The model also includes the geometric Brownian motion process as a special case ($\theta = 2$). The conditional density function for the case $\theta \neq 2$ can be found in Cox [1975], [1996], Emanuel and MacBeth [1982] and Schroder [1989]. The boundary behavior of the process is analyzed by Davydov and Linetsky [2001]. Valuation formulas for American options written on a CEV price process were derived by Kim and Yu [1996] and studied by Detemple and Tian [2002].

Figure 4.2 graphs the call exercise boundary in the CEV model. The plot illustrates the positive impact of the elasticity of variance, θ, on the boundary.

[9] Decreasing time-monotonicity implies that the prospects of a put option improve as time passes. This positive impact on the put payoff is curtailed by the negative impact on the discount factor. The respective strengths of these opposing forces will determine the properties of the put boundary.

This follows because an increase in θ increases the local variance of the process along the boundary. As $\theta \to 2$ the boundary converges to the boundary in the GBM model.

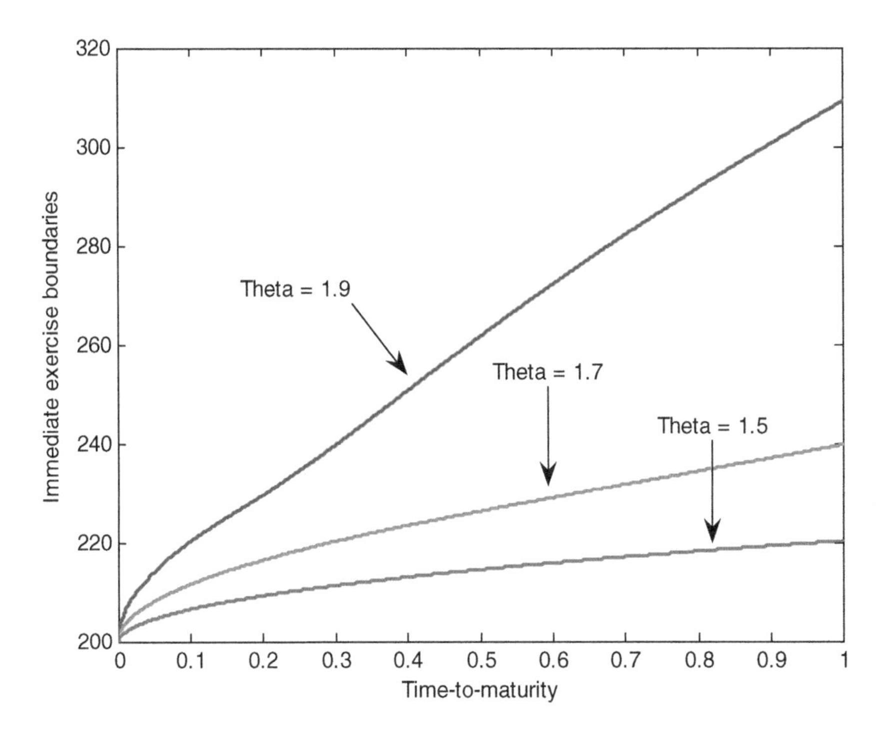

Figure 4.2: CEV model. Exercise boundary of an American call (y-axis) versus time to maturity (x-axis). The curves correspond to elasticities of variance $\theta = 1.9, 1.7$ and 1.5. The strike price is 100, the interest rate 6%, the dividend rate 3% and the volatility $\sigma = 0.2\sqrt{10}$. Computations use the integral equation method with 2,000 time steps.

4.7 Floating Strike Asian Options

Although floating strike Asian options have a complex payoff structure that depends on the strike process, they can sometimes be reduced to standard options written on a new underlying variable. This result, that follows from the change of numeraire approach of Geman, El Karoui and Rochet [1995], applies when the strike is computed using geometric averaging.

A floating strike Asian call option with geometric averaging pays off $(S - G)^+$ upon exercise where

$$G_t = \exp\left(\frac{1}{t}\int_0^t \log(S_u)\,du\right).$$

Assume that the underlying asset price S satisfies (4.1). Let $C^{fa}(S, G, t)$ denote the floating strike Asian option price. Taking the underlying asset as the new numeraire gives

$$C^{fa}(S, Y, t) = S_t \sup_{\tau \in \mathcal{S}_{t,T}} E_t^*\left[e^{-\delta(\tau - t)}(1 - Y_\tau)^+\right]$$

where $Y_\tau = G_\tau / S_\tau$ is the solution at time τ of the stochastic differential equation

$$dY_t = Y_t\left[\left(\delta - r - \frac{1}{t}\ln(Y_t)\right)dt + \sigma dz_t^*\right], \quad Y_0 = 1 \tag{4.12}$$

and z^* is a Brownian motion in the new numeraire (the underlying measure is Q^*, as defined in (2.23)). From this expression it follows that the floating strike Asian call with geometric averaging is equivalent to S_t put options, with unit strike, written on an asset that pays dividends at the rate $\delta_Y = r + \frac{1}{t}\ln(Y_t)$ in a financial market with interest rate δ. As the dividend rate is an increasing function of Y Proposition 39 (adapted to the put case) enables us to conclude that the exercise region is down-connected and has an upper boundary B_t^p.[10] Using this structure of the exercise set in combination with the EEP representation leads to the following formula,

Proposition 40 *Suppose that the underlying asset price follows the geometric Brownian motion process (4.1) and that the interest rate is constant. The value of an American floating strike Asian call option has the early exercise premium representation*

$$C^{fa}(S_t, G_t, t) = S_t\left[p(Y_t, t; \delta) + \pi^p(Y_t, t, B^p(\cdot); \delta)\right] \tag{4.13}$$

for $t \in [0, T]$, where $p(Y_t, t; \delta)$ represents the value of a European put option and $\pi^p(Y_t, t, B^p(\cdot); \delta)$ is the early exercise premium, in a financial market with interest rate δ and where the underlying asset price follows (4.12). Explicit formulas are

$$\begin{aligned} p(Y_t, t; \delta) &= e^{-\delta(T-t)}\left(N\left(\widehat{d}(Y_t, 1, t, T)\right)\right. \\ &\quad \left.-Y_t^{t/T} e^{q(t,T)} N\left(\widehat{d}(Y_t, 1, t, T) - \frac{v(t,T)}{T}\right)\right) \end{aligned} \tag{4.14}$$

[10] The dividend rate is also an increasing (decreasing) function of time if $Y_t < 1$ ($Y_t > 1$). As a result the solution of (4.12) fails to exhibit decreasing time-monotonicity. This prevents us from concluding that the exercise region is right-connected.

$$\begin{aligned}\pi^p\left(Y_t,t,B^p\left(\cdot\right);\delta\right) &= \int_t^T e^{-\delta(s-t)}\left(\delta N\left(\widehat{d}\left(Y_t,B^p_{s,}t,s\right)\right)\right. \\ &\left. -Y_t^{t/s}e^{q(t,s)}H\left(Y_t,B^p_s,t,s\right)\right)ds \end{aligned} \quad (4.15)$$

with

$$m\left(t,s\right)=-\frac{1}{2}\left(r-\delta+\frac{1}{2}\sigma^2\right)\left(s^2-t^2\right)$$

$$v\left(t,s\right)=\sigma\sqrt{\frac{s^3-t^3}{3}}$$

$$q\left(t,s\right)=\frac{m\left(t,s\right)}{s}+\frac{1}{2}\frac{v\left(t,s\right)^2}{s^2}$$

$$\begin{aligned}H\left(Y_t,B^p_t,t,s\right) &= \left(r+\widehat{e}\left(Y_t,1,t,s\right)\right)N\left(\widehat{d}\left(Y_t,B^p_s,t,s\right)-\frac{v\left(t,s\right)}{s}\right) \\ &-\frac{v\left(s,t\right)}{s^2}n\left(\widehat{d}\left(Y_t,B^p_s,t,s\right)-\frac{v\left(t,s\right)}{s}\right)\end{aligned}$$

$$\widehat{d}\left(Y_t,B,t,s\right)=\frac{s\log\left(B\right)-t\log\left(Y_t\right)-m\left(t,s\right)}{v\left(t,s\right)}$$

$$\widehat{e}\left(Y_t,1,t,s\right)=-\frac{v\left(t,s\right)}{s^2}\widehat{d}\left(Y_t,1,t,s\right)+\frac{v\left(t,s\right)^2}{s^3}.$$

The immediate exercise boundary solves the recursive integral equation

$$1-B^p_t=p\left(B^p_t,t;\delta\right)+\pi^p\left(B^p_t,t,B^p\left(\cdot\right);\delta\right)$$

subject to the boundary condition $B^p_{T_-}=\min\left\{1,\delta/\left(r+\frac{1}{T}\log\left(B^p_{T-}\right)\right)\right\}$. *At maturity* $B^p_T=1\geq B^p_{T_-}$.

The expressions for the components of the EEP formula in (4.14)-(4.15) reflect the fact that the underlying asset price is an exponential transform of a normally distributed random variable, i.e., $Y_t=\exp\left(x/t\right)$ where x is normally distributed with mean $m\left(0,t\right)$ and variance $v\left(0,t\right)^2$.

The analytical formulas in Proposition 40 were derived by Wu, Kwok and Yu [1999] using a partial differential equation approach.

4.8 American Forward Contracts

We complete this chapter with a study of American-style forward contracts, as described in section 3.5, in the standard model with constant coefficients. Let $V^f\left(S,t\right)$ be the value of the contract. The immediate exercise region, $\mathcal{E}^f\equiv\left\{\left(S,t\right)\in\mathbb{R}_+\times\left[0,T\right]:V^f\left(S,t\right)=S-K\right\}$, exhibits the following structure,[11]

[11] Exercise at maturity is not a matter of choice due to the obligatory nature of this contract. But given that $V^f(S,T)=S-K$ the set of prices for which this equality holds (i.e., the full space) is tagged on to the exercise region for times prior to maturity. This convention enables us to define the set $\mathcal{E}^f$ for all $t\in\left[0,T\right]$.

Proposition 41 *The immediate exercise region* $\mathcal{E}^f$ *has the following properties*
(i) $(S,t) \in \mathcal{E}$ *implies* $(S,s) \in \mathcal{E}$ *for all* $t \in [0,T]$ *and* $s \in [t,T]$.
(ii) $(S,t) \in \mathcal{E}$ *implies* $(\lambda S,t) \in \mathcal{E}$ *for all* $\lambda \geq 1$, *and for all* $t \in [0,T]$.
(iii) Suppose that $S < (r/\delta) K$. *Then* $(S,t) \notin \mathcal{E}$, *for all* $t \in [0,T)$.

The first two properties are identical to the corresponding ones for the call option contract. The third one differs in that it is only claimed that points (S,t) where the local gains $\delta S - rK < 0$ are negative belong to the continuation region. As will become clear the structure of the forward contract payoff implies that immediate exercise may well be optimal when $S < K$.

The counterpart of Proposition 32 is,

Proposition 42 *The price* $V^f(S,t)$ *of the American forward contract satisfies*
(i) $V^f(S,t)$ *is continuous on* $\mathbb{R}_+ \times [0,T]$.
(ii) $V^f(\cdot,t)$ *is non-decreasing and convex on* $\mathbb{R}_+$ *for all* $t \in [0,T]$.
(iii) $V^f(S,\cdot)$ *is non-increasing on* $[0,T]$ *for all* $S \in \mathbb{R}_+$.
(iv) $0 \leq \partial V^f(S,t)/\partial S \leq 1$ *on* $\mathbb{R}_+ \times [0,T]$; $\partial V^f(S,t)/\partial S = 1$ *for* $(S,t) \in \overset{o}{\mathcal{E}}$.

The arguments underlying these properties are similar to those employed to prove the corresponding results for the call option. As before convexity of the price (in (ii)) along with up-connectedness of the exercise region (property (ii) in Proposition 41) imply the bounds on the slope of the price in (iv). Note also that continuity of the price ensures that the exercise region is a closed set. It follows that the exercise boundary (i.e., the boundary of the immediate exercise region), $B^f \equiv \left\{B_t^f : t \in [0,T]\right\}$, belongs to $\mathcal{E}^f$. Up-connectedness of the exercise region implies that $B_t^f = \inf\left\{S : (S,t) \in \mathcal{E}^f\right\}$. Again, the exercise boundary is unique and depends on time alone. It has the following behavior,

Proposition 43 *The boundary* B^f *of the immediate exercise region of the American forward contract is continuous, non-increasing with respect to time and has limiting values* $\lim_{t\uparrow T} B_t^f = (r/\delta) K$ *and* $\lim_{T-t\uparrow\infty} B_t^f = B_{-\infty}^f \equiv (r/\delta) K (b+f) / (b+f-\sigma^2)$ *where* $b \equiv \delta - r + \frac{1}{2}\sigma^2$ *and* $f \equiv \left(b^2 + 2r\sigma^2\right)^{1/2}$. *At maturity* $B_T = 0$.

These properties are motivated along the same lines as for the call option. Note that the limiting values of the exercise boundary are those of the call boundary with the substitution of $(r/\delta) K$ in place of $\max\{K, (r/\delta) K\}$. This modification follows because the contract does not provide the choice of exercising, or not, at maturity. As a result the date T exercise threshold for the option, equal to K, ceases to play a role in the case of a forward contract.

To get a characterization of the price and the exercise boundary we specialize our earlier representation results.

Proposition 44 *Consider an American forward contract with payoff* $Y = S - K$ *and delivery date at or before* T. *Its value, in the market with GBM price process and constant interest rate, has the early exercise premium representation*

$$V^f(S_t,t) = v^f(S_t,t) + \pi\left(S_t,t,B^f(\cdot)\right), \quad \text{for } t \in [0,T],$$

where $v^f(S,t)$ *is the European forward contract value*

$$v^f(S_t,t) = S_t \exp(-\delta(T-t)) - K \exp(-r(T-t))$$

and $\pi_t^f \equiv \pi^f\left(S_t,t,B^f(\cdot)\right)$ *is the early exercise premium*

$$\begin{aligned}\pi_t^f &= \int_t^T \left(\delta S_t e^{-\delta(s-t)} N\left(d\left(S_t,B_s^f,s-t\right)\right)\right. \\ &\quad \left. -rKe^{-r(s-t)} N\left(d_1\left(S_t,B_s^f,s-t\right)\right)\right) ds\end{aligned}$$

with $d\left(S_t,B_s^f,s-t\right)$ *and* $d_1\left(S_t,B_s^f,s-t\right)$ *as defined in (4.4), (4.5). The immediate exercise boundary* B^f *solves the recursive integral equation* $B_t^f - K = v^f\left(B_t^f,t\right) + \pi^f\left(B_t^f,t,B^f(\cdot)\right)$ *subject to the limiting condition* $B_{T\_}^f = (r/\delta)K$.

As for an American call the forward contract with an early exercise provision derives its premium from the excess of dividends over the interest rate loss incurred. As a result its early exercise premium has the same structure as the one for the call option. There are two main differences with respect to the call price formula. The first one is that the European claim value appearing in the formula, $v^f(S_t,t)$, is the value of a standard (European) forward contract. The second is that the exercise boundaries differ. The mandatory delivery provision at the maturity date ensures that the forward price differs from the price of a call. The optimal exercise boundary reflects this basic difference in valuation. As the European-style forward contract may produce a negative payoff the American forward boundary is bounded above by the call boundary, $B_t^f \leq B_t$ for all $t \in [0,T]$.

Figure 4.3 illustrates a typical exercise boundary for the American forward contract. In the example displayed the limiting boundary value is $B_{T\_}^f = (r/\delta)K = 66.67$. For comparison purposes note that the call boundary converges to $B_{T\_} = \max\{1,r/\delta\}K = 100$. As time to maturity becomes large the boundaries of the forward contract and of the call converge to the common value $B_{-\infty}^f = B_{-\infty}$.

4.9 Appendix: Proofs

Proof of Proposition 31: Recall that $\mathcal{S}_{t,T}$ denotes the set of stopping times of the Brownian filtration with values in $[t,T]$. Using the GBM structure of the stock price process we can rewrite the stopping time problem as

$$C(S,t;T) = \sup_{\tau \in \mathcal{S}_{t,T}} \widetilde{E}\left[e^{-r(\tau-t)}(SN_{t,\tau}-K)^+\right]$$

where

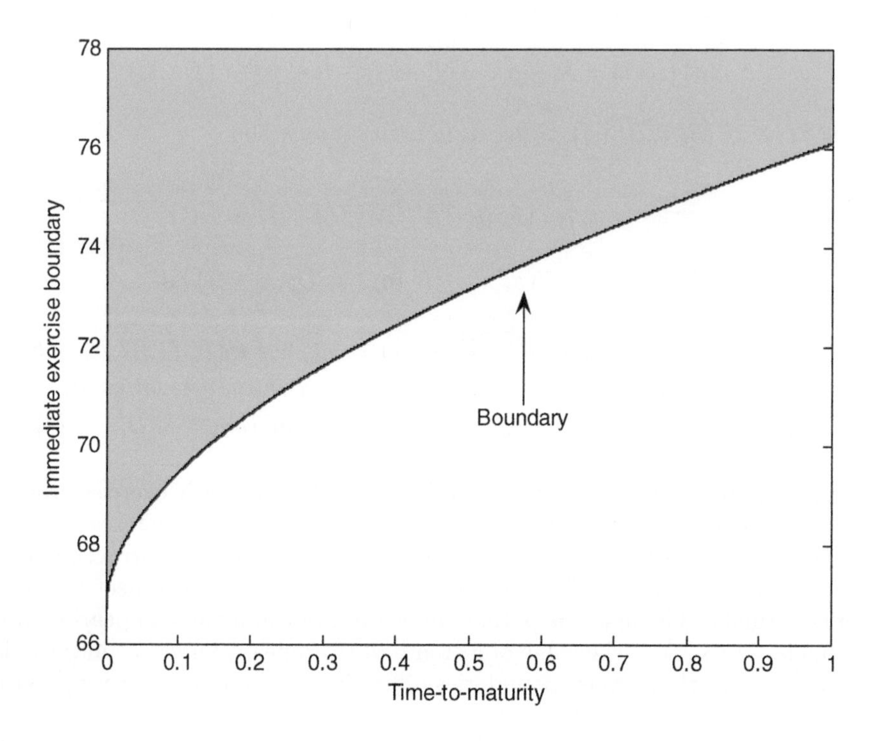

Figure 4.3: American Forward Contract: Immediate exercise boundary of an American Forward (y-axis) versus time to maturity (x-axis). The strike price is 100, the interest rate 4%, the dividend rate 6%, and the volatility 0.20. Maturity is 1 year. Computations use the integral equation method with 200 time steps.

$$N_{t,\tau} \equiv \exp\left(\left(r - \delta - \frac{1}{2}\sigma^2\right)(\tau - t) + \sigma\left(\tilde{z}_\tau - \tilde{z}_t\right)\right).$$

Note that the expectation above is unconditional due to the independence of Brownian increments. Moreover, as the filtration generated by the Brownian increment $\tilde{z}_s - \tilde{z}_t$, in the interval $[t, T]$, is identical to the Brownian filtration in $[0, T - t]$ we can replace the set of stopping times $\mathcal{S}_{t,T}$ by $\mathcal{S}_{0,T-t}$ in the optimization and write

$$C(S, t; T) = \sup_{\tau \in \mathcal{S}_{0,T-t}} \widetilde{E}\left[e^{-r\tau}\left(SN_\tau - K\right)^+\right]$$

where $N_\tau \equiv N_{0,\tau} = \exp\left(\left(r - \delta - \frac{1}{2}\sigma^2\right)\tau + \sigma\tilde{z}_\tau\right)$.

(i) Fix S and let $s \geq t$. The last expression above shows that the value at time s of an option with maturity date T is the same as the value at t of an option with maturity date $T-s+t$ and same strike price, i.e., $C(S,t;T-s+t) = C(S,s;T)$ (this follows immediately from $\mathcal{S}_{0,T-s+t-t} = \mathcal{S}_{0,T-s}$). Furthermore, as $\mathcal{S}_{0,T-s} \subset \mathcal{S}_{0,T-t}$, we obtain $C(S,t;T) \geq C(S,t;T-s+t) = C(S,s;T)$. By assumption, immediate exercise is optimal at t. Thus $(S-K)^{+} \geq C(S,s;T)$. If immediate exercise is suboptimal at (S,s) we also have $C(S,s;T) > (S-K)^{+}$, which contradicts the previous inequality.

(ii) Take $S^1 > S^2$ and suppose that $(S^2,t) \in \mathcal{E}$ while $(S^1,t) \notin \mathcal{E}$. Let τ_1 be the optimal stopping time at (S^1,t). We have the following sequence of relations

$$\begin{aligned}
C(S^1,t) &= \widetilde{E}\left[e^{-r\tau_1}\left(S^1 N_{\tau_1} - K\right)^{+}\right] \\
&= \widetilde{E}\left[e^{-r\tau_1}\left(S^2 N_{\tau_1} + \left(S^1 - S^2\right) N_{\tau_1} - K\right)^{+}\right] \\
&\leq \widetilde{E}\left[e^{-r\tau_1}(S^2 N_{\tau_1} - K)^{+}\right] + \widetilde{E}\left[e^{-r\tau_1}\left(S^1 - S^2\right) N_{\tau_1}\right] \\
&\leq C\left(S^2,t\right) + \left(S^1 - S^2\right)\widetilde{E}\left[e^{-r\tau_1} N_{\tau_1}\right] \\
&\leq C\left(S^2,t\right) + \left(S^1 - S^2\right) \\
&\leq S^2 - K + \left(S^1 - S^2\right) = S^1 - K.
\end{aligned}$$

The equality in the first line above results from the optimality of τ_1 at (S^1,t). The inequality in the third line uses $(a+b)^{+} \leq a^{+} + b^{+}$. The fourth line follows from the suboptimality of τ_1 at (S^2,t), the fifth from $S^1 - S^2 > 0$ and the supermartingale property of S, and the last line from the assumption that immediate exercise at (S^2,t) is optimal. Hence $C(S^1,t) \leq S^1 - K$, which contradicts the assumption that immediate exercise is suboptimal at (S^1,t). We conclude that $(S^1,t) \in \mathcal{E}$. This corresponds to the property stated, with $\lambda \equiv S^1/S^2 > 1$. The case $\lambda = 1$ is trivial.

(iii) Suppose that $0 < S < K$. As $P[S_v > K] > 0$, for some $v \in [t,T)$, immediate exercise is a suboptimal policy. Suppose that $K \leq S < (r/\delta)K$ and assume that immediate exercise is optimal, i.e., $C(S,t) = S - K$. Consider the portfolio consisting of 1 call option, 1 share of the stock held short and K dollars invested at the riskfree rate. Define the stopping time

$$\tau = \inf\left\{v \in [t,T] : S_v = (r/\delta)K\right\}$$

or $\tau = T$ if no such time exists. Suppose that we liquidate this portfolio at τ. The cash flows generated by this policy are

	time t	time τ
Buy Call	$-(S-K)$	$\max\left\{(S_\tau - K)^{+}, C(S_\tau,\tau)\right\}$
Sell Stock	$+S$	$-S_\tau - \int_t^\tau e^{r(\tau-v)}\delta S_v dv$
Invest K	$-K$	$Ke^{r(\tau-t)} = K + \int_t^\tau e^{r(\tau-v)} rK dv$
Total	0	$\max\left\{(K-S_\tau)^{+}, C(S_\tau,\tau) + K - S_\tau\right\}$ $- \int_t^\tau e^{r(\tau-v)}(\delta S_v - rK) dv$

Given that $rK - \delta S_v > 0$ for all $v < \tau$ we have constructed an arbitrage portfolio. As the existence of an equilibrium implies the absence of arbitrage opportunities it must be the case that $C(S,t) > S - K$, i.e., immediate exercise is a suboptimal policy.

(iv) Immediate exercise is optimal at the maturity date if and only if $S \geq K$. ■

Proof of Proposition 32:

(i) This result can be proved by using the continuity of the option payoff function and the continuity of the flow of the stochastic differential equation (4.1) relative to the initial values. Precise arguments can be found in Jaillet, Lamberton and Lapeyre [1990] (see their Proposition 2.2 and the corresponding proof). Continuity with respect to the first argument, the underlying asset price, also follows from the proof of property (iv) below.

(ii) Suppose $S^1 > S^2$ and define $S^\lambda \equiv \lambda S^1 + (1-\lambda) S^2$. As the call option payoff is non-decreasing and convex with respect to the asset price, we have, for any $\tau \in \mathcal{S}_{0,T-t}$,

$$(S^1 N_\tau - K)^+ \geq (S^2 N_\tau - K)^+$$

$$\lambda (S^1 N_\tau - K)^+ + (1-\lambda)(S^2 N_\tau - K)^+ \geq (S^\lambda N_\tau - K)^+ .$$

Taking present values and evaluating the payoffs at the optimal exercise policies associated with the two prices establishes that the option price is non-decreasing and convex.

(iii) This is a straightforward counterpart of Proposition 31 (i).

(iv) Consider (S^1,t) and (S^2,t) such that $S^1 > S^2$. For any stopping time $\tau \in \mathcal{S}_{0,T-t}$ we have

$$0 \leq (S^1_\tau - K)^+ - (S^2_\tau - K)^+ \leq S^1_\tau - S^2_\tau = (S^1 - S^2) N_\tau .$$

In particular this holds for the optimal stopping time τ_1 associated with (S^1,t). Hence, we can write

$$\begin{aligned}
0 &\leq C(S^1,t) - C(S^2,t) \\
&= \widetilde{E}\left[e^{-r\tau_1}(S^1 N_{\tau_1} - K)^+\right] - \widetilde{E}\left[e^{-r\tau_2}(S^2 N_{\tau_2} - K)^+\right] \\
&\leq \widetilde{E}\left[e^{-r\tau_1}(S^1 N_{\tau_1} - K)^+\right] - \widetilde{E}\left[e^{-r\tau_1}(S^2 N_{\tau_1} - K)^+\right] \\
&\qquad\qquad\qquad\qquad \text{(suboptimality of } \tau_1 \text{at } (S^2,t)\text{)} \\
&\leq \widetilde{E}\left[e^{-r\tau_1}(S^1 - S^2) N_{\tau_1}\right] \\
&= (S^1 - S^2)\, \widetilde{E}\left[e^{-r\tau_1} N_{\tau_1}\right] \\
&\leq (S^1 - S^2)
\end{aligned}$$

where the last inequality follows because $S^1 - S^2 > 0$ and because the discounted price of a dividend-paying asset is a Q-supermartingale. Dividing both sides by

$S^1 - S^2$ proves the statement (this argument also establishes the continuity of the option price with respect to S). ■

Proof of Proposition 33: Property (iii) of Proposition 32 implies that the boundary is non-increasing in time. This feature combined with the definition of the boundary implies that $B(t_-) \geq B(t) \geq B(t_+)$ where $B(t_-)$ and $B(t_+)$ are the left and right limits of the boundary at t.

Suppose now that $(S, v) \in \mathcal{E}$ for all $v > t$. As $\mathcal{E}$ is a closed set we have $\lim_{v \downarrow t} (S, v) \in \mathcal{E}$. Thus, $(S, t_+) \in \mathcal{E}$. In particular $(B(t_+), t_+) \in \mathcal{E}$ (take $S = B(t_+)$ in the previous argument). As $t_+ = t$, up-connectedness of the exercise region implies $B(t_+) \geq B(t)$. This combined with the reverse inequality (see above) establishes that $B(t_+) = B(t)$ (i.e., the exercise boundary is right continuous).

Left continuity of the boundary can be proved using the arguments in Jacka (1991) adapted to the call option case. Accordingly, suppose that $B(t_-) > B(t)$. Then there exists an open set $(B(t), B(t_-)) \equiv O$ such that $O \subset \mathcal{E}$ and for all $x \in O$ there exists a sequence $\{x^n\} \in \mathcal{C}$ with $x = \lim_{s \uparrow t} x^n$. Let us now use the fact that the option price is $\mathcal{C}^{2,1}(\mathbb{R}_+ \times [0, T])$ and satisfies the fundamental PDE (see Merton [1973a], Jacka [1991] and Jaillet, Lamberton and Lapeyre [1990])

$$C_t(S,t) + C_s(S,t) S(r-\delta) + \frac{1}{2} C_{ss}(S,t) S^2 \sigma^2 - rC(S,t) = 0$$

for all $(S, t) \in \mathcal{C}$. Evaluating this equation along the sequence $\{x^n\}$ shows that, in the limit (at the point $x = \lim_{s \uparrow t} x^n$ in the closure of the continuation region),

$$\frac{1}{2} C_{ss}(x,t) x^2 \sigma^2 = rC(x,t) - C_t(x,t) - C_s(x,t) x (r-\delta) = \delta x - rK.$$

As $B(t) \geq (r/\delta) K$ we conclude that $C_{ss}(x,t) \geq 0$ for all $x \in O$ and $C_{ss}(x,t) \geq \widehat{\varepsilon} > 0$ for x in a subset of O with positive Lebesgue measure. Integration by parts now gives

$$C(B(t_-), t) - C(B(t), t) - (B(t_-) - B(t)) = \int_{B(t)}^{B(t_-)} \int_{B(t)}^{y} C_{ss}(x,t)\, dx dy.$$

By definition of $\mathcal{E}$ the left-hand side of this expression is null. The right hand side, on the other hand, is no less than ε, for some $\varepsilon > 0$. This contradiction implies that the set O must have null Lebesgue measure, in other words that $B(t_-) = B(t)$.

To show that $B_{T_-} = \max\{K, (r/\delta) K\}$ note that property (iii) of Proposition 31 gives the lower bound $B_{T_-} \geq \max\{K, (r/\delta) K\}$. Suppose that $B_{T_-} > \max\{K, (r/\delta) K\}$ and consider a point $x \in (\max\{K, (r/\delta) K\}, B_{T_-})$. Then $C(x,t) > x - K$ for all $t \in [0, T)$. Thus, $\lim_{t \uparrow T} C(x,t) = C(x, T_-) > x - K$ (in fact $C(x, T_-) \geq x - K + \varepsilon$ for some $\varepsilon > 0$ when x is sufficiently below B_{T_-}). But, by the continuity of the price function, we also have $C(x, T_-) =$

$C(x,T) = x - K$ (as $x > K$), a contradiction. We conclude that the interval $\left(\max\{K, (r/\delta)K\}, B_{T_-}\right)$ has null Lebesgue measure, i.e., that $B_{T_-} = \max\{K, (r/\delta)K\}$.

To complete the proof we need to identify the boundary $B_{-\infty}$. This limiting value can be obtained from the EEP formula (4.2) in Theorem 34 taking $T = \infty$ (see Corollary 35). ■

Proof of Theorem 34: Proposition 31 implies $B \geq \max\{K, (r/\delta)K\}$. Hence $Y = (S-K)^+$ equals $S - K$ in $\mathcal{E}$. It follows that $dY_t = S_t\left[(r-\delta)dt + \sigma d\tilde{z}_t\right]$ in the exercise region, i.e., $dA_t^Y = S_t(r-\delta)dt$ on $\{S_t \geq B_t\}$. Theorem 21 then implies

$$\begin{aligned} C(S_t,t) &= c(S_t,t) + \widetilde{E}_t\left[\int_t^T e^{-r(v-t)}\left(r(S_v - K) - (r-\delta)S_v\right)1_{\{S_v \geq B_v\}}dv\right] \\ &= c(S_t,t) + \widetilde{E}_t\left[\int_t^T e^{-r(v-t)}\left(\delta S_v - rK\right)1_{\{S_v \geq B_v\}}dv\right]. \end{aligned}$$

Under the GBM assumption the expectation above can be computed explicitly. This leads to (4.2)-(4.3). The recursive equation for the optimal exercise boundary is obtained by imposing the boundary condition $C(B,t) = B - K$. ■

Proof of Corollary 35: When $T \uparrow \infty$ the immediate exercise boundary becomes time independent: $B = B_\infty$. It follows that

$$d(B_\infty, B_\infty, s-t) = \frac{\left(r - \delta + \frac{1}{2}\sigma^2\right)(s-t)}{\sigma\sqrt{s-t}}$$

and $d_1(B_\infty, B_\infty, s-t) = d(B_\infty, B_\infty, s-t) - \sigma\sqrt{s-t}$ are independent of B_∞. As the European call option value converges to 0 the recursive equation (4.6) becomes linear in B_∞ and has the unique solution $B_\infty = K(b+f)/\left(b+f-\sigma^2\right)$. The value of the option then follows from (4.2): the early exercise premium simplifies to $(B_\infty - K)(S/B_\infty)^{2\alpha/\sigma^2}$. ■

Proof of Proposition 36: The property is a special case of Theorem 26. ■

Proof of Theorem 37: In the exercise region the EEP representation states $S - K = c(S,t) + \pi(S,t,B(\cdot))$ for $S \geq B_t$. Twice differentiating with respect to S, and using the relation

$$e^{-r(s-t)}n\left(d(S,B_s,s-t) - \sigma\sqrt{s-t}\right) = e^{-\delta(s-t)}n\left(d(S,B_s,s-t)\right)\frac{S}{B_s},$$

gives

$$0 = c_{ss}(S,t) + \pi_{ss}(S,t,B(\cdot))$$

where

$$c_{ss}(S,t) = e^{-\delta(T-t)}n\left(d(S,K,T-t)\right)\frac{1}{S\sigma\sqrt{T-t}}$$

and

$$\begin{aligned}\pi_{ss}(S,t,B(\cdot)) &= \int_t^T \delta e^{-\delta(s-t)} n\left(d\left(S,B_s,s-t\right)\right) \frac{1}{S\sigma\sqrt{s-t}} ds \\ &\quad - \int_t^T e^{-\delta(s-t)} n\left(d\left(S,B_s,s-t\right)\right) d\left(S,B_s,s-t\right) \\ &\quad \times \left(\frac{\delta B_s - rK}{B_s}\right) \frac{1}{S\sigma^2(s-t)} ds.\end{aligned}$$

Given that the exercise set is closed we can take the limit as $S \to B_t$ to obtain

$$\begin{aligned}0 &= e^{-\delta(T-t)} n\left(d\left(B_t,K,T-t\right)\right) \frac{1}{\sigma\sqrt{T-t}} \\ &\quad + \int_t^T \delta e^{-\delta(s-t)} n\left(d\left(B_t,B_s,s-t\right)\right) \frac{1}{\sigma\sqrt{s-t}} ds \\ &\quad - \int_t^T e^{-\delta(s-t)} n\left(d\left(B_t,B_s,s-t\right)\right) d\left(B_t,B_s,s-t\right) \\ &\quad \times \left(\frac{\delta B_s - rK}{B_s}\right) \frac{1}{\sigma^2(s-t)} ds,\end{aligned}$$

and substituting

$$d\left(B_t,B_s,s-t\right) = \left(\log\left(B_t/B_s\right) + \left(r - \delta + \frac{1}{2}\sigma^2\right)(s-t)\right) / \sigma\sqrt{s-t}$$

we can write

$$\begin{aligned}&\log\left(B_t\right) \int_t^T e^{-\delta(s-t)} n\left(d\left(B_t,B_s,s-t\right)\right) \left(\frac{\delta B_s - rK}{B_s}\right) \frac{1}{\sigma^3 (s-t)^{3/2}} ds \\ &= e^{-\delta(T-t)} n\left(d\left(B_t,K,T-t\right)\right) \frac{1}{\sigma\sqrt{T-t}} \\ &\quad + \int_t^T \delta e^{-\delta(s-t)} n\left(d\left(B_t,B_s,s-t\right)\right) \frac{1}{\sigma\sqrt{s-t}} ds \\ &\quad - \int_t^T e^{-\delta(s-t)} n\left(d\left(B_t,B_s,s-t\right)\right) \frac{-\log\left(B_s\right) + \left(r-\delta+\frac{1}{2}\sigma^2\right)(s-t)}{\sigma\sqrt{s-t}} \\ &\quad \times \left(\frac{\delta B_s - rK}{B_s}\right) \frac{1}{\sigma^2(s-t)} ds.\end{aligned}$$

Solving for $\log(B_t)$ and using the definitions of $I(B_t, B(\cdot))$ and $J(B_t, B(\cdot))$ establishes the statements in the theorem. ■

Proof of Proposition 38: Taking the derivative of (4.2) and simplifying leads to the expression announced. Continuity follows from the continuity of $h(S_t, t)$ and $d(S_t, B_s, s-t)$ with respect to S, and the continuity of $N(x)$ and $n(x)$ with respect to their argument x. In the exercise region immediate exercise is optimal. It follows that $H(S,t) = 1$ for all $S > B_t, t \in [t,T)$. Continuity of the hedge ratio implies that $H(S,t) = 1$ on the boundary $S = B_t$. ■

Proof of Proposition 39: We first prove the existence of a non-increasing right-continuous function $B = \{B(t) : t \in [0,T]\}$ such that $\mathcal{E} = \{(S,t) \in \mathbb{R}_+ \times [0,T] : S \geq B(t)\}$. Our next two lemmas are instrumental,

Lemma 45 *Consider a claim with payoff* $0 \leq u(S) \leq S$ *and price* $V(S,t)$. *Suppose that*

(i) $u(S)$ *is increasing, and*

(ii) $u(S') - u(S) \leq S' - S$ *for all* $S' \geq S$.

Then, $V(S',t) - V(S,t) \leq S' - S$ *for all* $S' \geq S$.

Lemma 45 bounds the variation in the price function for a given change in the underlying price. The conditions on the payoff are standard: condition (ii) bounds the slope of the payoff by 1; both conditions hold for the call option payoff. Put-call symmetry ensures that put options are also covered by these conditions (see Schroder [1999] or Detemple [2001]).

Proof of Lemma 45: Consider two initial prices, S and S', such that $S' > S$. Let τ be the optimal stopping time at (S',t), i.e., $V(S',t) = \widetilde{E}_t\left[R'_{t,\tau}u(S'_\tau)\right]$ where $R'_{t,\tau} = \exp\left(-\int_t^\tau r(S'_u,u)\,du\right)$. As τ is suboptimal at (S,t) we have $V(S,t) \geq \widetilde{E}_t\left[R_{t,\tau}u(S_\tau)\right]$ with $R_{t,\tau} = \exp\left(-\int_t^\tau r(S_u,u)\,du\right)$. Thus,

$$\begin{aligned}
\Delta &\equiv V(S',t) - V(S,t) \\
&\leq \widetilde{E}_t\left[R'_{t,\tau}u(S'_\tau)\right] - \widetilde{E}_t\left[R_{t,\tau}u(S_\tau)\right] \\
&= \widetilde{E}_t\left[R'_{t,\tau}\left[u(S'_\tau) - u(S_\tau)\right]\right] + \widetilde{E}_t\left[\left(R'_{t,\tau} - R_{t,\tau}\right)u(S_\tau)\right]. \qquad (4.16)
\end{aligned}$$

The Comparison Theorem for solutions of Stochastic Differential Equations (Karatzas-Shreve [1988], Proposition 2.18) implies $Q\left[S_s \leq S'_s, \text{for all } s \in [t,T]\right] = 1$ (Q is the risk neutral measure). An approximation argument then shows that $Q\left[S_\tau \leq S'_\tau\right] = 1$ for all $\tau \in \mathcal{S}_{t,T}$.

As $R_{t,u}S_u + \int_t^u R_{t,v}\delta(S_v,v)\,S_v dv$ is a Q-martingale, Doob's optional sampling theorem gives

$$S_t = \widetilde{E}_t\left[R_{t,\tau}S_\tau\right] + \widetilde{E}_t\left[\int_t^\tau R_{t,v}\delta(S_v,v)\,S_v dv\right] \qquad (4.17)$$

$$S'_t = \widetilde{E}_t\left[R'_{t,\tau}S'_\tau\right] + \widetilde{E}_t\left[\int_t^\tau R'_{t,v}\delta(S'_v,v)\,S'_v dv\right]. \qquad (4.18)$$

Let $I \equiv \widetilde{E}_t \left[R'_{t,\tau} \left[u\left(S'_\tau\right) - u\left(S_\tau\right) \right] \right]$. We can write

$$\begin{aligned} I &\leq \widetilde{E}_t \left[R'_{t,\tau} \left(S'_\tau - S_\tau \right) \right] \\ &= S' - \widetilde{E}_t \left[\int_t^\tau R'_{t,v} \delta\left(S'_v, v\right) S'_v dv \right] - \widetilde{E}_t \left[R'_{t,\tau} S_\tau \right] \\ &= S' - \widetilde{E}_t \left[R_{t,\tau} S_\tau \right] - \widetilde{E}_t \left[\int_t^\tau R'_{t,v} \delta\left(S'_v, v\right) S'_v dv \right] + \widetilde{E}_t \left[R_{t,\tau} S_\tau \right] - \widetilde{E}_t \left[R'_{t,\tau} S_\tau \right] \\ &= S' - S - \widetilde{E}_t \left[\left(R'_{t,\tau} - R_{t,\tau} \right) S_\tau \right] \\ &\quad - \widetilde{E}_t \left[\int_t^\tau \left(R'_{t,\tau} \delta\left(S'_v, v\right) S'_v - R_{t,\tau} \delta\left(S_v, v\right) S_v \right) dv \right], \end{aligned}$$

where the first line follows from assumption (ii) in the lemma and the second and fourth lines use (4.17)-(4.18). Substituting this inequality in (4.16) and collecting terms gives the upper bound

$$\begin{aligned} \Delta &\leq S' - S - \widetilde{E}_t \left[\left(R'_{t,\tau} - R_{t,\tau} \right) \left(S_\tau - u\left(S_\tau\right) \right) \right] \\ &\quad - \widetilde{E}_t \left[\int_t^\tau \left(R'_{t,\tau} \delta\left(S'_v, v\right) S'_v - R_{t,\tau} \delta\left(S_v, v\right) S_v \right) dv \right]. \end{aligned} \tag{4.19}$$

Given that $S'_u \geq S_u$, for all $u \in [t, T]$ and $r(S, t)$ is decreasing in S, we have $r\left(S'_u, u\right) \leq r\left(S_u, u\right)$. Hence $\exp\left(-\int_t^\tau r\left(S'_u, u\right) du\right) \geq \exp\left(-\int_t^\tau r\left(S_u, u\right) du\right)$. As $S_\tau \geq u\left(S_\tau\right)$ we obtain

$$\widetilde{E}_t \left[\left(R'_{t,\tau} - R_{t,\tau} \right) \left(S_\tau - u\left(S_\tau\right) \right) \right] \geq 0. \tag{4.20}$$

On the other hand, $\delta\left(S'_v, v\right) S'_v \geq \delta\left(S_v, v\right) S_v$ by assumption and $R'_{t,\tau} \geq R_{t,\tau}$, so that

$$\widetilde{E}_t \left[\int_t^\tau \left(R'_{t,\tau} \delta\left(S'_v, v\right) S'_v - R_{t,\tau} \delta\left(S_v, v\right) S_v \right) dv \right] \geq 0. \tag{4.21}$$

Combining (4.20)-(4.21) with (4.19) gives $V(S', t) - V(S, t) \leq S' - S$, as claimed. ■

Lemma 46 *Consider a claim with non-decreasing payoff $u(S)$ such that $0 \leq u(S) \leq S$ and price $V(S, t)$. Suppose that the solution of (4.8) satisfies decreasing time-monotonicity. Then $V(S, t; T - (t' - t)) \geq V(S, t'; T)$ for all $t' \in [t, T]$.*

This lemma shows that the value of a claim with a fixed time to maturity cannot improve with the passage of time. This result can be traced to the depreciation, over time, of the discounted payoffs that can be achieved. The increasing structure of the payoff, the decreasing time-monotonicity and the properties of the interest rate combine to produce the property.

Proof of Lemma 46: Recall that the filtration generated by the Brownian process $\widetilde{z}_v - \widetilde{z}_t$ in $[t, T + t - t']$ is the same as the filtration generated by $\widetilde{z}_v - \widetilde{z}_{t'}$

in $[t', T]$. Let $U \equiv T - t'$ be the length of these time intervals and let $\widehat{z} = \{\widehat{z}_v : v \in [0, U]\}$ denote the Brownian motion that carries the common filtration. The set of stopping times of this filtration is $\widehat{\mathcal{S}}_{0,U}$.

With these definitions we can write

$$V(S, t; T - (t' - t)) = \sup_{\tau \in \mathcal{S}_{t,T-(t'-t)}} \widetilde{E}_t \left[R_{t,\tau} u \left(S_\tau^{(S,t)} \right) \right] = \sup_{\widehat{\tau} \in \widehat{\mathcal{S}}_{0,U}} \widehat{E} \left[\widehat{R}_{0,\widehat{\tau}} u \left(Z_{\widehat{\tau}} \right) \right].$$

The expression after the first equality is the usual value of the American claim where $S^{(S,t)}$ solves (4.8) subject to the initial condition $S_t = S$. To get the second equality we use the fact that the Brownian motion $\widehat{z}$ generates the same filtration and is such that $\widehat{z}_0 = 0$. The expectation $\widehat{E}$ is taken relative to this Brownian motion and the corresponding measure $\widehat{P}$. The process $Z = \{Z_v : v \in [0, U]\}$ solves

$$\frac{dZ_v}{Z_v} = \left(r\left(Z_v, v + t\right) - \delta\left(Z_v, v + t\right)\right) dv + \sigma\left(Z_v, v + t\right) d\widehat{z}_v; \qquad Z_0 = S$$

and the discount factor is

$$\widehat{R}_{0,\tau} = \exp\left(- \int_t^\tau r\left(S_v^{(S,t)}, v \right) dv \right) = \exp\left(- \int_0^{\widehat{\tau}} r\left(Z_v, v + t\right) dv \right).$$

Let $h \equiv t' - t$. Similar steps show that

$$V(S, t'; T) = \sup_{\widehat{\tau} \in \widehat{\mathcal{S}}_{0,U}} \widehat{E} \left[\widehat{R}_{0,\widehat{\tau}}^h u \left(Z_{\widehat{\tau}}^h \right) \right]$$

where $Z^h = \left\{ Z_v^h : v \in [0, U] \right\}$ solves the time-shifted equation

$$\frac{dZ_v^h}{Z_v^h} = \left(r\left(Z_v^h, v + h + t\right) - \delta\left(Z_v^h, v + h + t\right)\right) dv + \sigma\left(Z_v^h, v + h + t\right) d\widehat{z}_v; \; Z_0 = S$$

and

$$\widehat{R}_{0,\widehat{\tau}}^h = \exp\left(- \int_{t'}^\tau r\left(S_v^{(S,t')}, v \right) dv \right) = \exp\left(- \int_0^{\widehat{\tau}} r\left(Z_v^h, v + h + t\right) dv \right).$$

With these transformations the value functions become easy to compare. Decreasing time-monotonicity implies that $Z_v \geq Z_v^h$ for all $v \in [0, U]$ $(\widehat{P} - a.s.)$ and, as the payoff is a non-decreasing function, we obtain $u(Z_v) \geq u\left(Z_v^h\right)$ for all $v \in [0, U]$ $(\widehat{P} - a.s.)$. The properties of the interest rate (decreasing in the first argument and increasing in the second argument) also enable us to conclude that $\widehat{R}_{0,\widehat{\tau}}^h \leq \widehat{R}_{0,\widehat{\tau}}$. The lemma follows. ■

Proof of Proposition 39 (continued): To show that the exercise region is up-connected note that the call payoff satisfies the conditions of Lemma 45. Thus, $C(S', t) \leq C(S, t) + S' - S$ for $S' \geq S$. Suppose then that immediate exercise is optimal at (S, t) but not at (S', t). The optimality of exercise at (S, t), combined with the inequality above, implies $C(S', t) \leq S - K + S' - S = S' - K$. This contradicts the suboptimality of exercise at (S', t). We conclude that immediate exercise must also be optimal at (S', t). As this holds for any $(S, t) \in \mathcal{E}$ the exercise region is up-connected. The existence of a unique exercise boundary follows.

To show that the boundary is non-increasing (i.e., that the exercise region is right-connected) we proceed as follows. Suppose $(S, t) \in \mathcal{E}$ and $t' > t$. We want to show $(S, t') \in \mathcal{E}$. Suppose not, i.e., suppose that $C(S, t'; T) > (S - K)^+$. In this case we have

$$\begin{aligned} C(S,t;T) &\geq C(S,t;T-(t'-t)) && \text{(shorter maturity option)} \\ &\geq C(S,t';T) && \text{(decreasing time-monotonicity)} \\ &> (S-K)^+ && \text{(suboptimality of exercise at } t'\text{).} \end{aligned}$$

The inequality in the second line follows from Lemma 46. The last inequality is the assumption that immediate exercise is suboptimal at t'. Given that $(S, t) \in \mathcal{E}$ we have a contradiction.

The same arguments as in the proof of Proposition 33 can be used to prove the right continuity of the exercise boundary. ■

Proof of Proposition 40: The structure of the exercise set and the validity of the EEP representation follow from Proposition 39.

To establish (4.14)-(4.15) we pass to a new numeraire corresponding to the detrended asset price. Substituting the underlying price in the strike function gives

$$G_t = \exp\left(\frac{1}{t}\int_0^t \ln(S_u)\,du\right) = \exp\left(\frac{1}{2}\left(r - \delta - \frac{1}{2}\sigma^2\right)t + \sigma\frac{1}{t}\int_0^t \tilde{z}_u du\right)$$

and $Y_t = G_t / S_t \equiv \exp(x_t / t)$ with

$$x_t \equiv -\frac{1}{2}\left(r - \delta - \frac{1}{2}\sigma^2\right)t^2 + \sigma\left(\int_0^t \tilde{z}_u du - t\tilde{z}_t\right).$$

An application of Ito's lemma shows that

$$\int_0^t \tilde{z}_u du - t\tilde{z}_t = -\int_0^t u d\tilde{z}_u.$$

Substituting in x and using the translated (under the risk neutral measure) Brownian motion $dz_t^* \equiv -d\tilde{z}_t + \sigma dt$ gives

$$\begin{aligned} x_t &= -\frac{1}{2}\left(r - \delta - \frac{1}{2}\sigma^2\right)t^2 - \sigma\int_0^t u d\tilde{z}_u \\ &= -\frac{1}{2}\left(r - \delta + \frac{1}{2}\sigma^2\right)t^2 + \sigma\int_0^t u dz_u^*. \end{aligned}$$

Under the measure $dQ^* = \exp\left(-\frac{1}{2}\sigma^2 t + \sigma \tilde{z}_t\right) dQ$ the process z^* is a Brownian motion (by Girsanov's theorem). It follows that

$$x_s = x_t + m(t,s) + v(t,s) z, \qquad \text{for } s \geq t,$$

where $m(t,s) \equiv -\frac{1}{2}\left(r - \delta + \frac{1}{2}\sigma^2\right)\left(s^2 - t^2\right)$, $v(t,s) \equiv \sigma\sqrt{\left(s^3 - t^3\right)/3}$ and z is a random variable with standard normal distribution function.

The European option value is

$$p(Y_t, t; \delta) = e^{-\delta(T-t)}\left(E_t^*\left[1_{\{Y_T \leq 1\}}\right] - E_t^*\left[Y_T 1_{\{Y_T \leq 1\}}\right]\right).$$

The two probability elements are respectively,

$$E_t^*\left[1_{\{Y_T \leq 1\}}\right] = \int_{z \leq d(x_t,1,t,T)} n(z)\, dz = N\left(d\left(x_t, 1, t, T\right)\right)$$

and

$$\begin{aligned}
E_t^*\left[Y_T 1_{\{Y_T \leq 1\}}\right] &= \int_{z \leq d(x_t,1,t,T)} \exp\left(\frac{x_t + m(t,T) + v(t,T) z}{T}\right) n(z)\, dz \\
&= \int_{z \leq d(x_t,1,t,T)} \exp\left(\frac{x_t + m(t,T)}{T} + \frac{1}{2}\frac{v(t,T)^2}{T^2}\right) \\
&\quad \times \frac{1}{\sqrt{2\pi}} \exp\left(-\frac{1}{2}\left(z - \frac{v(t,T)}{T}\right)^2\right) dz \\
&= \exp\left(\frac{x_t}{T} + q(t,T)\right) N\left(d\left(x_t, 1, t, T\right) - \frac{v(t,T)}{T}\right)
\end{aligned}$$

where

$$d\left(x_t, 1, t, T\right) = \frac{T \ln(1) - x_t - m(t,T)}{v(t,T)} = -\frac{x_t + m(t,T)}{v(t,T)}$$

$$q(t,T) = \frac{m(t,T)}{T} + \frac{1}{2}\frac{v(t,T)^2}{T^2}.$$

The pricing formula (4.14) follows from the relations $\exp\left(x_t/T\right) = Y_t^{t/T}$ and

$$d\left(x_t, 1, t, T\right) = \frac{T \ln(1) - x_t - m(t,T)}{v(t,T)} = \widehat{d}\left(Y_t, 1, t, T\right).$$

For the early exercise premium note that, by Ito's lemma,

$$\begin{aligned}
dY &= Y\left[\frac{1}{t}\left(-\left(r - \delta + \frac{1}{2}\sigma^2\right) t dt + \sigma t dz_t^*\right) - \frac{1}{t^2} x_t dt + \frac{1}{2}\sigma^2 \frac{t^2}{t^2} dt\right] \\
&= Y\left[-\left(r - \delta + \frac{1}{t^2} x_t\right) dt + \sigma dz_t^*\right] \\
&\equiv Y\left[\left(\delta - \delta_Y(t)\right) dt + \sigma dz_t^*\right]
\end{aligned}$$

with $\delta_Y(t) \equiv r + x_t/t^2$. The EEP is given by

$$\pi^p(Y_t, t, B^p(\cdot); \delta) = e^{-\delta(s-t)} \int_t^T \left(E_t^*\left[\delta 1_{\{Y_s \le B_s^p\}}\right] - E_t^*\left[\delta_Y(s) Y_s 1_{\{Y_s \le B_s^p\}}\right]\right) ds.$$

The first probability element is

$$E_t^*\left[\delta 1_{\{Y_s \le B_s^p\}}\right] = \delta N\left(d\left(x_t, B_s^p, t, s\right)\right).$$

The second one simplifies as follows,

$$\begin{aligned}
& E_t^*\left[\delta_Y(s) Y_s 1_{\{Y_s \le B_s^p\}}\right] \\
= & \int_{z \le d(x_t, B_s^p, s, t)} \left(r + \frac{x_t + m(t,s) + v(t,s) z}{s^2}\right) \\
& \times \exp\left(\frac{x_t + m(t,s) + v(t,s) z}{s}\right) n(z)\, dz \\
= & \left(r + \frac{x_t + m(t,s)}{s^2} + \frac{v(t,s)^2}{s^3}\right) \exp\left(\frac{x_t}{s} + q(t,s)\right) \\
& \times N\left(d\left(x_t, B_s^p, t, s\right) - \frac{v(t,s)}{s}\right) \\
& + \frac{v(t,s)}{s^2} \exp\left(\frac{x_t}{s} + q(t,s)\right) \int_{z \le d(x_t, B_s^p, t, s)} \left(z - \frac{v(t,s)}{s}\right) \\
& \times \frac{1}{\sqrt{2\pi}} \exp\left(-\frac{1}{2}\left(z - \frac{v(t,s)}{s}\right)^2\right) dz
\end{aligned}$$

where the integral in the last term equals $-n\left(d\left(x_t, B_s^p, t, s\right) - \frac{v(t,s)}{s}\right)$. Combining all these expressions produces the formulas in (4.15). The recursive equation for the immediate exercise boundary is obtained using standard arguments. The boundary condition follows from the dividend $\delta_Y(T_-) = r + \frac{1}{T}\ln\left(B_{T_-}^p\right)$ paid by Y as $t \to T$. ■

Proof of Proposition 41: Properties (i) and (iii) are standard. For (ii) note that

$$V^f(\lambda S, t) = \sup_{\tau \in \mathcal{S}_{t,T}} \widetilde{E}\left[e^{-r(\tau - t)}\left(\lambda S N_{t,\tau} - K\right)\right] \le V^f(S,t) + (\lambda - 1) S$$

for $\lambda \ge 1$, where τ stands for the optimal exercise time. A standard argument can now be applied to conclude that the claim holds. ■

Proof of Proposition 42: The arguments in Jaillet, Lamberton and Lapeyre [1990], Proposition 2.2, can be invoked to prove the continuity in (i). Property (ii) is a consequence of the affine, increasing payoff structure. Property (iii) is the counterpart of (i) in Proposition 41. The proof of (iv) is an adaptation of the proof of (iv) in Proposition 32. ■

Proof of Proposition 43: The proof parallels the proof of Proposition 33 for call options. The only difference is the "exercise" boundary at maturity, equal to $B_T = 0$. This value is implied by the obligatory nature of the contract at the expiration date. ■

Proof of Proposition 44: An application of Theorem 21 to the payoff function $S - K$ leads to the formulas indicated. ■

Chapter 5

Barrier and Capped Options

This chapter examines the valuation of American barrier and capped options. Barrier options, such as down-and-out calls, are considered first (section 5.1). The case of capped options is treated next (section 5.2). Contracts with constant and growing caps are analyzed in the Black-Scholes setting with constant coefficients, of chapter 4. Results are also reported for capped options with stochastic caps, that are written on nondividend-paying assets with stochastic volatility. Insights about the optimal exercise policy when the underlying asset price follows a diffusion process are also provided (section 5.3).

5.1 Barrier Options

5.1.1 Definitions and Literature

A *barrier* option is an option whose payoff depends on whether the underlying asset price reaches a pre-specified level, the barrier, during a certain time period. The most common examples are knock-in and knock-out options. A *knock-in* option comes into existence, while a *knock-out* option expires, when the barrier is reached. Knock-in and knock-out options each come in two varieties (down and up) depending on the position of the barrier relative to the initial asset price. If the barrier is below the underlying price at inception, the option is a *down-and-in* or a *down-and-out* option; if it lies above, the option is an *up-and-in* or an *up-and-out* option. All these contracts exist in the form of puts and calls. For instance, a down-and-in put is a put option that comes into existence when the underlying asset price decreases to a pre-specified barrier; a down-and-out call is a call option that expires when the barrier is reached. Knock-out options can involve a payoff at expiration. If they do, the payment is called a *rebate*.

Vanilla barrier options, such as those described above, are sometimes called "one touch" barrier options to emphasize the fact that an event materializes at the first touch of the pre-specified barrier. This contractual provision can be contrasted with more exotic barrier and event specifications that have begun to appear in derivatives markets and that will be dealt with in chapter 7.

European-style barrier options have been studied by Cox and Rubinstein [1985], Rubinstein and Reiner [1991] and Boyle [1992]. Hedging aspects are addressed by Bermin [2002], [2003]. American-style contracts are examined by Gao, Huang and Subrahmanyam [2000].

5.1.2 Valuation

Our analysis is carried out in the context of the Black-Scholes market setting. Consider an American down-and-out call option with maturity date T, exercise price K and barrier H with $K > H$. In the event that the barrier is reached at or before maturity the option payoff is null (the rebate is assumed to be null). On the complementary event the holder has the right to exercise the option and collects $(S - K)^+$ if he/she chooses to do so.

The down-and-out call option is a simple example of a path-dependent contract because the payoff depends on the trajectories followed by the underlying asset price. In spite of this apparent complexity the valuation of this American-style contract is straightforward. It rests on a simple extension of the EEP representation to American-style contracts with random maturity. Formulas of this type emerge in several valuation problems involving derivatives with trigger levels (such as the capped options examined further in this chapter).

Theorem 47 *(EEP representation) Consider an American down-and-out call option with maturity date T, exercise price K and barrier $K > H$. Let B^{doc} be the optimal exercise boundary, $C^{do}(S, t, H, B^{doc}(\cdot))$ the value of the contract, and $c^{do}(S, t, H)$ the value of a down-and-out European call with identical characteristics (T, K, H). For $H < S < B^{doc}$ and $t \in [0, T]$, the value of the American down-and-out call is given by*

$$C^{do}(S, t, H, B^{doc}(\cdot)) = c^{do}(S, t, H) + \pi^{do}(S, t, H, B^{doc}(\cdot)) \tag{5.1}$$

where

$$c^{do}(S, t, H) = \widetilde{E}_t\left[e^{-r(T-t)}(S_T - K)^+ 1_{\{\tau_H > T\}}\right]$$

$$\pi^{do}(S, t, H, B^{doc}(\cdot)) \equiv \widetilde{E}_t\left[\int_t^{\tau_H \wedge T} e^{-r(v-t)}(\delta S_v - rK)\, 1_{\{S_v \geq B_v^{doc}\}} dv\right]$$

and $\tau_H \equiv \inf\{v \in [t, T] : S_v = H\}$ denotes the first hitting time of H in $[t, T]$, or $\tau_H \equiv \infty$ if no such time exists in $[t, T]$. The optimal exercise boundary solves the recursive integral equation

$$B^{doc}(t) - K = C^{do}(B^{doc}(t), t, H, B^{doc}(\cdot)) \tag{5.2}$$

subject to the boundary condition $B^{doc}(T_-) = K \vee \frac{r}{\delta}K$.

This decomposition of the American down-and-out call option is similar to the early exercise premium representation for standard American options (Theorem 34). The first component is the European down-and-out call price. The second component, the early exercise premium, again represents the present value of the benefits from early exercise. For the contract under consideration the benefits are collected as long as the contract is alive, i.e., as long as the underlying asset price exceeds the barrier H. When the barrier is attained (on the event $\{\tau_H \leq T\}$) the option is retired and loses all value.

As usual, the EEP representation (5.1) leads to an equation for the exercise boundary (see (5.2)). The intuition here is standard. As immediate exercise is optimal on the boundary the value given by (5.1) must equal the exercise payoff when both are evaluated at $S = B^{doc}$.

The components of the EEP representation can be expressed in more explicit form by taking advantage of the availability of a closed form solution for the joint distribution of the asset price and the hitting time τ_H.

Corollary 48 *The EEP components of the American down-and-out call option value are*

$$\begin{aligned} c^{do}(S,t,H) &= Se^{-\delta(T-t)}\left[N\left(-d^-(K)+\sigma\sqrt{T-t}\right)\right. \\ &\quad \left. -\lambda^{2\frac{b}{\sigma^2}-2}N\left(-d^+(K)+\sigma\sqrt{T-t}\right)\right] \\ &\quad -Ke^{-r(T-t)}\left[N\left(-d^-(K)\right)-\lambda^{2\frac{b}{\sigma^2}}N\left(-d^+(K)\right)\right] \end{aligned} \tag{5.3}$$

and

$$\begin{aligned} \pi^{do}\left(S,t,H,B^{doc}(\cdot)\right) &= \int_t^T \delta Se^{-\delta(v-t)}\left[N\left(-d^-\left(B_v^{doc}\right)+\sigma\sqrt{v-t}\right)\right. \\ &\quad \left.-\lambda^{2\frac{b}{\sigma^2}-2}N\left(-d^+\left(B_v^{doc}\right)+\sigma\sqrt{v-t}\right)\right]dv \\ &\quad -\int_t^T rKe^{-r(v-t)}\left[N\left(-d^-\left(B_v^{doc}\right)\right)\right. \\ &\quad \left.-\lambda^{2\frac{b}{\sigma^2}}N\left(-d^+\left(B_v^{doc}\right)\right)\right]dv \end{aligned} \tag{5.4}$$

where

$$d^{\pm}(x) = \frac{\pm\log(\lambda)-\log(H)+\log(x)+b(v-t)}{\sigma\sqrt{v-t}}$$

and $\lambda = \frac{S}{H}$, $b = \delta - r + \frac{1}{2}\sigma^2$. *In (5.3) terms in* $d^{\pm}(x)$ *are evaluated at* $v = T$.

For the up-and-out put similar computations can be performed to obtain

Theorem 49 *(EEP representation) Consider an American up-and-out put option with maturity date* T*, exercise price* K *and barrier* $K < H$*. Let* B^{uop} *be the optimal exercise boundary,* $P^{uo}(S,t,H,B^{uop}(\cdot))$ *the value of the contract,* *and* $p^{uo}(S,t,H)$ *the value of an up-and-out European put with identical characteristics* (T,K,H). *For* $H > S > B^{uop}$ *and* $t \in [0,T]$, *the value of the American up-and-out put is given by*

$$P^{uo}(S,t,H,B^{uop}(\cdot)) = p^{uo}(S,t,H) + \pi^{uop}(S,t,H,B^{uop}(\cdot)) \tag{5.5}$$

where the European up-and-out put and the early exercise premium are given by

$$\begin{aligned} p^{uo}(S,t,H) &= Ke^{-r(T-t)}\left[N\left(d^{-}(K)\right) - \lambda^{2\frac{b}{\sigma^2}} N\left(d^{+}(K)\right)\right] \\ &\quad -Se^{-\delta(v-t)}\left[N\left(d^{-}(K) - \sigma\sqrt{v-t}\right)\right. \\ &\quad \left.-\lambda^{2\frac{b}{\sigma^2}-2} N\left(d^{+}(K) - \sigma\sqrt{v-t}\right)\right] \end{aligned}$$

$$\begin{aligned} \pi^{uop}(S,t,H,B^{uop}(\cdot)) &= \int_t^T rKe^{-r(v-t)}\left[N\left(d^{-}(B_v^{uop})\right)\right. \\ &\quad \left.-\lambda^{2\frac{b}{\sigma^2}} N\left(d^{+}(B_v^{uop})\right)\right] dv \\ &\quad -\int_t^T \delta Se^{-\delta(v-t)}\left[SN\left(d^{-}(B_v^{uop}) - \sigma\sqrt{v-t}\right)\right. \\ &\quad \left.-\lambda^{2\frac{b}{\sigma^2}-2} N\left(d^{+}(B_v^{uop}) - \sigma\sqrt{v-t}\right)\right] dv. \end{aligned}$$

The optimal exercise boundary solves the recursive integral equation

$$K - B^{uop}(t) = P^{uo}(B^{uop}(t), t, H, B^{uop}(\cdot)) \tag{5.6}$$

subject to the boundary condition $B^{uop}(T_{-}) = K \wedge \frac{r}{\delta}K$.

Inspection of the formulas for the down-and-out call and the up-and-out put reveals the following relationship

Proposition 50 *(American barrier option put-call symmetry) Consider an American up-and-out put option with characteristics* (T,K,H), *written on an asset whose price satisfies (4.1). The symmetry relation* $P^{uo}(S,K,H,r,\delta,T) = C^{do}(K,S,KS/H,\delta,r,T)$ *holds. The exercise boundaries are related by* $B^{uop} = SK/B^{doc}$.

This proposition extends the put-call symmetry property for standard American options (Theorem 3.13) to the case of barrier options. An important difference, relative to plain vanilla contracts, is that the barrier must also be adjusted when passing from the put to the call. The exercise boundaries on the other hand stand in the usual relationship. This result, that can be proved by using the change of numeraire outlined in section 2.7 (i.e., passing to the equivalent probability measure Q^* in (2.23)), is a special case of the general put-call symmetry for random maturity options (see Section 4 in Detemple [2001]).

5.2 Capped Options

5.2.1 Definitions, Examples and Literature

A capped option is a contract whose payoff is limited by some pre-specified amount. For example a capped call option with cap L and strike K pays off $(S \wedge L - K)^+$ upon exercise; a capped put option with cap L and strike K pays off $(K - S \vee L)^+$.[1] Capped options can be European- or American-style. They can also incorporate automatic exercise provisions. For example, a capped call option with maturity date T and automatic exercise at the cap is a call with random maturity given by the first time at which the underlying asset price reaches the cap, or the maturity date if no such time exists prior to T. This contract is a knock-out barrier option with rebate equal to the option payoff at the trigger date. Finally, the cap provision can be a constant, a function of time, or a stochastic process. Examples of capped options with time-dependent caps include contracts with discrete monitoring dates.

Several contracts with cap provisions have been issued by firms or financial institutions in the last two decades. One example is the MILES contract (Mexican Index-Linked Euro Security). This security is a package comprised of several securities including an American call option on the dollar value of the Mexican stock index. The contract is unusual in that it has both a time-dependent cap and a restriction on the exercise period.

Other examples are the capped options on the S&P 100 and S&P 500 indices that were introduced by the Chicago Board of Options Exchange (CBOE) in November 1991. These capped index options, which are called CAPS, combine a European exercise feature (the holder of the security cannot choose to exercise until the maturity of the contract) with an automatic exercise provision. The automatic exercise provision is triggered if the index value exceeds the cap at the close of any trading day (see Flesaker [1992] for a critical analysis of these options). Additional examples of European capped options include the range forward contract, collar loans, barrier options, indexed notes and index currency option notes (see Boyle and Turnbull [1989] and Rubinstein and Reiner [1991] for details).

European capped options are studied by Boyle and Turnbull [1989]. American-style contracts are examined by Broadie and Detemple [1995], [1997a], [1999] in the Black-Scholes setting and Detemple and Tian [2002] in a more general class of diffusion models.

5.2.2 Constant Cap

In this section and the next we suppose that the market satisfies the Black-Scholes assumptions (model with constant coefficients) with positive interest rate. In this setting consider an American capped call option with maturity date T, exercise price K and constant cap L with $L > K$. Upon exercise the

[1] The call payoff is capped by $L - K$; for the put option the maximum payoff is $K - L$. The constant L corresponds to the cap on the underlying asset price.

contract pays $(S \wedge L - K)^+$. Let B^L and $C^L(S,t)$ denote the optimal exercise boundary and the price of the capped option, respectively. The optimal exercise boundary is characterized in Theorem 51 and illustrated in Figure 5.1

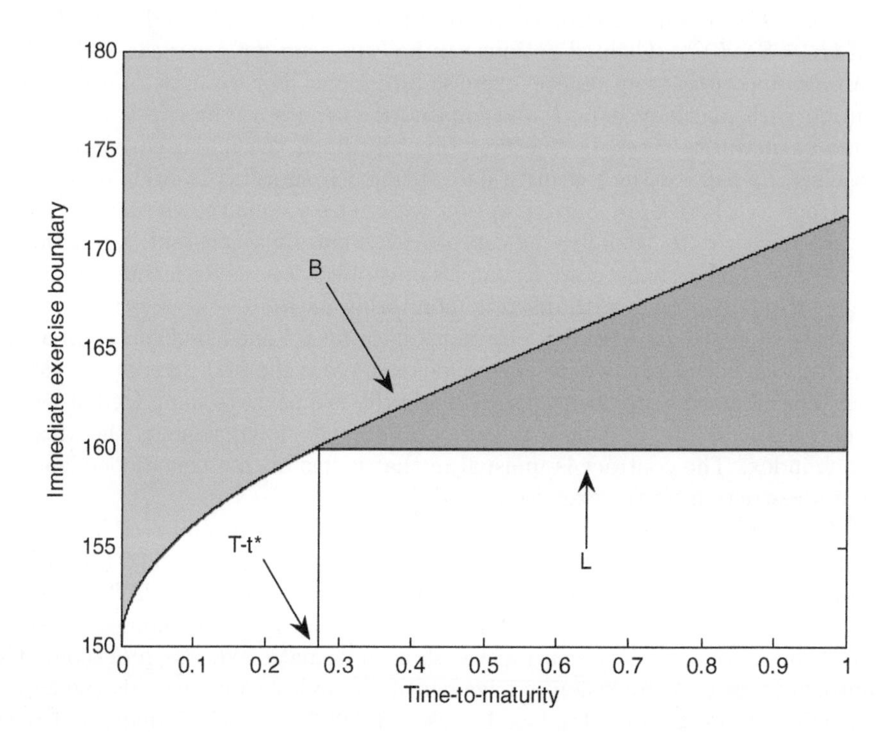

Figure 5.1: GBM model. Exercise boundary of a capped call option (y-axis) versus time to maturity (x-axis). The strike price is 100, the cap 160, the interest rate 6% and the dividend rate 4%. Computations use the integral equation method with 2,000 time steps.

Theorem 51 *Consider an American-style capped call option with maturity date* T*, exercise price* K *and constant cap equal to* L *with* $L > K$*. The optimal exercise boundary* B^L *is given by*

$$B^L = L \wedge B$$

where B *denotes the optimal exercise boundary of an American uncapped call option with same maturity date and exercise price.*

The exercise boundary of the capped call option has a very simple structure indeed! Intuition for this result can be gathered by examining the various configurations of the asset price S, the cap L and the exercise boundary B that can materialize. When S is above the cap waiting to exercise is suboptimal due to the time value of money. When S is below the cap and $B(t) \wedge L = B(t)$ the holder of the capped option can implement the optimal exercise policy and match the payoff of the uncapped American option with identical strike and maturity date (see Figure 5.1). This policy will therefore be optimal for the capped option as well. From these two observations one concludes that the exercise boundary for the capped option B^L is always bounded above by $B \wedge L$ and is equal to B at the first time t^* at which $B(t) = L$. When S lies below the cap and $B(t) \wedge L = L$ (i.e., before t^*) the argument is more subtle. In this situation it can be shown that the holder of the capped option can in fact duplicate the optimal exercise policy and the payoff of an uncapped option with shorter time-to-maturity for which immediate exercise is suboptimal (see the proof in the Appendix).

With the optimal exercise policy fully identified, the valuation of the contract becomes easy to perform. Let t^* be the solution of the equation

$$B(t,T) = L$$

if an interior solution, with value in $[0,T]$, exists. If $B(t,T) < L$ for all $t \in [0,T]$ set $t^* = 0$. If $B(t,T) > L$ for all $t \in [0,T]$ set $t^* = T$. The next theorem provides a valuation formula for the American capped call option.

Theorem 52 *Consider an American-style capped call option with maturity date T, exercise price K and constant cap equal to L $(L > K)$. For $S \geq L \wedge B$ the option value is $(S \wedge L) - K$. For $S < L \wedge B$ and $t \geq t^*$ the option value is $C^L(S,t) = C(S,t)$. For $S < L \wedge B$ and $t < t^*$ the option is worth $C^L(S,t)$ given by*

$$(L-K)\widetilde{E}_t\left[e^{-r(\tau_L - t)}1_{\{\tau_L < t^*\}}\right] + \widetilde{E}_t\left[e^{-r(t^*-t)}C(S_{t^*}, t^*)\,1_{\{\tau_L \geq t^*\}}\right] \tag{5.7}$$

where $\tau_L \equiv \inf\{v \in [t,T] : S_v = L\}$ denotes the first hitting time of L in $[t,T]$ and $\tau_L \equiv T$ if no such time exists in $[t,T]$. The representation formula in (5.7) can be simplified by computing the expectations explicitly

$$\begin{aligned} C^L(S,t) &= (L-K)\left(\lambda^{2\phi/\sigma^2}N(d_0) + \lambda^{2\alpha/\sigma^2}N\left(d_0 + 2f\sqrt{t^*-t}/\sigma^2\right)\right) \\ &\quad + e^{-r(t^*-t)}\int_0^L C(x,t^*)\,u(x,t,t^*)\,dx \end{aligned} \tag{5.8}$$

where

$$u(x,t,t^*) = \frac{n\left(d_1^-(x)\right) - \lambda^{1-2(r-\delta)/\sigma^2}n\left(d_1^+(x)\right)}{x\sigma\sqrt{t^*-t}} \tag{5.9}$$

$$d_0 = \frac{\log(\lambda) - f(t^*-t)}{\sigma\sqrt{t^*-t}} \tag{5.10}$$

$$d_1^{\pm}(x) = \frac{\pm \log(\lambda) - \log(L) + \log(x) + b(t^* - t)}{\sigma\sqrt{t^* - t}} \tag{5.11}$$

and $b = \delta - r + \frac{1}{2}\sigma^2$, $f = \sqrt{b^2 + 2r\sigma^2}$, $\phi = \frac{1}{2}(b - f)$, $\alpha = \frac{1}{2}(b + f)$ *and* $\lambda = S/L$.

An alternative decomposition that draws on Theorems 51 and 52 relates the value of the American capped option to the value of a capped option with automatic exercise at the cap.

Theorem 53 *(Early exercise premium representation) Let* $C^{ae}(S, t, L)$ *be the value of a capped option with automatic exercise at the cap (see formula (5.13) below). For* $S < L \wedge B$ *and* $t \in [0, T]$*, the value of the American capped option is given by*

$$C^L(S, t) = C^{ae}(S, t, L) + \widetilde{E}_t \left[\int_t^{\tau_L} e^{-r(v-t)} (\delta S_v - rK) 1_{\{S_v \geq B_v\}} dv \right] \tag{5.12}$$

where $\tau_L \equiv \inf\{v \in [t, T] : S_v = L\}$ *denotes the first hitting time of* L *in* $[t, T]$*, and* $\tau_L \equiv T$ *if no such time exists in* $[t, T]$.

This decomposition of the American option value is similar to the early exercise premium representation for standard American options (Theorem 34). It differs in that it relates the value of the option contract to the value of a contract that may be automatically exercised before the maturity date (the standard representation uses the value of a European option with exercise at the maturity date as the benchmark). This is the first component on the right hand side of (5.12): a call with automatic exercise at the cap. The second component in (5.12) measures the premium achieved by optimally exercising prior to reaching the cap. The premium is the present value of net benefits in the event of exercise, consisting of the dividends collected on the underlying asset net of the interest costs incurred.

When dividends are sufficiently low the exercise boundary of the uncapped option B exceeds the cap L at all times. In this case the valuation formulas (5.7) and (5.12) simplify significantly. Straightforward deductions show that,

Corollary 54 *(American capped call valuation with low dividends) Suppose that* $\delta \leq rK/L$*. For* $S < L$ *and* $t \in [0, T]$*, the value of the American capped call option equals the value of the corresponding capped call option with automatic exercise at the cap*

$$\begin{aligned} C^L(S, t) &= C^{ae}(S, t, L) \\ &= (L - K)\left(\lambda^{2\phi/\sigma^2} N(d_0) + \lambda^{2\alpha/\sigma^2} N\left(d_0 + 2f\sqrt{\tau}/\sigma\right)\right) \\ &\quad + Se^{-\delta\tau}\left(N\left(d_1^-(L) - \sigma\sqrt{\tau}\right) - N\left(d_1^-(K) - \sigma\sqrt{\tau}\right)\right) \\ &\quad - \lambda^{-2(r-\delta)/\sigma^2} Le^{-\delta\tau}\left(N\left(d_1^+(L) - \sigma\sqrt{\tau}\right) - N\left(d_1^+(K) - \sigma\sqrt{\tau}\right)\right) \\ &\quad - Ke^{-r\tau}\left(N\left(d_1^-(L)\right) - N\left(d_1^-(K)\right)\right. \\ &\qquad \left. - \lambda^{1-2(r-\delta)/\sigma^2}\left(N\left(d_1^+(L)\right) - N\left(d_1^+(K)\right)\right)\right). \end{aligned} \tag{5.13}$$

In (5.13) the expressions for d_0 and $d_1^{\pm}(x)$ are the same as in (5.10)-(5.11) but with $\tau = T - t$ replacing $t^* - t$. The expressions for b, f, ϕ and α are the same as in Theorem 52.

Remark 5 *The value of a European capped call option with strike price K, cap L and maturity T (the option with payoff $(S_T \wedge L - K)^+$ at date T) is given by*

$$\begin{aligned} C^e(S,t,L) &= Se^{-\delta(T-t)}\left(N\left(d_1^-(L) - \sigma\sqrt{T-t}\right) - N\left(d_1^-(K) - \sigma\sqrt{T-t}\right)\right) \\ &\quad -Ke^{-r(T-t)}\left(1 - N\left(d_1^-(K)\right)\right) + Le^{-r(T-t)}\left(1 - N\left(d_1^-(L)\right)\right). \end{aligned}$$

The European capped option value can serve as a benchmark to measure the gains from early exercise (prior to maturity) embedded in the American capped option value. The early exercise premium is particularly simple to compute in the case of low dividends (formula (5.13)).

Remark 6 *If $L \uparrow \infty$ the European capped call option value $C^e(S,t,L)$ converges to the Black-Scholes formula adjusted for dividends (see corollary 15).*

5.2.3 Capped Options with Growing Caps

Consider now the class of American capped options whose caps grow at a constant rate $g \geq 0$. Suppose that

$$L_t = L_0 e^{gt}, \ t \in [0,T] \tag{5.14}$$

where we assume that $L_0 > K$. Let t^* denote the solution to the equation $B(t,T) = L_t$, if an interior solution in $[0,T]$ exists. If $B(t,T) < L_t$ for all $t \in [0,T]$ set $t^* = 0$. If $B(t,T) > L_t$ for all $t \in [0,T]$ set $t^* = T$. In order to describe the optimal exercise region it is necessary to consider the class of exercise policies defined next and illustrated in Figure 5.2.

Definition 55 *((t_e, t^*, t_f) Exercise Policy). Let t_e and t_f satisfy $0 \leq t_e \leq t_f \leq T$ and $t_e \leq t^* \leq T$. Define the stopping times*

$$\begin{aligned} \tau_1 &= \inf\{v \in [t_e, t_f \vee t^*] : S_v = L_v\} \text{ or if no such } v \text{ exists set } \tau_1 = T \\ \tau_2 &= t_f \text{ if } S_{t_f} \geq L_{t_f} \text{ otherwise set } \tau_2 = T \\ \tau_3 &= \inf\{v \in [t^*, T] : S_v = B_v\} \text{ or if no such } v \text{ exists set } \tau_3 = T. \end{aligned}$$

An exercise policy is a (t_e, t^*, t_f)-policy if the option is exercised at the stopping time $\tau_1 \wedge \tau_2 \wedge \tau_3$.

A (t_e, t^*, t_f)-policy is an exercise policy that is completely described by the three parameters indicated. Moreover, one of the parameters (t^*) is already determined as the solution to an equation, while the other two (t_e and t_f) are arbitrary constants, up to the restrictions $0 \leq t_e \leq t_f \leq T$ and $t_e \leq t^* \leq T$.

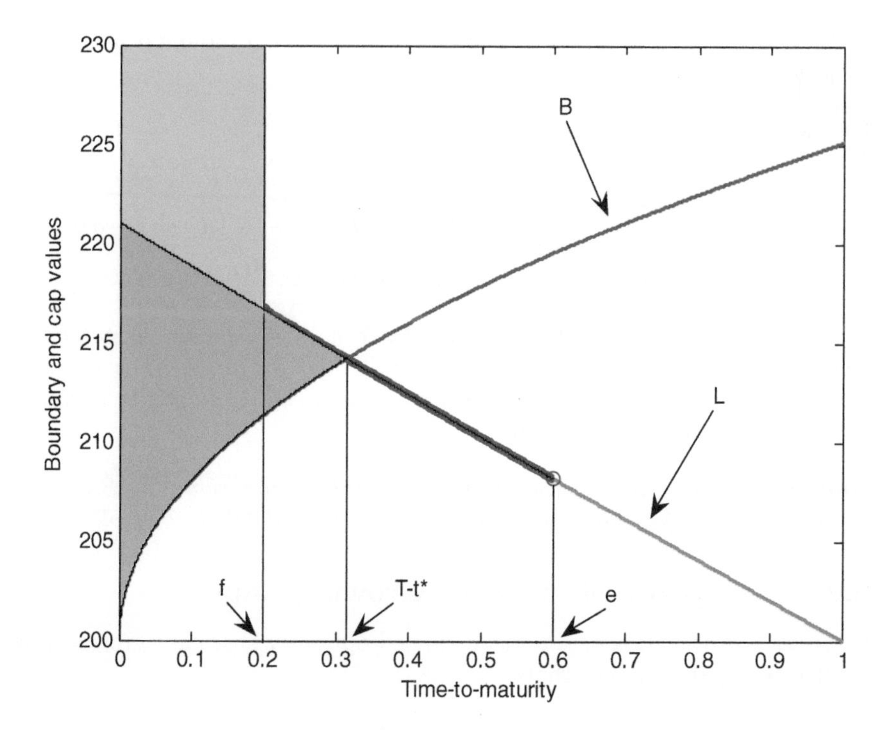

Figure 5.2: A (t_e, t^*, t_f)-policy for a capped call option with growing cap. Time-to-maturity is on the x-axis: point e corresponds to $T - t_e$, point f is $T - t_f$. The option strike is $K = 100$, the initial cap $L_0 = 200$, the cap's growth rate is $g = 10\%$, the interest rate $r = 6\%$, the dividend rate $\delta = 3\%$ and the volatility coefficient $\sigma = 0.2$. Computations use the integral equation with 2000 time steps.

A (t_e, t^*, t_f)-policy mandates immediate exercise along the cap if the time-to-maturity is less than $T - t_e$. In the interval $[0, T - t^*]$ exercise is immediate if the underlying asset price lies in between the exercise boundary B and the cap L. For times-to-maturity less than $T - t_f$ immediate exercise is also implemented if the asset price lies above the cap.

Our first theorem identifies the optimal exercise policy.

Theorem 56 *Consider an American capped call option with exercise price K, maturity date T and cap given by equation (5.14). Then the optimal exercise policy is a (t_e, t^*, t_f)-policy.*

This result is quite surprising because it shows that the optimal stopping time belongs to the class of (t_e, t^*, t_f)-policies introduced in definition 55. As we shall see shortly, the parameter t_f (in addition to t^*) is already determined by the structure of the capped option payoff, the cap process, the underlying asset price process and the interest rate. So $t_e \in [0, t^*]$ is the only parameter that remains to be determined: pricing the American capped call option has been reduced to the identification of a single parameter, t_e. This boils down to a univariate optimization problem that can be resolved by using the valuation formula given in Theorem 57 below.

Intuition for the optimality of a (t_e, t^*, t_f)-policy is as follows. First, note that immediate exercise is optimal for the capped option in the region $B < S < L$ because it is optimal for the uncapped option and the payoffs are identical. Similarly, when $t \geq t^*$ and $S < B(t)$ the capped option holder can implement the exercise policy of the uncapped option and collect the same payoff. Second, note that immediate exercise is also optimal if $t \geq t_f$ and $S > L_t$. In this region the discounted payoff $\psi_t \equiv e^{-rt}(L_t - K)$ decreases almost surely and, as a result, waiting reduces value. The reverse argument applies when $t < t_f$ and $S > L_t$. Here the discounted payoff ψ_t increases over time which means that the policy of exercising at $t_f \wedge \tau_L$, where τ_L is the first hitting time of the cap L, increases value. Third, when $t < t^*$ and $S < L_t$ one can find an uncapped option with shorter time-to-maturity whose exercise boundary falls below the cap at all subsequent times. Implementation of this exercise policy dominates immediate exercise for the capped option holder. Lastly, along the cap (i.e., when $S = L$ and $t < t^*$) the benefits of waiting (which dominate above the cap) are curtailed by the risk of falling below the cap. When the maturity date is sufficiently close (i.e., for t close to t^*) this risk is enough to prompt immediate exercise. With longer horizons the appreciation in the discounted value of the cap can be sufficient for the optimality of a waiting policy.

The value of a (t_e, t^*, t_f)-policy is easy to calculate. The price of the American-style capped contract naturally follows.

Theorem 57 *(Valuation of American capped option with growing cap) Define*

$$t_f \equiv \arg \max_{t \in [0,T]} \left\{ e^{-rt}(L_t - K) \right\}.$$

The value of the American capped option with growing cap is given by

$$C^L(S, 0) = \max_{t_e} \left\{ C^L(t_e, t^*, t_f) : t_e \in [0, t^* \wedge t_f] \right\} \tag{5.15}$$

where

$$C^L(t_e, t^*, t_f) = \widetilde{E}_t \left[e^{-r(t_e - t)} \left\{ C^u 1_{\{S_{t_e} > L_{t_e}\}} + C^d 1_{\{S_{t_e} \leq L_{t_e}\}} \right\} \right] \tag{5.16}$$

and C^u and C^d are the values at time t_e in the events $\{S_{t_e} > L_{t_e}\}$ and $\{S_{t_e} \leq L_{t_e}\}$, respectively. Explicit formulas for C^u and C^d are given in Broadie and Detemple [1995].

Consider a fixed parameter t_e along with the parameters t^* and t_f defined above. The value of the (t_e, t^*, t_f)-policy is given by $C^L(t_e, t^*, t_f)$ in formula (5.16). It follows immediately that the value of the American-style capped option is obtained by maximizing this value relative to the free parameter t_e. Even though this optimization problem does not have a simple closed form solution it can be resolved numerically, by using standard methods.

5.2.4 Stochastic Cap, Interest Rate and Volatility

This section considers a fairly general class of American capped options written on nondividend-paying assets in a setting with stochastic interest rate and stochastic return volatility. The asset price S satisfies (under the Q-measure)

$$dS_t = S_t \left[r_t dt + \sigma_t d\tilde{z}_t \right], t \in [0, T]; \quad S_0 \text{ given} \tag{5.17}$$

where volatility follows a stochastic process $\sigma \equiv \{\sigma_t, \mathcal{F}_t : t \in [0, T]\}$ that is progressively measurable, bounded above and bounded away from zero (P-a.s.) and the interest rate $r \equiv \{r_t, \mathcal{F}_t : t \in [0, T]\}$ is progressively measurable, strictly positive and bounded.

The capped call option has payoff $(S \wedge L - K)^+$, where L satisfies

$$dL_t = L_t g_t dt, t \in [0, T], \quad L_0 \text{ given.} \tag{5.18}$$

We assume that the growth rate of the cap, g, is a progressively measurable process such that $L_t > K$ for all $t \in [0, T]$ and such that the following growth condition holds,

$$(g_t - r) L_t + r_t K < 0, t \in [0, T]. \tag{5.19}$$

The model (5.17)–(5.19) for the underlying asset price and for the cap is fairly general. It allows for a stochastic return volatility, a stochastic interest rate as well as a stochastic growth rate for the cap. The factor underlying the stochastic behavior of these quantities is the same Brownian motion that affects the asset price: the market is complete. The cap's growth rate may take positive as well as negative values as long as condition (5.19) is satisfied. This condition is a restriction on the growth rate of the cap that is clearly satisfied if the cap is constant or decreasing. It is satisfied even when the growth rate of the cap is positive as long as it is not too large. Broadie and Detemple [1999] consider the special case with constant interest rate.

For this model the following result holds,

Theorem 58 *Consider an American capped call option with stochastic cap given by (5.18)–(5.19) when the interest rate is a positive, bounded and progressively measurable process, and the underlying asset price satisfies (5.17). The optimal exercise boundary is a stochastic process given by $B^L = L$. If $S \geq L$ immediate exercise is optimal and $C^L(S, t) = L - K$. If $S < L$ the optimal exercise policy is described by the stopping time τ_L where $\tau_L \equiv \inf\{v \in [t, T] : S_v = L_v\}$, or $\tau_L \equiv T$ if no such time exists. For $S < L$ and for all $t \in [0, T]$, the value of the capped option is*

$$\begin{aligned} C^L(S,t) &= \widetilde{E}\left[\exp\left(-\int_t^{\tau_L} r_v dv\right)(L_{\tau_L} - K) 1_{\{\tau_L < T\}}\right] \\ &\quad + \widetilde{E}\left[\exp\left(-\int_t^{T} r_v dv\right)(S_T - K)^+ 1_{\{\tau_L \geq T\}}\right]. \end{aligned} \tag{5.20}$$

Although the valuation problem under consideration involves path dependence (as the cap, the interest rate and the volatility of the underlying asset are adapted processes) it nevertheless has an explicit solution as far as the optimal exercise policy is concerned. The intuition for this result parallels the intuition for the optimality of exercise at t_f above the cap when the cap grows at a constant rate (section 5.2.3). When the underlying asset price exceeds the cap immediate exercise is optimal if the discounted payoff is a strictly decreasing function of time. In the case under consideration the discounted payoff equals $\psi_t \equiv \exp\left(-\int_0^t r_v dv\right)(L_t - K)$ when $S_t \geq L_t$. It follows that $d\psi_t = \exp\left(-\int_0^t r_v dv\right)\left[(g_t - r_t) L_t + r_t K\right] dt$, which is negative under condition (5.19). Hence in all circumstances (independently of the next realizations of the Brownian motion z) waiting to exercise is a dominated policy. When the underlying asset is below the cap, on the contrary, immediate exercise is suboptimal: given that the asset does not pay dividends there is no incentive to exercise. Combining these two insights shows that it is optimal to exercise at the first time at which the asset price reaches the cap. The value of the contract is the value of this exercise policy.

5.3 Diffusion Processes

Let us now revert to the diffusion model of section 4.6 along with the accompanying assumptions. In this context we reexamine the optimal exercise decision for a capped call option with constant cap.

Our main result, that draws on Detemple and Tian [2002], extends Theorem 51.

Theorem 59 *Consider an American capped call option with constant cap L and suppose that the underlying asset price follows the diffusion process (4.8), where the coefficients satisfy the assumptions stated. The immediate exercise region of the capped call option is $\mathcal{E}(L) = \{(S,t) \in \mathbb{R}_+ \times [0,T) : S \geq B(t) \wedge L\}$ where $B(t)$ is the exercise boundary of the corresponding uncapped option with identical characteristics (T, K).*

Quite remarkably, the structure of the exercise region is the same as for the GBM price process: the capped option's exercise boundary is the lesser of the cap and the boundary of an uncapped option with the same maturity date and strike price (Figure 5.3 provides an illustration in the CEV model). Thus, as before, the exercise boundary of the capped option is known once the boundary of the uncapped option has been identified.

Intuition for this result can be provided along the same lines as before. The same arguments apply for each of the three possible configurations of (S, L, B). We can complement these intuitions by also noticing that the optimal exercise boundary of the capped option is a non-increasing of time (i.e., a non-decreasing

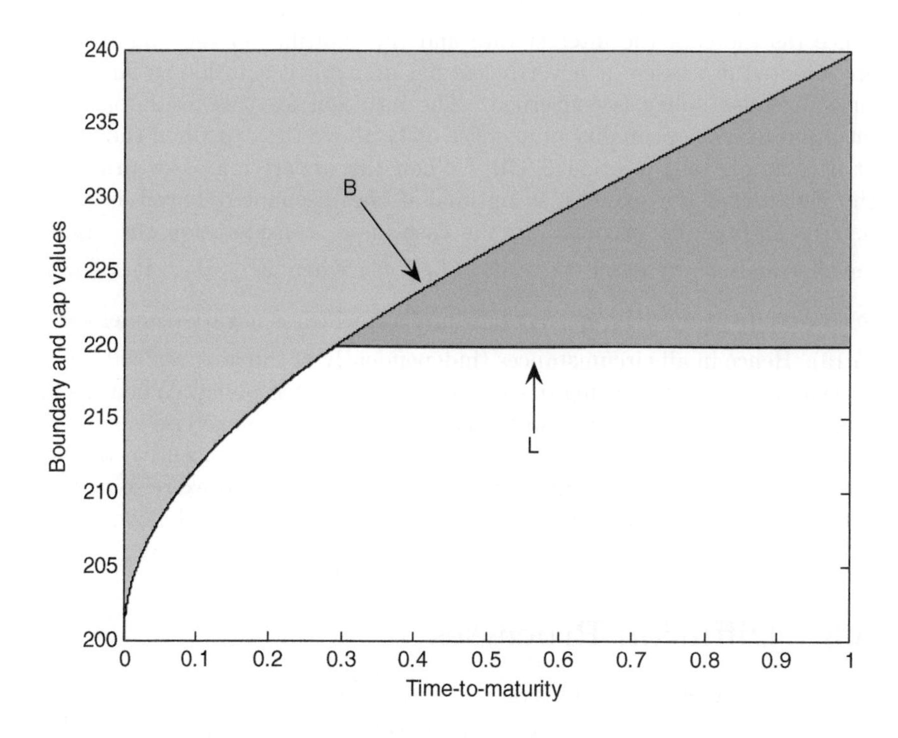

Figure 5.3: CEV model. Exercise boundary for a capped call with constant cap (y-axis) versus time to maturity (x-axis). The option strike is $K = 100$ and the cap $L = 220$. The interest rate is $r = 6\%$, the dividend rate $\delta = 3\%$, the volatility coefficient $\sigma = 0.2\sqrt{10}$ and the elasticity of variance $\theta = 1.7$. Computations use the integral equation method with 2000 time steps.

function of time-to-maturity). As B^L is bounded above by $B(t) \wedge L$ and equals $B(t^*) = L$ at time t^* it must be that $B^L = B \wedge L$.

5.4 Appendix: Proofs

Proof of Theorem 47: Let $C^{do}(S,t)$ denote the down-and-out call price and let $\mathcal{E}^{doc} \equiv \left\{(S,t) \in \mathbb{R}_+ \times [0,T) : C^{do}(S,t) = (S-K)^+\right\}$ be the immediate exercise region. As $K > H$ there is no incentive to exercise when $K > S > H$. Similarly, exercising in the region $H < S < (r/\delta) K$ is a dominated policy as it would induce a local loss equal to $\delta S - rK < 0$. It can also be verified that $\mathcal{E}^{doc}$ has properties that are similar to $\mathcal{E}$, the exercise region of a standard American call. In particular, the region is up- and right-connected (in the time dimension). We conclude that $\mathcal{E}^{doc} \equiv \left\{(S,t) \in \mathbb{R}_+ \times [0,T) : S \geq B^{doc}(t)\right\}$

for some non-increasing function of time $B^{doc}(\cdot)$. The function B^{doc} is the immediate exercise boundary. It is continuous and converges to $(1 \vee (r/\delta)) K$ as $t \to T$.

As the down-and-out call expires at the first hitting time τ_H of the level H and as the payoff $(S_{\tau_H} - K)^+ = 0$, the contract can be viewed as an American option with random maturity τ_H. The early exercise premium representation

$$\begin{aligned} C^{do}(S,t,H,B^{doc}(\cdot)) &= \widetilde{E}_t\left[e^{-r(T\wedge\tau_H - t)}\left(S_{T\wedge\tau_H} - K\right)^+\right] \\ &+\widetilde{E}_t\left[\int_t^{T\wedge\tau_H} e^{-r(v-t)}\left(\delta S_v - rK\right)1_{\{S_v \geq B_v^{doc}\}}dv\right] \end{aligned}$$

follows by a straightforward extension of the results in chapter 3. On the event $\{\tau_H \leq T\}$ the payoff $(S_{T\wedge\tau_H} - K)^+ = (S_{\tau_H} - K)^+ = 0$. On the complementary event $(S_{T\wedge\tau_H} - K)^+ = (S_T - K)^+$. The formula in the theorem follows.

The recursive integral equation for the boundary (5.2) is obtained from (5.1) using the fact that immediate exercise is optimal at the boundary, i.e., at $S = B^{doc}$. ■

In order to demonstrate Corollaries 48, 54 and Theorem 49 we need the auxiliary lemmas stated below. These lemmas describe distributions and truncated moment generating functions involving the first passage times of sets with exponential boundaries, when the asset price follows a geometric Brownian motion process. The formulas provided can be inferred from the joint law of a Brownian motion and its running maximum (see Karatzas and Shreve [1988], p. 95).

Lemma 60 *Suppose that S satisfies the stochastic differential equation $dS = S[(r-\delta)dt + S\sigma dz_t]$. For $A \leq B$ define the conditional probability,*

$$U(S,A,B,t,T) = P\left[S_T \leq A \text{ and } S_v < B \text{ for } v \in [t,T) \middle| S_t = S\right].$$

Then

$$U(S,A,B,t,T) = N(d(A)) - \left(\frac{S}{B}\right)^{1-2(r-\delta)/\sigma^2} N(d(A,B))$$

when $S \leq B$ and $t \in [0,T]$, where

$$\begin{aligned} d(A) &= \frac{\log(A) - \log(S) + b(T-t)}{\sigma\sqrt{T-t}} \\ d(A,B) &= \frac{\log(S) + \log(A) - 2\log(B) + b(T-t)}{\sigma\sqrt{T-t}} \end{aligned}$$

and $b = \delta - r + \frac{1}{2}\sigma^2$.

Lemma 61 *Suppose that $L_t = L_0 e^{gt}$ and that τ_L represents the first time in $[t,T]$ at which $S = L$ (the first passage time of the level L). Then, for $t \in [0,T]$,* *the conditional probability* $P[\tau_L \geq T | S_t = S]$ *is given by*

$$\begin{cases} N\left(-\overline{d}_2^{+}(\lambda_t, T-t)\right) - (\lambda_t)^{1+2(g+\delta-r)/\sigma^2} N\left(-\overline{d}_2^{-}(\lambda_t, T-t)\right) & \text{if } S \leq L_t \\ N\left(\overline{d}_2^{+}(\lambda_t, T-t)\right) - (\lambda_t)^{1+2(g+\delta-r)/\sigma^2} N\left(\overline{d}_2^{-}(\lambda_t, T-t)\right) & \text{if } S \geq L_t \end{cases}$$

where

$$d_2^{\pm}(\lambda_t, T-t) = \frac{\pm \log(\lambda_t) - \overline{b}(T-t)}{\sigma\sqrt{T-t}}$$

and $\lambda_t = S/L_t$, $\overline{b} = g + \delta - r + \frac{1}{2}\sigma^2$.

Lemma 62 *The conditional density of the first passage time* τ_L *is, for* $y \geq t$,

$$\gamma(y; S, t) = \begin{cases} -n\left(\overline{d}_2^{+}(\lambda_t, y-t)\right) \frac{\log(\lambda_t)}{\sigma(y-t)y^3/2} & \text{if } S \leq L_t \\ n\left(\overline{d}_2^{+}(\lambda_t, y-t)\right) \frac{\log(\lambda_t)}{\sigma(y-t)^3/2} & \text{if } S \geq L_t. \end{cases}$$

Lemma 63 *The truncated conditional moment generating function of* τ_L *is (at date 0)*

$$\int_0^t e^{(a-r)y}\gamma(y; S_0, 0)\, dy$$

$$= \begin{cases} \lambda_0^{2\frac{\phi(a)}{\sigma^2}} N(d_0(a)) + \lambda_0^{2\frac{\alpha(a)}{\sigma^2}} N\left(d_0(a) + 2f(a)\frac{\sqrt{t}}{\sigma}\right) & \text{if } \lambda_0 \leq 1 \\ \lambda_0^{2\frac{\phi(a)}{\sigma^2}} N(-d_0(a)) + \lambda_0^{2\frac{\alpha(a)}{\sigma^2}} N\left(-d_0(a) - 2f(a)\frac{\sqrt{t}}{\sigma}\right) & \text{if } \lambda_0 \geq 1 \end{cases}$$

where

$$d_0(a) = \frac{\log(\lambda_0) - f(a)t}{\sigma\sqrt{t}}$$

with $\lambda_0 = S_0/L_0$, $\overline{b} = g + \delta - r + \frac{1}{2}\sigma^2$, $f(a) = \left(\overline{b}^2 + 2(r-a)\sigma^2\right)^{\frac{1}{2}}$, $\phi(a) = \frac{1}{2}\left(\overline{b} - f(a)\right)$ *and* $\alpha(a) = \frac{1}{2}\left(\overline{b} + f(a)\right)$.

Proof of Corollary 48: Using lemma 60 and a change of variable leads to

$$P[S_T \geq A \text{ and } S_v > B \text{ for } v \in [t, T) | S_t = S]$$
$$= N(-d(A)) - \left(\frac{S}{B}\right)^{2b/\sigma^2} N(-d(A, B))$$

for $A \geq B$. The European option value is

$$c^{do}(S, t, H) = \int_H^\infty \exp(-r(T-t))(x-K)^+ \overline{u}(x; S, H, t, T)\, dx$$

where $\overline{u}(x; S, H, t, T)$ is the conditional density of the distribution above. Performing the integration gives the expression for the European down-and-out call. A similar computation produces the value of the early exercise premium. ■

Proof of Theorem 49: The proof parallels the proofs of Theorem 47 and Corollary 48. ■

Proof of Proposition 50: Put-call symmetry can be deduced from the formulas of Theorems 47 and 49. Alternatively, the general symmetry result of chapter 2 can be specialized to produce the results stated. ■

Proof of Theorem 51: The proof proceeds by considering all the possible configurations for the asset price S.
Case (i): Suppose first that $S \geq L$. In this situation immediate exercise is optimal because the exercise payoff is $(S \wedge L - K)^+ = L - K$, the maximum payoff attainable, and the interest rate is strictly positive.

Case (ii): Suppose that $B \leq S < L$. As $(S \wedge L - K)^+ \leq S - K$ the inequality

$$C^L(S,t) \leq C(S,t)$$

always holds. In the region under consideration immediate exercise is optimal for the holder of the uncapped option. Thus $C^L(S,t) \leq (S-K)^+ = (S-K)$. Given that immediate exercise is a feasible policy for the capped option and yields a payoff equal to $(S \wedge L - K)^+ = (S-K)^+ = (S-K)$, we conclude that immediate exercise is optimal for the capped option as well (if not there exists a waiting policy that dominates immediate exercise for the capped option, hence for the uncapped option - a contradiction when $S \geq B$).

Case (iii): Suppose that $S < B \wedge L$. We must show that immediate exercise is suboptimal. Consider first the case $L > \max\{(r/\delta)K, K\}$. Let $B(t,T)$ be the exercise boundary for an uncapped option with exercise price K and maturity date T. Recall that $B(t,T)$ is a non-increasing continuous function of time converging to $K \vee (r/\delta)K$ when t goes to T. Hence, in the case under consideration, we can always find a shorter maturity $T_0 \leq T$, such that $S < B(t,T_0) < L$. Clearly the policy of exercising at the first hitting time in $[t,T_0)$ of the boundary $\{B(v,T_0), v \in [t,T_0)\}$, or at T_0 if no such time exists, is feasible for the holder of the capped option. This policy has the same payoff as the uncapped option with shorter maturity date T_0. We conclude that

$$C(S,t,T_0) \leq C^L(S,t).$$

Because immediate exercise is suboptimal for the shorter maturity uncapped option when $S < B(t,T_0)$ we must have $(S-K)^+ < C(S,t,T_0) \leq C^L(S,t)$. It follows that immediate exercise is also suboptimal for the capped option. Consider next the case $L \leq (r/\delta)K$. Let τ denote the minimum of T and of the first hitting time of the set $[L,\infty)$. The policy of exercising at τ dominates immediate exercise because $\delta S_v - rK < 0$ for $v \in [t,\tau)$. ■

Proof of Theorem 52: The proof follows immediately from Theorem 51. ■

Proof of Theorem 53: The structure of the optimal exercise boundary shows that the set of exercise policies can be restricted, without loss, to the set of bounded stopping times $\tau \leq \tau_L$. The representation formula in the theorem follows. ■

Proof of Corollary 54: When $\delta \leq rK/L$ the immediate exercise boundary is given by the cap L. The formula follows from the previous theorem. ■

Proof of Theorem 56: The following cases arise.
Case (i): Suppose first that $B \leq S < L$. Then the same argument as in the proof of Theorem 51, case (ii) applies and demonstrates that immediate exercise is an optimal policy.

Case (ii): Consider now the case $S < B \wedge L$ and suppose that $(r/\delta) K > K$. If $L_t \geq (r/\delta) K$ the argument in the proof of Theorem 51, case (iii) applies. If $L_t < (r/\delta) K$ the policy of exercising at the stopping time τ equal to the first hitting time of the set $[(r/\delta) K \wedge L, \infty)$ or T if no such time exists dominates immediate exercise because $\delta S_v - rK < 0$ for $v \in [t, \tau)$. In the case $(r/\delta) K \leq K$ we have $L_t > K$ for all $t \in [0, T]$ (by assumption) and the argument of Theorem 51, case (iii), again applies.

Case (iii): Suppose now that $S > L$. It can be verified that the discounted payoff function $\psi_t \equiv e^{-rt} (L_t - K)$ has a unique maximum at

$$t_f = \arg \max_{t \in [0,T]} e^{-rt} (L_t - K).$$

If t_f is in the interior of $[0, T]$ the function ψ_t is strictly increasing for $t < t_f$ and strictly decreasing for $t > t_f$. If $t_f = 0$ the discounted payoff ψ_t is strictly decreasing for $t > t_f$; if $t_f = T$ it is strictly increasing for $t < t_f$. Hence, if $t \geq t_f$ immediate exercise dominates any waiting policy. If $t < t_f$ the policy of exercising at the first hitting time of L or at t_f dominates immediate exercise.

Case (iv): Finally, suppose that immediate exercise is optimal at some time $t < t^*$ when $S = L$. Then it is optimal to exercise at all $v \in [t, t^*]$ when $S_v = L_v$. Suppose not, i.e., suppose that there exists $u \in [t, t^*]$ such that $S_u = L_u$ and $C^L (S_u, u; T) > L_u - K$ (for clarity we indicate the maturity dates of the options under consideration). At t we have

$$\begin{aligned} L_t - K &= C^L (S_t, t; T) \\ &\geq C^L (S_t, t; T - (u - t)) \quad \text{(shorter maturity option)} \\ &= C^H (S_t, u; T) \quad (H \text{ is } L \text{ translated by } t - u : H(u) = L(t)) \\ &\geq C^L (S_u, u; T) - (L_u - L_t) \quad \text{(see Lemma 64 below)}. \end{aligned}$$

(See Figure 5.4 for an illustration of L and its translation H.) If immediate exercise is suboptimal at u then $C^L (S_u, u; T) > L_u - K$ (recall that $S_u = L_u$) so that $L_t - K > (L_u - K) - (L_u - L_t) = L_t - K$, a contradiction. We conclude that immediate exercise is optimal along the cap, for all $v \in [t, t^*]$, if it is optimal at time t (i.e., the exercise region is a connected segment along the cap). ■

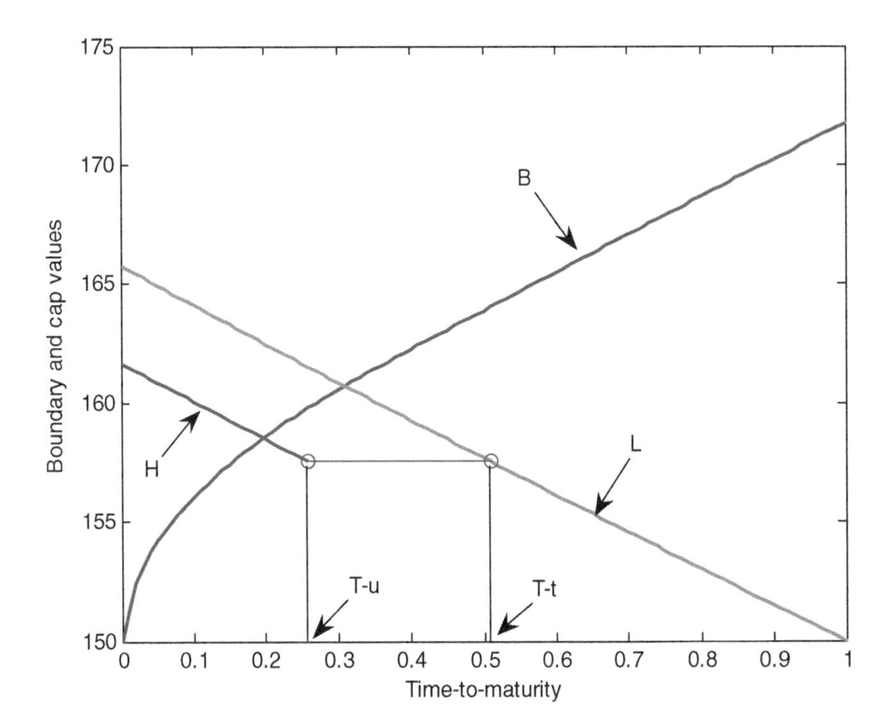

Figure 5.4: Boundary and cap (y-axis) versus time to maturity (x-axis). The exercise boundary of an uncapped call B, the cap function L and its translation H are plotted against time to maturity. The option strike is $K = 100$, the initial cap $L_0 = 150$ and the maturity date $T = 1$. The interest rate is $r = 6\%$, the dividend rate $\delta = 4\%$ and the volatility coefficient $\sigma = 0.2$. Computations use the integral equation method with 100 time steps.

Lemma 64 *Suppose that the underlying asset price S satisfies (4.1). Consider two American capped call options written on S, with common maturity date T and exercise price K, and respective caps L and H satisfying (5.14) with initial condition $L_0 > H_0$. Let $S_0^1 = L_0$ and $S_0^2 = H_0$. Then $C^L\left(S_0^1, 0\right) \leq C^H\left(S_0^2, 0\right) + L_0 - H_0$.*

Proof of Lemma 64: For any stopping time $\tau \in \mathcal{S}_{0,T}$ we have

$$\begin{aligned} 0 &\leq \left(S_\tau^1 \wedge L_\tau - K\right)^+ - \left(S_\tau^2 \wedge H_\tau - K\right)^+ \\ &\leq S_\tau^1 \wedge L_\tau - S_\tau^2 \wedge H_\tau \\ &= S_0^1 N_{0,\tau} \wedge L_0 e^{g\tau} - S_0^2 N_{0,\tau} \wedge H_0 e^{g\tau} \end{aligned}$$

where $N_{0,\tau}$ is the exponential process defined in the proof of proposition 31. As $S_0^1 = L_0$ and $S_0^2 = H_0$ the expression on the last line is $\left(S_0^1 - S_0^2\right)\left(N_{0,\tau} \wedge e^{g\tau}\right)$, which is bounded above by $\left(S_0^1 - S_0^2\right) N_{0,\tau}$. This upper bound for the payoff holds, in particular, for the optimal stopping time τ_1 associated with $\left(S_0^1, 0\right)$. Hence, we can write

$$\begin{aligned}
0 &\leq C^L\left(S_0^1, 0\right) - C^H\left(S_0^2, 0\right) \\
&= \widetilde{E}\left[e^{-r\tau_1}\left(S_0^1 N_{0,\tau_1} \wedge L_0 e^{g\tau_1} - K\right)^+\right] \\
&\quad - \widetilde{E}\left[e^{-r\tau_2}\left(S_0^2 N_{0,\tau_2} \wedge H_0 e^{g\tau_2} - K\right)^+\right] \\
&\leq \widetilde{E}\left[e^{-r\tau_1}\left(S_0^1 - S_0^2\right) N_{0,\tau_1}\right] \qquad \text{(suboptimality of } \tau_1 \text{ at } \left(S_0^2, 0\right) \\
&\qquad\qquad \text{and upper bound)} \\
&\leq S_0^1 - S_0^2 \qquad (Q - \text{supermartingale property of } R_{0,t} S_t).
\end{aligned}$$

By assumption $S_0^1 = L_0$ and $S_0^2 = H_0$. This proves the claim. ■

Proof of Theorem 57: The valuation formula follows from the exercise policy described in Theorem 56. ■

Proof of Theorem 58: We must show the optimality of stopping at the first hitting time of the cap. The valuation formula (5.20) is the value under that exercise policy.

(i) Suppose first that $S < L$ and assume that immediate exercise is optimal. Consider the investment policy described below along with the exercise policy τ_L defined in the theorem

	time t	time $\tau_L < T$	time $\tau_L \geq T$
Buy Call	$-(S-K)$	$L_{\tau_L} - K$	$(S_T - K)^+$
Sell Stock	$+S$	$-S_{\tau_L}$	$-S_T$
Invest K	$-K$	$K \exp\left(\int_t^{\tau_L} r_v dv\right)$	$K \exp\left(\int_t^T r_v dv\right)$
Total	0	$K \exp\left(\int_t^{\tau_L} r_v dv\right)$	$K \exp\left(\int_t^T r_v dv\right)$
		$-K$	$-K 1_{\{S_T \geq K\}} - S_T 1_{\{S_T < K\}}$

As the payoff on the event $\{\tau_L \geq T\}$ is bounded below by

$$-K 1_{\{S_T < K\}} + K\left(\exp\left(\int_t^T r_v dv\right) - 1_{\{S_T \geq K\}}\right) = K\left(\exp\left(\int_t^T r_v dv\right) - 1\right)$$

and as $r > 0$ the policy outlined represents an arbitrage opportunity. The absence of arbitrages in equilibrium implies that immediate exercise is a suboptimal policy.

(ii) Consider now the case $S \geq L$. By Ito's lemma the discounted payoff $\psi_t \equiv \exp\left(-\int_0^t r_v dv\right)(L_t - K)$ satisfies

$$d\psi_t = \exp\left(-\int_0^t r_v dv\right)\left[(g_t - r_t) L_t + r_t K\right] dt, \quad t \in [0, T].$$

Condition (5.19) implies that the process ψ is decreasing (P-a.s.). The optimality of immediate exercise follows as any waiting policy leads to a decrease in the discounted payoff. ■

Proof of Theorem 59: Let t^* be the smallest solution of the equation $B(t) = L$; or $t^* = T$ if $L \leq B(T)$; or $t^* = 0$ if $L \geq B(0)$. If the boundary is only right-continuous and the cap passes in between two limit points, i.e., $L \in [B(t), \inf\{B_s : s < t\}]$, at some discontinuity point t, let t^* be this unique point t.

By the same arguments as in the proof of Theorem 51 we conclude that $B^L(t) \leq B(t) \wedge L$, with equality for $t \in [t^*, T)$ (i.e., $B^L(t) = B(t)$ for $t \in [t^*, T)$). To show that the upper bound is attained for $t \in [0, t^*)$ we will use the fact that the function $t \to B^L(t)$ is non-increasing. Define the difference $\Lambda(t) \equiv B(t) \wedge L - B^L(t)$ and note that $\Lambda(t) \geq 0$ for $t \in [0, T)$ while $\Lambda(t) = 0$ for $t \in [t^*, T)$. These properties combined with $B^L(t)$ non-increasing and L constant imply that $\Lambda(t) = 0$ for $t \in [0, t^*)$. ■

Chapter 6

Options on Multiple Assets

This chapter is devoted to American options written on multiple underlying assets. We first provide background information about multiasset options (section 6.1) and describe the financial environment (section 6.2). We then study, in succession, options on the maximum of two assets (section 6.3), spread options (section 6.4), options on the average of two assets (section 6.5) and options on the minimum of two assets (section 6.6). For each of these contracts the optimal exercise region is characterized and valuation formulas are provided. Extensions to derivatives on $n > 2$ assets are presented in Appendix A.

6.1 Definitions, Examples and Literature

An increasing number of contractual rewards and obligations, held by economic entities, involve options that are written on multiple underlying assets. Some of the payoffs that have become fairly common include options on the maximum of several asset prices (max-options), options on the minimum of several prices (min-options), options on the difference between two prices (spread options) and options on an average of prices (portfolio options). Contingent claims with these characteristics can be found on financial exchanges, in over-the-counter transactions, in cash flows generated by investment decisions of firms and in executive compensation plans.

The simplest example of a multiasset contingent claim is an index option. A contract of this type is typically written on a portfolio of exchange-traded assets and has an exercise payoff depending on an average of prices. For instance, S&P 100 Index Options, that have traded on the CBOE since March 1983, are American-style options written on the Standard & Poor's 100 Index, a capitalization-weighted index of 100 stocks. S&P 100 Index LEAPS (Long-term Equity AnticiPation Securities) are long-dated American-style options, on the same index. Options on the difference between asset prices, spread options, can also be found on several exchanges. Crack Spread Options, traded on the New York Mercantile Exchange (NYMEX), are essentially contracts on the difference between raw material and finished product prices. The Gasoline Crack Spread Option is written on the spread between the NYMEX New York Harbor unleaded gasoline futures price and the NYMEX crude oil futures price. The Heating Oil Crack Spread Option is on the spread between the NYMEX New York Harbor heating oil futures price and the NYMEX crude oil futures price. Both contracts are American-style.

Derivatives paying the maximum or minimum of several prices or the difference between two prices appear in a variety of contexts, often outside the realm of exchanges. Corporate bonds and managerial contracts commonly include 'sweeteners', taking the form of multiasset payoffs, designed to enhance the attractiveness of an issue or to provide adequate incentives. Embedded American max-options are routinely found in capital budgeting decisions confronting firms. They are typically associated with assets that can be put to multiple, but exclusive, uses. Real estate development companies, for instance, often face a choice among several exclusive alternatives regarding land use. A particular type of spread options, options to exchange one asset for another, is also common. It characterizes certain employment switching decisions and is the hallmark of stock tender offers. Switching options are also standard in capital budgeting decisions.

In most cases the assets underlying these claims pay dividends or produce cash flows. As shown in chapters 3 and 4, it is optimal to exercise an American-style claim written on a single dividend-paying asset, under appropriate circumstances. Intuition suggests that the same ought to be true for multiasset claims. Clearly, if the local gains are large enough, early exercise will be optimal when the right conditions prevail. The determination of these conditions, however, is not as straightforward as in the single asset case. The presence of multiple prices, that jointly determine the exercise region, is an obvious complication. More subtle differences appear in terms of intuition. As we shall see, some of the basic insights from the single asset case do not carry over when several prices affect the payoff. In addition, small variations in contractual structures can lead to important changes in optimal exercise policies. The multivariate nature of these claims ultimately results in complex decision-making. Precise descriptions of the exercise regions become useful for accurate decision-making, as well as for pricing and risk management purposes.

The early literature on multiasset options has primarily focused on European-style contracts. European options to exchange one asset for another were analyzed by Margrabe [1978]. Valuation formulas for European put and call options on the maximum or minimum of two assets can be found in Johnson [1981] and Stulz [1982]. The case of several assets is treated by Johnson [1987]. In comparison, American options on multiple dividend-paying assets have only received attention in more recent years. Tan and Vetzal [1994] perform numerical simulations to identify the immediate exercise region for some types of exotic options. Geltner, Riddiough and Stojanovic [1996] also provide insights about the exercise region for a perpetual option on the best of two assets in the context of land use choice. The case of perpetual options on the minimum of two nondividend-paying assets is investigated by Villeneuve [1997]. The results presented in this chapter are mainly drawn from Broadie and Detemple [1997a] and Detemple, Feng and Tian [2003] who provide detailed descriptions of the exercise regions and valuation formulas associated with common multiasset options.

6.2 The Financial Market

Through most of this chapter we consider derivative securities written on a pair of assets, with prices S^1 and S^2. Under the risk neutral measure, the price of asset $i, i = 1, 2$, evolves according to

$$dS_t^i = S_t^i \left[(r - \delta_i) \, dt + \sigma_i d\widetilde{z}_t^i \right] \tag{6.1}$$

where $r \geq 0$ is the interest rate, $\delta_i \geq 0$ is the dividend yield and $\sigma_i > 0$ is the volatility of the asset's return. The coefficients (r, δ_i, σ_i) are constants. The processes $\widetilde{z}^i, i = 1, 2$, are standard Brownian motions, under the risk neutral measure. Their correlation coefficient ρ is constant and satisfies the non-degeneracy condition $\mid \rho \mid < 1$.

Appendix A will allow for a more general framework with $n > 2$ assets. Prices will also be modelled as geometric Brownian motions with coefficients $(\delta_i, \sigma_i), i = 1, ..., n$.

6.3 Call Options on the Maximum of 2 Prices

We start with an analysis of an American call option written on the maximum of the two asset prices S^1, S^2 (a max-call option). Let T be the maturity date of the option. The option payoff, if exercised at some time $t \leq T$, is $(S_t^1 \vee S_t^2 - K)^+$, where $S_t^1 \vee S_t^2 \equiv \max(S_t^1, S_t^2)$. Let $C^x(S_t^1, S_t^2, t)$ be the option value at (S_t^1, S_t^2, t). Drawing on the results in chapter 3 value can be expressed as

$$C^x(S^1, S^2, t) = \sup_{\tau \in \mathcal{S}_{t,T}} \widetilde{E}_t \left[e^{-r(\tau - t)} (S_\tau^1 \vee S_\tau^2 - K)^+ \right],$$

where $\mathcal{S}_{t,T}$ is the class of stopping times of the filtration generated by the underlying Brownian motion processes $(\widetilde{z}_1, \widetilde{z}_2)$. The immediate exercise region is the set

$$\mathcal{E}^{c,x} \equiv \left\{ (S^1, S^2, t) \in \mathbb{R}_+^2 \times [0, T] : C_t^x(S^1, S^2) = (S^1 \vee S^2 - K)^+ \right\}$$

of price-date points (S^1, S^2, t) where the option price matches the immediate exercise payoff. Its complement is the continuation region

$$\mathcal{C}^{c,x} \equiv \left\{ (S^1, S^2, t) \in \mathbb{R}_+^2 \times [0, T] : C_t^x(S^1, S^2) > (S^1 \vee S^2 - K)^+ \right\}.$$

The t-section of $\mathcal{E}^{c,x}$, denoted by $\mathcal{E}^{c,x}(t) \equiv \left\{ (S^1, S^2) \in \mathbb{R}_+^2 : (S^1, S^2, t) \in \mathcal{E}^{c,x} \right\}$, is the set of price pairs (S^1, S^2) in the exercise region at the fixed time t.

6.3.1 Exercise Region of a Max-Call Option

In order to provide perspective it is useful to recall some of the fundamental properties of the exercise regions of American-style vanilla call options. Three features, in particular, deserve to be highlighted. The first one is the up-connectedness of the call exercise region: $(S,t) \in \mathcal{E}$ implies $(\lambda S,t) \in \mathcal{E}$ for all $\lambda \geq 1$. The second is the optimality of immediate exercise for sufficiently large asset value: when $\delta > 0$ there exists a constant M such that $(S,t) \in \mathcal{E}$ for all $S \geq M$. The last one is the convexity of the time section of the exercise region with respect to the asset price.

These properties, at first sight, appear fairly generic and it is tempting to conjecture that an American max-call will give rise to similar behavior.

Conjecture 65 *Let $S \equiv (S^1, S^2)$ be the price vector. The exercise region of the max-call has the following properties*

(i) $(S,t) \in \mathcal{E}^{c,x}$ implies $(\lambda_1 S^1, \lambda_2 S^2, t) \in \mathcal{E}^{c,x}$ for all $\lambda_1 \geq 1$ and $\lambda_2 \geq 1$.

(ii) Suppose that $\delta_1 > 0$ and $\delta_2 > 0$. Then there exist two constants M_1 and M_2 such that $(S,t) \in \mathcal{E}^{c,x}$ for all $S^1 \geq M^1$ and all $S^2 \geq M^2$.

(iii) $(S,t) \in \mathcal{E}^{c,x}$ and $\left(\widetilde{S},t\right) \in \mathcal{E}^{c,x}$ implies $(S(\lambda),t) \in \mathcal{E}^{c,x}$ for all $\lambda \in [0,1]$ where $S(\lambda) \equiv \lambda S + (1-\lambda)\widetilde{S}$.

Conjecture (i) is prompted by the fact that the exercise payoff increases when either one or both prices increase. If it is optimal to exercise at initial prices there should be even more incentives to exercise at higher prices. Indeed, the local gains collected in the event of early exercise increase as prices increase. The intuition for (ii) stems from the bounded nature of the exercise boundary in the single asset case with dividends. By analogy, it ought to be optimal to exercise, in the two asset case, when both prices exceed a threshold. Convexity of the max-call payoff function motivates (iii).

Quite surprisingly, these seemingly intuitive properties are not valid. As will become clear, attention must be restricted to subregions of the space $\mathcal{E}^{c,x}$ in order to obtain features that are similar to those for $\mathcal{E}$. The set

$$\mathcal{G}_i \equiv \left\{(S^1,S^2,t) \in \mathbb{R}_+^2 \times [0,T] : S^i = S^1 \vee S^2\right\},$$

where the price of asset i is the maximum of the two prices, plays a crucial role. Taking its intersection with the exercise region gives $\mathcal{E}_i^{c,x} = \mathcal{E}^{c,x} \cap \mathcal{G}_i$. This is the subset of the immediate exercise region in which S^i is the maximum of the two prices. Our next proposition shows that the first two conjectures above are invalid and establishes the relevance of the subregions $\mathcal{E}_i^{c,x}, i = 1,2$.

Proposition 66 *(Diagonal behavior) $S^1 = S^2 > 0$ and $t < T$ implies $(S^1,S^2,t) \notin \mathcal{E}^{c,x}$.*

Proposition 66 shows that immediate exercise before maturity is never optimal when the underlying asset prices are equal (i.e., along the diagonal of the 2-dimensional price space). It follows that the diagonal $S^1 = S^2$, at $t < T$, belongs to the continuation region and that the exercise region, at $t < T$, can be written as the union of two sets with empty intersection.

Some intuition for the suboptimality of immediate exercise along the diagonal can be provided along the following lines. Suppose that asset prices are the same. To simplify matters also assume that both dividend yields equal the interest rate and that prices are independent. Then, the probability of an increase in any one of the underlying asset prices over the next increment of time is roughly equal to 1/2. But the probability of an increase in the maximum of the two prices is about 3/4. The max-option holder has therefore a much better chance of improving his/her payoff by waiting, than the holder of an option on a single underlying asset. These improved odds are sufficient to induce the optimality of a waiting strategy. When correlation differs from zero the probabilities of the events described above are modified. Yet, as long as correlation is not perfect the odds of an increase in the maximum of two prices are better than the odds of an increase in any one of the underlying prices. This enhances the value of waiting for the max-option.

Additional motivation can be provided by noticing that the policy of delaying exercise up to some fixed time $s > t$ has value

$$\begin{aligned}
& \widetilde{E}_t\left[e^{-r(s-t)}\left(S_s^1 \vee S_s^2 - K\right)^+\right] \\
= \ & \widetilde{E}_t\left[e^{-r(s-t)}\left(S_s^1 - K + \left(S_s^2 - S_s^1\right)^+\right)^+\right] \\
\geq \ & \widetilde{E}_t\left[e^{-r(s-t)}\left(S_s^1 - K\right)\right] + \widetilde{E}_t\left[e^{-r(s-t)}\left(S_s^2 - S_s^1\right)^+\right] \\
= \ & S^1 e^{-\delta_1(s-t)} - Ke^{-r(s-t)} + \widetilde{E}_t\left[e^{-r(s-t)}\left(S_s^2 - S_s^1\right)^+\right]
\end{aligned}$$

where the first component of the lower bound on the last line is the value of a forward contract on asset 1 with delivery price K and the second component is the value of a European option to exchange asset 2 for asset 1, both with maturity date s. When s approaches t, the value of the forward contract converges, at a finite rate, to $S^1 - K$. At the same time the value of the exchange option decreases to zero at a rate that increases to infinity in the limit (see Figure 6.1 for an illustration). The value of the package consisting of the forward contract and the exchange option is therefore strictly positive for a range of times greater than t. The existence of an exercise time s, strictly larger than t, that dominates immediate exercise follows.

The subregions $\mathcal{E}_i^{c,x}, i = 1, 2$, of the exercise region display many of the properties satisfied by the exercise region in the single asset case. The next proposition summarizes some of the salient features of these sets. We recall the vector notation $S = \left(S^1, S^2\right)$ and $\widetilde{S} = \left(\widetilde{S}^1, \widetilde{S}^2\right)$. We also use $B^1(t)$ to denote the optimal exercise boundary for a call option written on asset 1 alone.

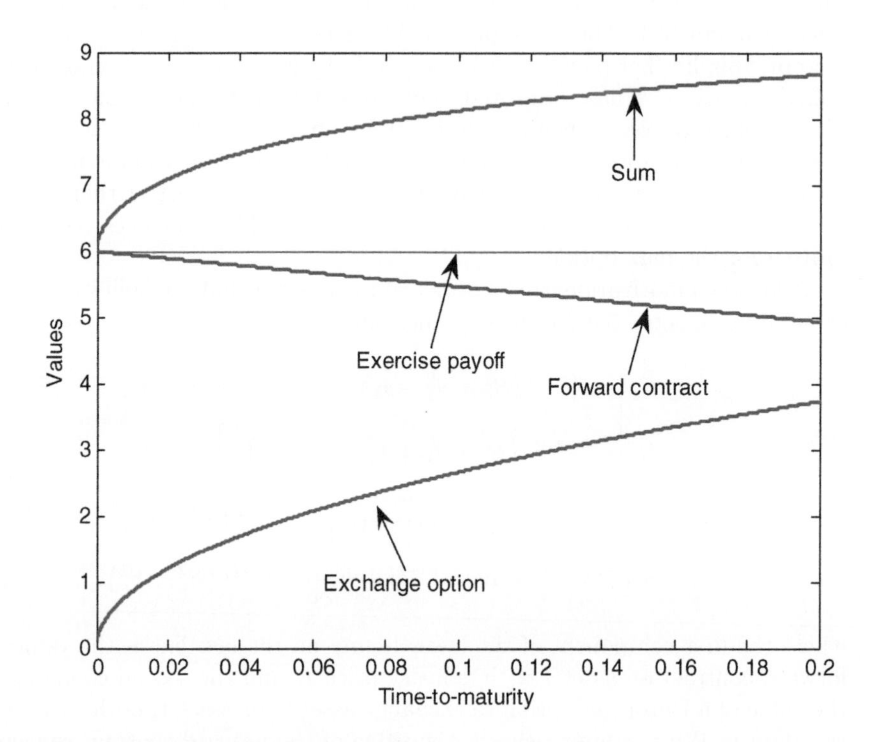

Figure 6.1: This figure displays the behavior of the lower bound for the value of a max-call option at a point on the diagonal. The values of the forward contract, the European exchange option and the sum of the two (portfolio) are graphed against time-to-maturity. Parameter values are $S^1 = S^2 = 106$, $K = 100$, $\delta_1 = \delta_2 = 6\%$, $\sigma_1 = \sigma_2 = 0.20$, $\rho = 0.5$, $r = 1\%$. The immediate exercise value is 6. Time-to-maturity varies between 0 and 0.2.

Proposition 67 *The immediate exercise region for the max-option,* $\mathcal{E}^{c,x} = \mathcal{E}_1^{c,x} \cup \mathcal{E}_2^{c,x}$*, satisfies the following properties.*

(i) $(S,t) \in \mathcal{E}^{c,x}$ *implies* $(S,s) \in \mathcal{E}^{c,x}$ *for all* $t \leq s \leq T$.

(ii) $(S^1, S^2, t) \in \mathcal{E}_1^{c,x}$ *implies* $(\lambda S^1, S^2, t) \in \mathcal{E}_1^{c,x}$ *for all* $\lambda \geq 1$.

(iii) $(S^1, S^2, t) \in \mathcal{E}_1^{c,x}$ *implies* $(S^1, \lambda S^2, t) \in \mathcal{E}_1^{c,x}$ *for all* $0 \leq \lambda \leq 1$.

(iv) $(S^1, 0, t) \in \mathcal{E}_1^{c,x}$ *if and only if* $S^1 \geq B^1(t)$.

(v) $(S,t) \in \mathcal{E}_1^{c,x}$ *and* $\left(\widetilde{S}, t\right) \in \mathcal{E}_1^{c,x}$ *implies* $(S(\lambda), t) \in \mathcal{E}_1^{c,x}$ *for all* $0 \leq \lambda \leq 1$.

Results (ii), (iii), (iv) and (v) also hold for the subregion $\mathcal{E}_2^{c,x}$.

The first property captures the standard notion that the exercise region expands as the maturity date approaches. This feature follows from the classic observation that the set of exercise opportunities shrinks when the time-to-maturity decreases. Given that a short-dated option cannot be more valuable than a long-dated option, immediate exercise must be an optimal policy for the short-dated contract if it is optimal for the long-dated one. Property (ii) is a connectedness property of the subregion $\mathcal{E}_1^{c,x}$. It shows that immediate exercise remains optimal when S^1 increases, if it is optimal at the initial point (S^1, S^2, t) (hence establishes right-connectedness of the set $\mathcal{E}_1^{c,x}$). The intuition rests on the price bound

$$C^x\left(\lambda S^1, S^2, t\right) \leq C^x\left(S^1, S^2, t\right) + \lambda S^1 - S^1.$$

If immediate exercise is optimal at (S^1, S^2, t) and S^1 is the largest of the two prices (i.e. $(S^1, S^2, t) \in \mathcal{E}_1^{c,x}$) the upper bound equals $\lambda S^1 - K$, which represents the immediate exercise payoff at $(\lambda S^1, S^2, t)$. It follows that immediate exercise must also be optimal at $(\lambda S^1, S^2, t)$. Property (iii) is also a connectedness property. It shows that the subregion $\mathcal{E}_1^{c,x}$ is connected when S^2 decreases.

For property (iv) it suffices to notice that the price process S^2 has an absorbing barrier at zero. At the point $(S^1, 0, t)$ the call option on the maximum of two prices becomes a call option on a single asset: $C^x(S^1, 0, t) = C(S^1, t)$. Its exercise region naturally corresponds to the exercise region of the call on asset 1. Property (iv) shows that the subregions $\mathcal{E}_1^{c,x}, \mathcal{E}_2^{c,x}$ are non-empty, for $t < T$, when dividends are positive. In the light of Proposition 66 this establishes that part (iii) of Conjecture 65 fails.

The last property, property (v), establishes the convexity of the subregion $\mathcal{E}_1^{c,x}$ with respect to prices. This feature is implied by the convexity of the payoff function with respect to the vector $S = (S^1, S^2)$ and the multiplicative nature of the geometric Brownian motion process (6.1) (i.e., the fact that $S_\tau^i = S^i N_{t,\tau}^i$ for $\tau \in \mathcal{S}_{t,T}$, where $N_{t,\tau}^i \equiv \exp\left(\left(r - \delta_i - \frac{1}{2}\sigma_i^2\right)(\tau - t) + \sigma_i\left(z_\tau^i - z_t^i\right)\right)$).

The results in Propositions 66 and 67 put a fair amount of structure on the t-sections $\mathcal{E}_i^{c,x}(t) \equiv \left\{(S^1, S^2) \in \mathbb{R}_+^2 : (S^1, S^2, t) \in \mathcal{E}_i^{c,x}\right\}$ of the subregions $\mathcal{E}_i^{c,x}, i = 1, 2$. The suboptimality of immediate exercise along the diagonal implies that $\mathcal{E}^{c,x}(t) = \mathcal{E}_1^{c,x}(t) \cup \mathcal{E}_2^{c,x}(t)$ with $\mathcal{E}_1^{c,x}(t) \cap \mathcal{E}_2^{c,x}(t) = \varnothing$ for $t < T$. The price-convexity of $\mathcal{E}_i^{c,x}$ shows that the t-section $\mathcal{E}_i^{c,x}(t)$ is convex, for all $t \in [0, T]$. The next proposition, in addition, establishes that the t-sections diverge from the diagonal as prices increase, before the maturity date. To state this result define the open cone

$$R(\lambda_1, \lambda_2) \equiv \left\{ (S^1, S^2) \in \mathbb{R}_+^2 : \lambda_2 S^1 < S^2 < \lambda_1 S^1 \right\}$$

for $0 \leq \lambda_2 < \lambda_1$. The set $R(\lambda_1, \lambda_2)$ is the cone with vertex at the origin and sides determined by the price ratios λ_1, λ_2.

Proposition 68 *(Divergence of the t-sections of the exercise region). For each $t < T$ there exists a pair (λ_1, λ_2) with $0 \leq \lambda_2 < 1 < \lambda_1$ such that $\mathcal{E}^{c,x}(t) \cap R(\lambda_1, \lambda_2) = \emptyset$.*

Proposition 66 shows that the diagonal belongs to the continuation region, prior to maturity. Proposition 68 establishes a stronger property: it shows the existence of an open cone, containing the diagonal, that does not intersect the t-section $\mathcal{E}^{c,x}(t)$ (hence belongs to the continuation region). Given that the sides of the cone diverge from the diagonal when prices increase, the boundaries of the t-sections will also eventually diverge when prices become sufficiently large (a formal definition of the boundaries of the sets $\mathcal{E}_i^{c,x}(t)$ is provided in the next section).

Figures 6.2-6.4 illustrate the structure of the exercise region. Figure 6.2 shows the behavior of the boundaries near maturity. In this example the boundaries of the t-sections converge to piecewise linear functions as t approaches T: for the set $\mathcal{E}_1^{c,x}(t)$ the limiting value of its boundary is the diagonal for $S^2 \geq K$ and the strike K for $S^2 < K$. Figures 6.3 and 6.4 show the t-sections of the exercise region 3 months and 6 months before maturity. The suboptimality of immediate exercise along the diagonal (Proposition 66) and the divergence of the t-sections as prices increase (Proposition 68) appear clearly at all maturities. The shrinkage of the exercise region, as maturity recedes, is also apparent (Proposition 67(i)).

6.3.2 Valuation of Max-Call Options

We now examine the valuation of the contract. The price function $C^x(S^1, S^2, t)$ exhibits the following features.

Proposition 69 *The value of the American max-call option, $C^x(S^1, S^2, t)$ has the following properties:*

(i) $C^x(S^1, S^2, t)$ is continuous on $\mathbb{R}_+^2 \times [0, T]$.

(ii) $C^x(\cdot, \cdot, t)$ is non-decreasing on $\mathbb{R}_+^2$ for all $t \in [0, T]$.

(iii) $C^x(S^1, S^2, \cdot)$ is non-increasing on $[0, T]$ for all $(S^1, S^2) \in \mathbb{R}_+^2$.

(iv) $C^x(\cdot, \cdot, t)$ is convex on $\mathbb{R}_+^2$, for all $t \in [0, T]$.

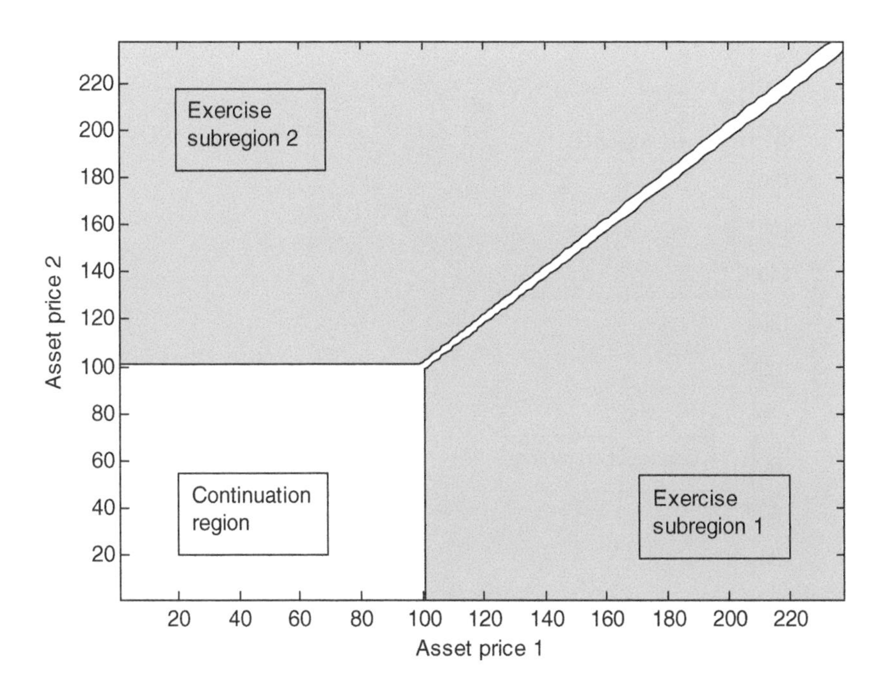

Figure 6.2: This figure displays the exercise region of a max-call option one day prior to maturity. Parameter values are $\delta_1 = \delta_2 = 6\%$, $\sigma_1 = \sigma_2 = 0.20$, $\rho = 0$, $r = 6\%$ and $K = 100$. Computations use a bivariate version of the binomial model with 400 time steps.

The continuity of the payoff function $\left(S^1 \vee S^2 - K\right)^+$ with respect to the price pair $\left(S^1, S^2\right)$ combines with the continuity of the flow generated by stochastic differential equations of the form (6.1) (i.e., the continuity of $S^i_\tau = S^i N^i_{t,\tau}$, $\tau \in \mathcal{S}_{t,T}$, with respect to the initial position S^i) to produce property (i). The non-decreasing structure of the price function in (ii) is also due to the non-decreasing behaviors of the payoff and of the flow of the SDE (6.1). The intuition underlying property (iii) rests on the loss of value associated with a reduced set of exercise opportunities (see property (i) of Proposition 67). The convexity of the payoff function is the source of (iv).

The continuity of the price function implies that the continuation region $\mathcal{C}^{c,x}$ is an open set while the immediate exercise region $\mathcal{E}^{c,x}$ is a closed set. Based on the structure of the exercise set we can define the boundary functions

$$B_1^{c,x}\left(S^2, t\right) = \inf\left\{S^1 \in \mathbb{R}_+ : \left(S^1, S^2\right) \in \mathcal{E}_1^{c,x}(t)\right\}$$

$$B_2^{c,x}\left(S^1, t\right) = \inf\left\{S^2 \in \mathbb{R}_+ : \left(S^1, S^2\right) \in \mathcal{E}_2^{c,x}(t)\right\}$$

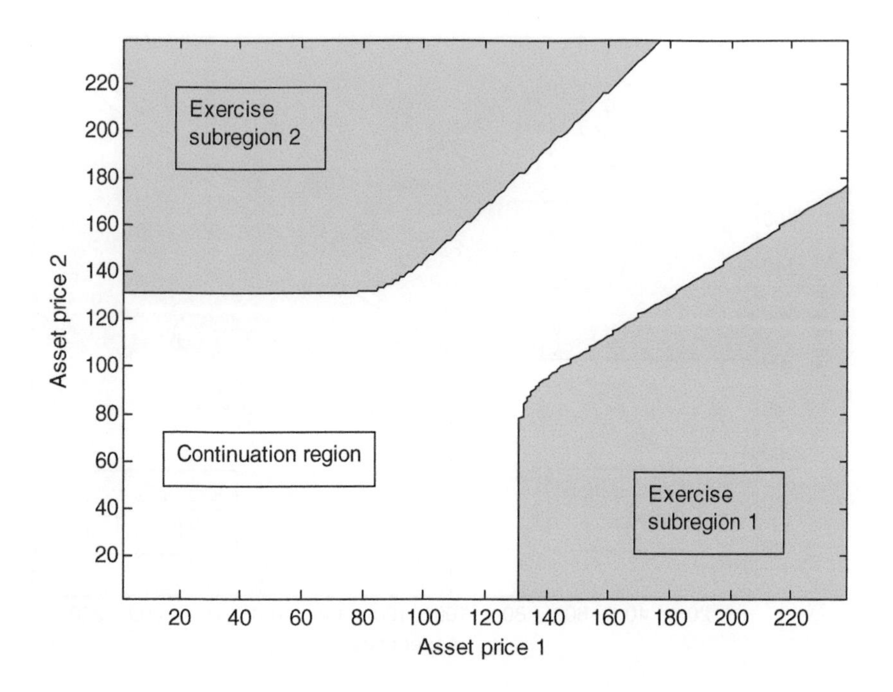

Figure 6.3: This figure displays the exercise region of a max-call option 6 months before maturity. Parameter values are $\delta_1 = \delta_2 = 6\%$, $\sigma_1 = \sigma_2 = 0.20$, $\rho = 0$, $r = 6\%$ and $K = 100$. Computations use a bivariate version of the binomial model with 400 time steps.

for $t \in [0, T]$. The function $B_1^{c,x}(S^2, t)$ represents the boundary of the t-section $\mathcal{E}_1^{c,x}(t)$; $B_2^{c,x}(S^1, t)$ is the boundary of $\mathcal{E}_2^{c,x}(t)$. The t-section of the exercise region can be rewritten as

$$\mathcal{E}^{c,x}(t) = \left\{ (S^1, S^2) \in \mathbb{R}_+^2 : S^1 \geq B_1^{c,x}(S^2, t) \text{ or } S^2 \geq B_2^{c,x}(S^1, t) \right\}.$$

The optimal exercise time is the first time at which one of the two boundaries is crossed

$$\tau = \inf \left\{ t \in [0, T] : S_t^1 \geq B_1^{c,x}(S_t^2, t) \text{ or } S_t^2 \geq B_2^{c,x}(S_t^1, t) \right\}.$$

If such a time does not exist set $\tau = T$ (with the usual convention that the payoff is null in that event).

Properties of the boundaries can be inferred from the results in Propositions 66, 67 and 68. The function $B_1^{c,x}(S^2, t)$, for instance, can be shown to be non-decreasing and convex in S^2, non-increasing in t, bounded below by S^2 (the

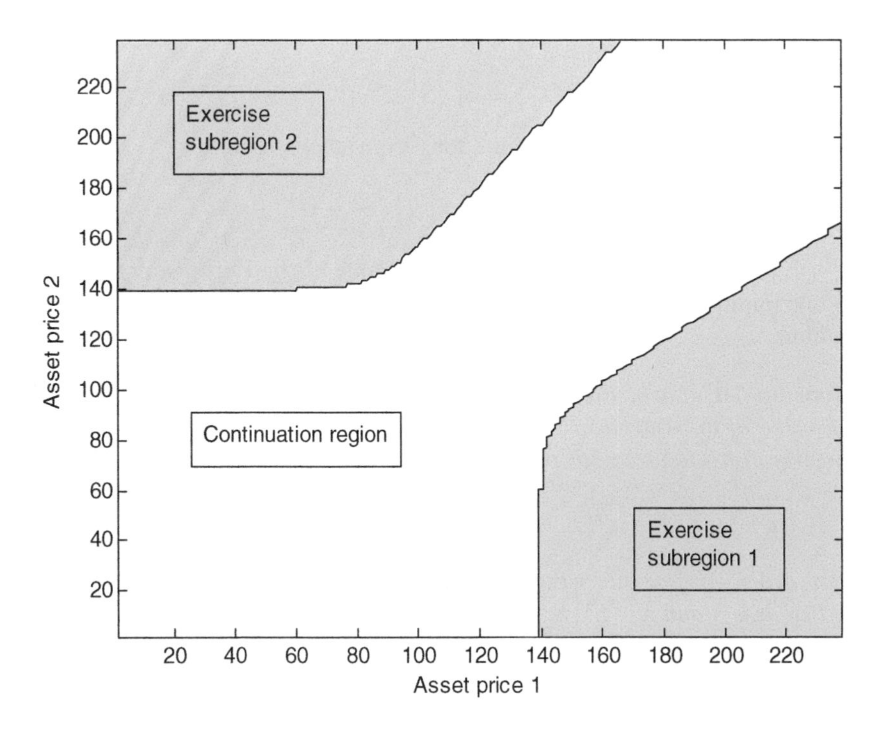

Figure 6.4: This figure displays the exercise region of a max-call option 1 year before maturity. Parameter values are $\delta_1 = \delta_2 = 6\%$, $\sigma_1 = \sigma_2 = 0.20$, $\rho = 0$, $r = 6\%$ and $K = 100$. Computations use a bivariate version of the binomial model with 400 time steps.

diagonal) and such that $B_1^{c,x}(0,t) = B^1(t)$. It has limiting value

$$B_1^{c,x}\left(S^2, T_-\right) = \begin{cases} S^2 & \text{if } S^2 \geq \max\left(K, \frac{r}{\delta_1}K\right) \\ \max\left(K, \frac{r}{\delta_1}K\right) & \text{if } S^2 < \max\left(K, \frac{r}{\delta_1}K\right). \end{cases}$$

As for the single asset case $B_1^{c,x}\left(S^2, \cdot\right)$ has a discontinuity at T if the interest rate exceeds the dividend rate, $r > \delta_1$.

The results in chapter 3 show that the max-option price has an early exercise premium representation. Based on the characterization of the optimal stopping time above we introduce the functions

$$c_t^x\left(S^1, S^2\right) = \widetilde{E}_t\left[e^{-r(T-t)}\left(\max\left(S_T^1 \vee S_T^2\right) - K\right)^+\right]$$

$$\pi_1^x\left(S^1, S^2, t; B_1^{c,x}\right) = \widetilde{E}_t\left[\int_t^T e^{-r(v-t)}\left(\delta_1 S_v^1 - rK\right) 1_{\left\{S_v^1 \geq B_1^{c,x}\left(S_v^2, v\right)\right\}}\right] dv$$

$$\pi_2^x\left(S^1,S^2,t;B_2^{c,x}\right)=\widetilde{E}_t\left[\int_t^T e^{-r(v-t)}\left(\delta_2 S_v^2-rK\right)1_{\left\{S_v^2\geq B_2^{c,x}\left(S_v^1,v\right)\right\}}\right]dv$$

$$\pi^x\left(S^1,S^2,t;B_1^{c,x},B_2^{c,x}\right)=\pi_1^x\left(S^1,S^2,t;B_1^x\right)+\pi_2^x\left(S^1,S^2,t;B_2^x\right)$$

where $\pi_i^x\left(S^1,S^2,t;B_i^{c,x}\right)$ is defined for a continuous surface

$$\left\{B_i^{c,x}\left(S^j,v\right):v\in[0,T],S^j\in\mathbb{R}^+,j\neq i\right\},$$

$i,j=1,2$. The function $c_t^x\left(S^1,S^2\right)$ is the value of the European call option on the maximum of the two prices and $\pi^x\left(S^1,S^2,t;B_1^{c,x},B_2^{c,x}\right)$ is the early exercise premium.

Theorem 70 *(Early exercise premium representation for max-call options). The value of an American call option on the maximum of two asset prices has the early exercise premium representation*

$$C_t^x\left(S^1,S^2;B_1^{c,x},B_2^{c,x}\right)=c_t^x\left(S^1,S^2\right)+\pi_t^x\left(S^1,S^2,B_1^{c,x},B_2^{c,x}\right)\tag{6.2}$$

where $c_t^x\left(S^1,S^2\right)$ is the value of a European call option on the maximum of the two assets and $\pi^x\left(S^1,S^2,t;B_1^{c,x},B_2^{c,x}\right)$ is the early exercise premium. The exercise boundaries $B_1^{c,x},B_2^{c,x}$ solve the system of coupled non-linear integral equations

$$B_1^{c,x}\left(S^2,t\right)-K=C_t^x\left(B_1^{c,x}\left(S^2,t\right),S^2;B_1^{c,x},B_2^{c,x}\right)\tag{6.3}$$

$$B_2^{c,x}\left(S^1,t\right)-K=C_t^x\left(S^1,B_2^{c,x}\left(S^1,t\right);B_1^{c,x},B_2^{c,x}\right)\tag{6.4}$$

subject to the boundary conditions

$$\lim_{t\uparrow T}B_1^{c,x}\left(S^2,t\right)=B^1\left(T_-\right)\vee S^2,\quad \lim_{t\uparrow T}B_2^{c,x}\left(S^1,t\right)=B^2\left(T_-\right)\vee S^1\tag{6.5}$$

$$B_1^{c,x}\left(0,t\right)=B^1\left(t\right),\quad B_2^x\left(0,t\right)=B^2\left(t\right).\tag{6.6}$$

At maturity $B_1^{c,x}\left(S^2,T\right)=K\vee S^2\leq B^1\left(T_-\right)\vee S^2$ and $B_2^{c,x}\left(S^1,t\right)=K\vee S^1\leq B^2\left(T_-\right)\vee S^1$.

Formula (6.2) decomposes the price of the American max-option into a European option component and an early exercise premium. As usual the EEP component is the present value of the local gains realized by exercising prior to the maturity date of the contract. For the max-option the local benefits of exercise can further be decomposed into two parts. Each of these corresponds to the benefits provided by the maximum of the two prices in the event of exercise ($\delta_i S_v^i-rK$ in the event $S_v^i\geq B_i^{c,x}\left(S_v^j,v\right),j\neq i$).

Applying the EEP formula at boundary points leads to a system of equations for the boundary components $B_1^{c,x},B_2^{c,x}$. The equations obtained have the standard recursive structure. They are also coupled: the integral equation for $B_1^{c,x}$ depends on $B_2^{c,x}$ and conversely. This follows from the fact that the exercise premium, at any given point (S^1, S^2, t), depends on both boundary surfaces integrated over future dates and prices.

More explicit formulas can be provided for the European option and the exercise premium components, by using the geometric Brownian motion assumption (see Johnson (1981) and Stulz (1982) for the European option part). The resulting integral equations, along with the relevant boundary conditions, can in principle serve as a basis for the numerical implementation of the model (see chapter 8 for a description of the integral equation method for the single asset case).

6.3.3 Dual Strike Max-Options

Dual strike max-options generalize max-options. A dual strike max-call pays off $(\max(S^1 - K_1, S^2 - K_2))^+$, i.e., it involves a strike price associated with each one of the underlying asset prices. It collapses to a max-call when the strike prices are equal.

Let $\mathcal{E}^{c,d}$ be the immediate exercise region of a dual strike max-call option. For each i define the subregion

$$\mathcal{E}_i^{c,d} = \mathcal{E}^{c,d} \cap \left\{ (S^1, S^2, t) \in \mathbb{R}_+^2 \times [0, T] : S^i - K_i = \max(S^1 - K_1, S^2 - K_2) \right\},$$

$i = 1, 2$. The set $\mathcal{E}_i^{c,d}$ is the subset of the exercise region in which the spread $S^i - K_i$ is greatest. The exercise region displays the following properties.

Proposition 71 *The immediate exercise region of the dual strike max-call $\mathcal{E}^{c,d}$ has the following properties*

(i) $(S, t) \in \mathcal{E}^{c,d}$ implies $(S, s) \in \mathcal{E}^{c,d}$ for all $t \leq s \leq T$.

(ii) $(S^1, S^2, t) \in \mathcal{E}_1^{c,d}$ implies $(\lambda S^1, S^2, t) \in \mathcal{E}_1^{c,d}$ for all $\lambda \geq 1$.

(iii) $(S^1, S^2, t) \in \mathcal{E}_1^{c,d}$ implies $(S^1, \lambda S^2, t) \in \mathcal{E}_1^{c,d}$ for all $0 \leq \lambda \leq 1$.

(iv) $(S^1, 0, t) \in \mathcal{E}_1^{c,d}$ if and only if $S^1 \geq B^1(t)$.

(v) If $S^2 = S^1 + K_2 - K_1$ and $t < T$, then $(S^1, S^2, t) \notin \mathcal{E}^{c,d}$.

(vi) $(S, t) \in \mathcal{E}_1^{c,d}$ and $(\widetilde{S}, t) \in \mathcal{E}_1^{c,d}$ implies $(S(\lambda), t) \in \mathcal{E}_1^{c,d}$ for all $0 \leq \lambda \leq 1$.

Results (ii), (iii), (iv) and (vi) also hold for the subregion $\mathcal{E}_2^{c,d}$.

The dual strike call option displays properties that are similar to those characterizing the call option on the maximum of two prices. Prior to maturity, the t-section of the exercise region is the union of two disjoint sets. These subregions lie on each side of the line $S^2 = S^1 + K_2 - K_1$ (the translated diagonal): immediate exercise along the translated diagonal is a suboptimal policy before the maturity date.

An EEP formula for the price of the dual strike max-option can also be written as in Theorem 70.

6.3.4 Put Options on the Minimum of 2 Prices

The American put option on the minimum of two assets, or min-put option, is also closely related to the call option on the maximum of two assets. The min-put option payoff is $(K - S^1 \wedge S^2)^+$ at exercise, where $S^1 \wedge S^2 \equiv \min(S^1, S^2)$. Let $\mathcal{E}^{p,m}$ be the immediate exercise region and define the subregions

$$\mathcal{E}_i^{p,m} = \mathcal{E}^{p,m} \cap \{(S^1, S^2, t) \in \mathbb{R}_+^2 \times [0, T] : S^i = S^1 \wedge S^2\},$$

for $i = 1, 2$. Also let $\mathcal{E}_i^p$ denote the immediate exercise region of a standard put option on asset i, $i = 1, 2$. The exercise region satisfies

Proposition 72 *The immediate exercise region for the min-put option, $\mathcal{E}^{p,m}$, has the following properties*

(i) $(S, t) \in \mathcal{E}^{p,m}$ implies $(S, s) \in \mathcal{E}^{p,m}$ for all $t \leq s \leq T$.

(ii) $(S^1, S^2, t) \in \mathcal{E}_1^{p,m}$ implies $(\lambda S^1, S^2, t) \in \mathcal{E}_1^{p,m}$ for all $0 \leq \lambda \leq 1$.

(iii) $(S^1, S^2, t) \in \mathcal{E}_1^{p,m}$ implies $(S^1, \lambda S^2, t) \in \mathcal{E}_1^{p,m}$ for all $\lambda \geq 1$.

(iv) $(S^1, S^2, t) \in \mathcal{E}_1^{p,m}$ implies $(S^1, t) \in \mathcal{E}_1^p$ (i.e., $\mathcal{E}_1^{p,m} \subseteq \mathcal{E}_1^p$).

(v) $(S^1, 0, t) \in \mathcal{E}_1^{p,m}$.

(vi) Suppose $(S^1, t) \in \mathcal{E}_1^p$. Then there exists S^{2*} such that $(S^1, S^2, t) \in \mathcal{E}_1^{p,m}$ for all $S^2 \geq S^{2*}$.

(vii) If $S^1 = S^2 \neq 0$ and $t < T$, then $(S^1, S^2, t) \notin \mathcal{E}^{p,m}$.

(viii) $(S, t) \in \mathcal{E}_1^{p,m}$ and $(\widetilde{S}, t) \in \mathcal{E}_1^{p,m}$ implies $(S(\lambda), t) \in \mathcal{E}_1^{p,m}$ for $\lambda \in [0, 1]$.

Results (ii)-(vi) and (viii) also hold for the subregion $\mathcal{E}_2^{p,m}$.

The properties of the min-put option parallel those of the max-call option and can be proved using similar arguments. Formulas in the spirit of those in Theorem 70 can also be provided.

6.3.5 Economic Implications

The analysis performed shows that an American call on the maximum of two assets displays unusual properties. The most surprising aspect is perhaps the fact that waiting to exercise is the optimal policy before the maturity date of the contract if the underlying assets have identical prices. Moreover, under these circumstances, waiting is optimal even if the underlying asset prices or their associated dividend yields are very large. This property of the max-call option stands in sharp contrast with the exercise behavior associated with a standard call option written on a single underlying asset.

The diagonal behavior of the optimal exercise policy of the max-call option has interesting implications for capital budgeting and investment decisions involving a choice between multiple exclusive alternatives. Interpreted in that context the result in Proposition 66 shows that economic entities may optimally delay certain investments when the present values of the alternatives under consideration are identical. Given that projects often involve assets that can be put to multiple exclusive uses this motive for delaying investments is likely to arise in a variety of situations. This insight complements traditional theories of investment behavior under uncertainty (see Dixit and Pindyck [1994]).

A complementary property is the divergence of the exercise region: the distance between the diagonal and the exercise boundaries increases as prices increase (i.e., moving up along the diagonal). Thus, when asset prices are large, immediate exercise may be suboptimal even if the asset prices differ by significant amounts. Interpreted in the context of investment theory this suggests that the adoption of a project among a set of exclusive alternatives may be optimally delayed even if present values are large and differ significantly.

6.4 American Spread Options

A *call spread option* is a contingent claim written on the spread between two prices, with payoff $\left(\max\left(S^2 - S^1, 0\right) - K\right)^+ = \left(S^2 - S^1 - K\right)^+$ upon exercise. When $K = 0$ the spread option becomes an option to exchange asset 1 for asset 2.

Let $C^s\left(S^1, S^2, t\right)$ be the value of the call spread option at the point $\left(S^1, S^2, t\right)$. Relying on the results in chapter 3 we can write

$$C^s\left(S^1, S^2, t\right) = \sup_{\tau \in \mathcal{S}_{t,T}} \widetilde{E}_t\left[e^{-r(\tau - t)}\left(S_\tau^2 - S_\tau^1 - K\right)^+\right].$$

The immediate exercise region is the set

$$\mathcal{E}^{c,s} \equiv \left\{\left(S^1, S^2, t\right) \in \mathbb{R}_+^2 \times [0, T] : C_t^s\left(S^1, S^2\right) = \left(S^2 - S^1 - K\right)^+\right\}.$$

Its t-section is $\mathcal{E}^{c,s}(t)$.

6.4.1 Exercise Region and Valuation

Our next proposition provides key properties of the exercise region.

Proposition 73 *The immediate exercise region for a call spread option, $\mathcal{E}^{c,s}$, has the following properties*

(i) $\left(S^1, S^2, t\right) \in \mathcal{E}^{c,s}$ *and* $t < T$ *implies* $S^2 > S^1 + K$.

(ii) $(S, t) \in \mathcal{E}^{c,s}$ *implies* $(S, s) \in \mathcal{E}^{c,s}$ *for all* $t \leq s \leq T$.

(iii) $\left(S^1, S^2, t\right) \in \mathcal{E}^{c,s}$ *implies* $\left(S^1, \lambda S^2, t\right) \in \mathcal{E}^{c,s}$ *for all* $\lambda \geq 1$.

(iv) $\left(S^1, S^2, t\right) \in \mathcal{E}^{c,s}$ *implies* $\left(\lambda S^1, S^2, t\right) \in \mathcal{E}^{c,s}$ *for all* $0 \leq \lambda \leq 1$.

(v) $\left(0, S^2, t\right) \in \mathcal{E}^{c,s}$ *if and only if* $S^2 \geq B^2(t)$.

(vi) $(S, t) \in \mathcal{E}^{c,s}$ *and* $\left(\widetilde{S}, t\right) \in \mathcal{E}^{c,s}$ *implies* $(S(\lambda), t) \in \mathcal{E}^{c,s}$ *for all* $0 \leq \lambda \leq 1$.

Most of these properties are intuitive adaptations of those characterizing max-options (or dual strike options). For property (i) note that the payoff is non-positive if $S^2 \leq S^1 + K$.

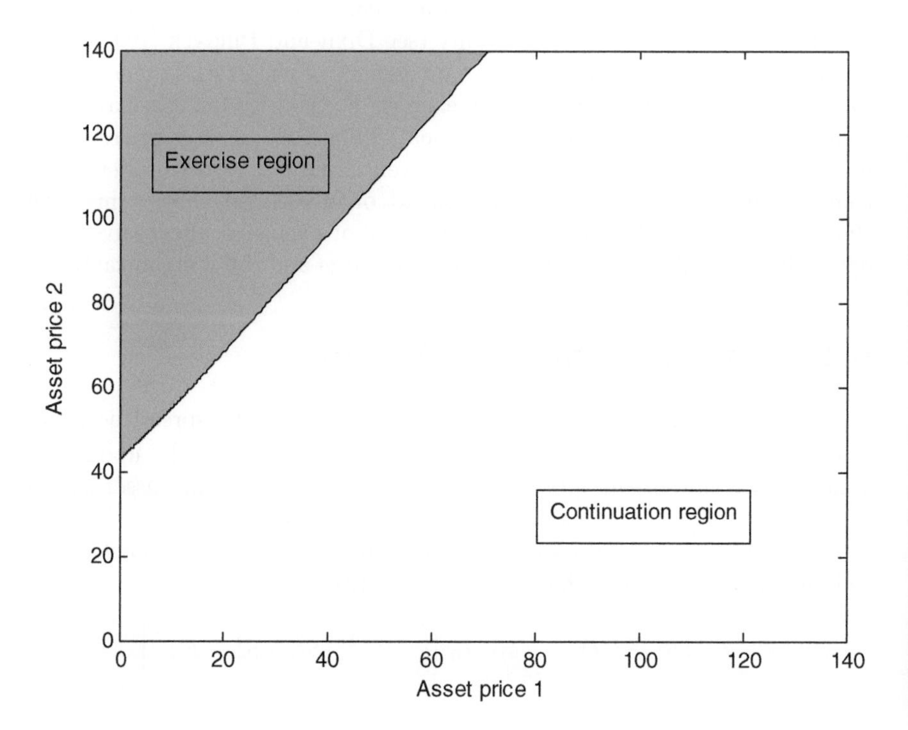

Figure 6.5: This figure displays the exercise region (the upper shaded region) of a spread-call option 6 months before maturity. Parameter values are $\delta_1 = 4\%, \delta_2 = 6\%$, $\sigma_1 = 0.30$, $\sigma_2 = 0.10$, $\rho = 0$, $r = 2\%$ and $K = 40$. Computations use a bivariate version of the binomial model with 400 time steps.

Proposition 73 implies that the t-section of the exercise region, $\mathcal{E}^{c,s}(t)$, is a convex set that lies above the line $S^2 > S^1 + K$. An illustration is provided in Figure 6.5. In this example the exercise boundary displays very mild curvature.

The price of the spread option can also be written so as to emphasize the gains from early exercise. Consider the functions

$$c_t^s\left(S^1, S^2\right) = \widetilde{E}_t\left[e^{-r(T-t)}\left(S_T^2 - S_T^1 - K\right)^+\right]$$

$$\pi_t^s\left(S^1, S^2; B^{c,s}\right) = \widetilde{E}_t\left[\int_t^T e^{-r(v-t)}\left(\delta_2 S_v^2 - \delta_1 S_v^1 - rK\right) 1_{\{S_v^2 \geq B^{c,s}(S_v^1, v)\}} dv\right]$$

where $\left\{B^{c,s}\left(S^1, v\right) : v \in [0, T], S^1 \in \mathbb{R}^+\right\}$ is a continuous surface. The function $c_t^s\left(S^1, S^2\right)$ is the value of a European call spread option, $\pi_t^s\left(S^1, S^2; B^{c,s}\right)$ is the early exercise premium and $B^{c,s}\left(S^1, v\right)$ represents the immediate exercise boundary. The following decomposition of the spread option price holds.

Theorem 74 *(Early exercise premium representation for spread options). The value of an American call spread option has the early exercise premium representation*

$$C_t^s\left(S^1, S^2; B^{c,s}\right) = c_t^s\left(S^1, S^2\right) + \pi_t^s\left(S^1, S^2; B^{c,s}\right)$$

where $c_t^s\left(S^1, S^2\right)$ is the value of a European call spread option and $\pi_t^s\left(S^1, S^2; B^{c,s}\right)$ is the early exercise premium. The exercise boundary $B^{c,s}$ solves the recursive integral equation

$$B^{c,s}\left(S^1, t\right) - K = c_t^s\left(S^1, B^{c,s}\left(S^1, t\right)\right) + \pi_t^s\left(S^1, B^{c,s}\left(S^1, t\right); B^{c,s}\right)$$

subject to the boundary conditions

$$\begin{aligned} \lim_{t\uparrow T} B^{c,s}\left(S^1, t\right) &= \max\left(\frac{\delta_1}{\delta_2}S^1 + \frac{r}{\delta_2}K, S^1 + K\right) \\ B^{c,s}(0, t) &= B^2(t). \end{aligned}$$

At maturity $B^{c,s}\left(S^1, T\right) = S^1 + K \leq B^{c,s}\left(S^1, T_-\right)$.

6.4.2 Options to Exchange One Asset for Another

When the strike is null, $K = 0$, the call spread option reduces to an option to exchange one asset for another, with exercise payoff $\left(S^2 - S^1\right)^+$. Using the price ratio $S^e \equiv S^2/S^1$ enables us to write $\left(S^2 - S^1\right)^+ = S^1\left(S^e - 1\right)^+$. The exchange option payoff is the same as the payoff of an option written on a single asset with price S^e and unit strike, adjusted by the random quantity S^1.

These observations suggest that the change of numeraire approach, reviewed in section 2.7, will be effective in pricing this contract. Given that price processes are of the form (6.1), it follows immediately from an adaptation of Theorem 8, with numeraire $X_t^{ca} = S_t^1 \exp(\delta_1 t)$, that

$$C^e\left(S^1, S^2, t\right) = S^1 \sup_{\tau\in\mathcal{S}_{t,T}} E_t^*\left[\exp\left(-\delta_1(\tau - t)\right)\left(S_\tau^e - 1\right)^+\right]$$

where

$$dS_t^e = S_t^e\left[\left(\delta_1 - \delta_2\right)dt + \sigma_e dz_t^e\right] \tag{6.7}$$

with $\sigma_e = \sqrt{\sigma_1^2 - 2\sigma_1\rho\sigma_2 + \sigma_2^2}$ and where z^e is a Brownian motion under the pricing measure Q^* whose Q-density is $\eta_T^* = \exp\left(-\frac{1}{2}\sigma_1^2 T + \sigma_1\widetilde{z}_T\right)$. The American exchange option is therefore equivalent to S^1 American calls with unit strike, written on a single underlying asset with dividend yield δ_2 and volatility σ_e, in a financial market with interest rate δ_1.

Based on this equivalence one obtains the following result,

Theorem 75 *(Early exercise premium representation for exchange options). The value of the American option to exchange asset one for asset two is*

$$C^e\left(S^1, S^2, t; B^e\right) = S_t^1 C\left(S_t^e, t; B^e\right) \tag{6.8}$$

where $C\left(S_t^e, t; B^e\right) = c\left(S_t^e, t\right) + \pi\left(S_t^e, t; B^e\right)$ *is the value of an American option with unit strike on an asset with price* S_t^e*, dividend yield* δ_2 *and volatility* σ_e*, in a market with interest rate* δ_1*. More precisely,* $c\left(S_t^e, t\right)$ *is the European option value*

$$\begin{aligned} c\left(S_t^e, t\right) &= S_t^e e^{-\delta_2(T-t)} N\left(d\left(S_t^e, 1, T-t, \delta_1, \delta_2, \sigma_e\right)\right) \\ &\quad - e^{-\delta_1(T-t)} N\left(d\left(S_t^e, 1, T-t, \delta_1, \delta_2, \sigma_e\right) - \sigma_e\sqrt{T-t}\right) \end{aligned} \tag{6.9}$$

and $\pi_t \equiv \pi\left(S_t^e, t; B^e(\cdot)\right)$ *the early exercise premium*

$$\begin{aligned} \pi_t &= \int_t^T \delta_2 S_t^e e^{-\delta_2(v-t)} N\left(d\left(S_t^e, B_v^e, v-t, \delta_1, \delta_2, \sigma_e\right)\right) dv \\ &\quad - \int_t^T \delta_1 e^{-\delta_1(v-t)} N\left(d\left(S_t^e, B_v^e, v-t, \delta_1, \delta_2, \sigma_e\right) - \sigma_e\sqrt{v-t}\right) dv \end{aligned} \tag{6.10}$$

where

$$d\left(S_t^e, B_v^e, v, \delta_1, \delta_2, \sigma_e\right) = \frac{\log\left(S_t^e / B_v^e\right) + \left(\delta_1 - \delta_2 + \frac{1}{2}\sigma_e^2\right) v}{\sigma_e\sqrt{v}}. \tag{6.11}$$

The optimal exercise boundary B^e *solves the recursive integral equation* $B_t^e - 1 = c\left(B_t^e, t\right) + \pi\left(B_t^e, t; B^e\right)$ *subject to the boundary condition* $B_{T-}^e = \left(\delta_1/\delta_2\right) \vee 1$*.*

Formulas (6.8)-(6.11) and the discussion preceding the proposition show that the American exchange option is symmetric to a standard option priced in a modified financial market. This symmetry property was first discovered by Rubinstein [1991] in the context of a binomial tree setting.

The proposition shows that immediate exercise is optimal when the asset price, expressed in the new numeraire, exceeds the boundary, i.e., $S_t^e \geq B^e(t)$. Equivalently, it is optimal to exercise when $S_t^2 \geq B^e(t) S_t^1$. In the plane $\left(S^1, S^2\right)$ the exercise region corresponds to a cone issued from the origin whose sides are the ray with slope $B^e(t)$ and the vertical axis. This exercise region coincides with the limit of the exercise region for the spread option as the strike converges to zero.

6.4.3 Exchange Options with Proportional Caps

A capped exchange option has a payoff equal to $\left(S^2 - S^1\right)^+ \wedge \widehat{L} S^1$ where $\widehat{L} > 0$ (equivalently we can write $\left(S^2 \wedge L S^1 - S^1\right)^+$ with $L = \widehat{L} + 1$).

Note that the payoff function for this contract is non-convex. It also lacks smoothness in a region where the exercise value is strictly positive (i.e. when

$S^2 - S^1 = \widehat{L}S^1$). It follows that the standard form of the early exercise premium representation, as in Theorem 75, does not apply. As for the case of capped call options, it is nevertheless possible to identify the optimal exercise boundary and to deduce a straightforward valuation formula. In fact the usual change of numeraire reduces this contract to a standard capped call in a modified financial market.

Theorem 76 *The value of an American capped exchange option with proportional cap (with payoff $(S^2 - S^1)^+ \wedge \widehat{L}S^1$ where $\widehat{L} > 0$) is given by*

$$C^{ec}\left(S^1, S^2, t; B^{ec}\right) = S_t^1 C^L\left(S_t^e, t; B^L\right) \tag{6.12}$$

where $C^L\left(S_t^e, t; B^L\right)$ is the value of an American-style capped call option, written on a single asset with price S_t^e, dividend δ_2 and volatility σ_e (see (6.7)), with strike 1 and cap on the underlying's price $L = \widehat{L} + 1$, in a market with interest rate δ_1. Accordingly, immediate exercise is optimal when $S^2 \geq B^{ec}S^1$ where the boundary B^{ec} is given by

$$B^{ec} = B^L \equiv B \wedge L, \tag{6.13}$$

i.e., the boundary is the minimum of the exercise boundary B for a standard uncapped option on the asset price S^e and the cap L. The value of the capped option $C^L\left(S_t^e, t; B^L\right)$ is obtained from Theorems 52 and 53 with the appropriate substitutions for the coefficients.

The change of numeraire method proves again very effective in this context. By expressing the payoff in the new unit of account one sees immediately that the contract reduces to the familiar form studied in chapter 5. It follows that our earlier results apply, with a suitable relabeling of the parameters, and this leads to the formulas in Theorem 76.

6.5 Options on an Average of 2 Prices

An option on a portfolio of assets is an example of a contract whose payoff depends on an average of prices. Several averaging methods can be employed. We examine the cases of geometric averaging (section 6.5.1) and arithmetic averaging (section 6.5.2).

6.5.1 Geometric Averaging

For generality we consider a larger class of contracts with payoffs

$$\left(\left(S^1\right)^{\gamma_1}\left(S^2\right)^{\gamma_2} - K\right)^+ \text{ for some } \gamma_1, \gamma_2 > 0 \text{ and } \gamma_1 + \gamma_2 = \gamma.$$

These are call options on a geometric sum of assets. Special members of this class include calls on the power of a product of prices ($\gamma_1 = \gamma_2 = \gamma/2$), calls on a product of prices ($\gamma_1 = \gamma_2 = 1$), calls on a geometric average of assets ($\gamma = 1$) and calls on an equally-weighted geometric average of assets ($\gamma_1 = \gamma_2 = 1/2$).

A simple argument establishes that these options are effectively written on a single underlying asset. Defining $Y \equiv (S^1)^{\gamma_1} (S^2)^{\gamma_2}$ and applying Ito's lemma gives

$$dY_t = Y_t [(r - \delta_Y) dt + \sigma_Y dz_t] \tag{6.14}$$

where

$$\delta_Y = (1 - \gamma) r + \gamma_1 \delta_1 + \gamma_2 \delta_2 - \frac{1}{2} (\sigma_Y^2 - \gamma_1 \sigma_1^2 - \gamma_2 \sigma_2^2)$$

$$\sigma_Y = \sqrt{(\gamma_1 \sigma_1)^2 + 2\gamma_1 \gamma_2 \sigma_1 \rho \sigma_2 + (\gamma_2 \sigma_2)^2}$$

$$dz_t = \frac{\gamma_1 \sigma_1 d\tilde{z}_t^1 + \gamma_2 \sigma_2 d\tilde{z}_t^2}{\sigma_Y}.$$

One concludes that Y follows a geometric Brownian motion process with implicit dividend yield δ_Y and volatility coefficient σ_Y. As (6.14) does not depend on the original prices S_1, S_2, we conclude that the geometric sum option is identical to an option written on Y alone. To proceed, let us assume that $\delta_Y \geq 0$ (if the implicit dividend δ_Y is negative there are no incentives to exercise early). Also let $B_t (\delta_Y, \sigma_Y^2)$ be the optimal exercise boundary of an American call on Y and let $C_t (Y)$ be its value.

Theorem 77 *The optimal exercise boundary for an American geometric sum option is*

$$B^{gs} (S^1, t) = \left(\frac{B_t}{(S^1)^{\gamma_1}} \right)^{1/\gamma_2} \tag{6.15}$$

where $B_t = B_t (\delta_Y, \sigma_Y^2)$ is the exercise boundary for the single asset call written on the process Y given in (6.14). The option value is $C^{gs} (S^1, S^2, t) = C(Y, t)$ where $C(Y, t)$ is the value of the American call on Y.

As announced by the discussion preceding the proposition the option on a geometric sum is identical to an option written on asset Y alone. The decision to exercise can be based directly on Y: immediate exercise is optimal at the first time at which Y equals the boundary B. Alternatively, the decision can be based on the two underlying assets. If asset 2 is taken as the reference asset then immediate exercise is optimal at the first time at which S_2 reaches the boundary $B^{gs} (S^1, t)$ described in the proposition. As could have been expected, this boundary is a non-linear transform of B, where the degree of non-linearity depends on the parameter γ_2. The relation also depends on the price of asset 1 and the associated power γ_1.

6.5.2 Arithmetic Averaging

Consider an American option written on a weighted average of asset prices with constant weights. The underlying asset in this contract can be viewed as a portfolio of two assets. The analysis of this type of contract is useful as it also helps to shed light on the structure of the exercise region for a more complicated contract, the min-option (see section 6.6)). The payoff on a call option upon exercise is $\left(wS^1 + (1-w)\, S^2 - K\right)^+$, where $w \in (0,1)$. The choice $w = 1/2$ corresponds to a call on the arithmetic average of two prices. The next proposition describes the optimal exercise region. We recall that B^i denotes the exercise boundary of a call option on asset i, $i = 1, 2$.

Proposition 78 *Let $\mathcal{E}^{c,w}$ be the optimal exercise region for a call option on the weighted average of 2 prices where $w \in (0,1)$ is constant. Then*
(i) $\left(0, S^2, t\right) \in \mathcal{E}^{c,w}$ if and only if $S^2 \geq \frac{1}{1-w} B^2(t)$.
(ii) $\left(S^1, 0, t\right) \in \mathcal{E}^{c,w}$ if and only if $S^1 \geq \frac{1}{w} B^1(t)$.
(iii) $\left(S^1, S^2, t\right) \in \mathcal{E}^{c,w}$ implies $\left(\lambda^1 S^1, \lambda^2 S^2, t\right) \in \mathcal{E}^{c,w}$ with $\lambda^1 \geq 1$, $\lambda^2 \geq 1$.
(iv) $(S,t) \in \mathcal{E}^{c,w}$ and $\left(\widetilde{S}, t\right) \in \mathcal{E}^{c,w}$ implies $(S(\lambda), t) \in \mathcal{E}^{c,w}$ for all $\lambda \in [0,1]$.
(v) $\left(S^1, S^2, t\right) \in \mathcal{E}^{c,w}$ implies $\left(S^1, S^2, s\right) \in \mathcal{E}^{c,w}$ for $T \geq s \geq t$.

Most of these properties are intuitive. Property (iii) establishes that immediate exercise remains optimal when both prices increase. This feature follows from the upper bound on the payoff function

$$
\begin{aligned}
& \left(w\lambda^1 S^1 + (1-w)\,\lambda^2 S^2 - K\right)^+ \\
\leq\ & \left(wS^1 + (1-w)\, S^2 - K\right)^+ + w\left(\lambda^1 - 1\right) S^1 + (1-w)\left(\lambda^2 - 1\right) S^2,
\end{aligned}
$$

that holds for $\lambda^1, \lambda^2 \geq 1$. The payoff bound implies the price bound

$$
C^w\left(\lambda^1 S^1, \lambda^2 S^2, t\right) \leq C^w\left(S^1, S^2, t\right) + w\left(\lambda^1 - 1\right) S^1 + (1-w)\left(\lambda^2 - 1\right) S^2.
$$

As immediate exercise is optimal at $\left(S^1, S^2, t\right)$ the price bound becomes $w\lambda^1 S^1 + (1-w)\,\lambda^2 S^2 - K$, the exercise value at the point $\left(\lambda^1 S^1, \lambda^2 S^2, t\right)$. Immediate exercise must therefore be an optimal policy at the point $\left(\lambda^1 S^1, \lambda^2 S^2, t\right)$.

Figure 6.6 illustrates the exercise region for a weighted average option. A valuation formula is given next.

Theorem 79 *(Early exercise premium representation for weighted average options). The value of the American call option on a weighted (arithmetic) average of two assets is*

$$
C^w\left(S^1, S^2, t; B^{c,w}\right) = c^w\left(S^1, S^2, t\right) + \pi^w\left(S^1, S^2, t; B^{c,w}\right)
$$

where $c^w\left(S^1, S^2, t\right)$ is the value of the European option on the weighted average

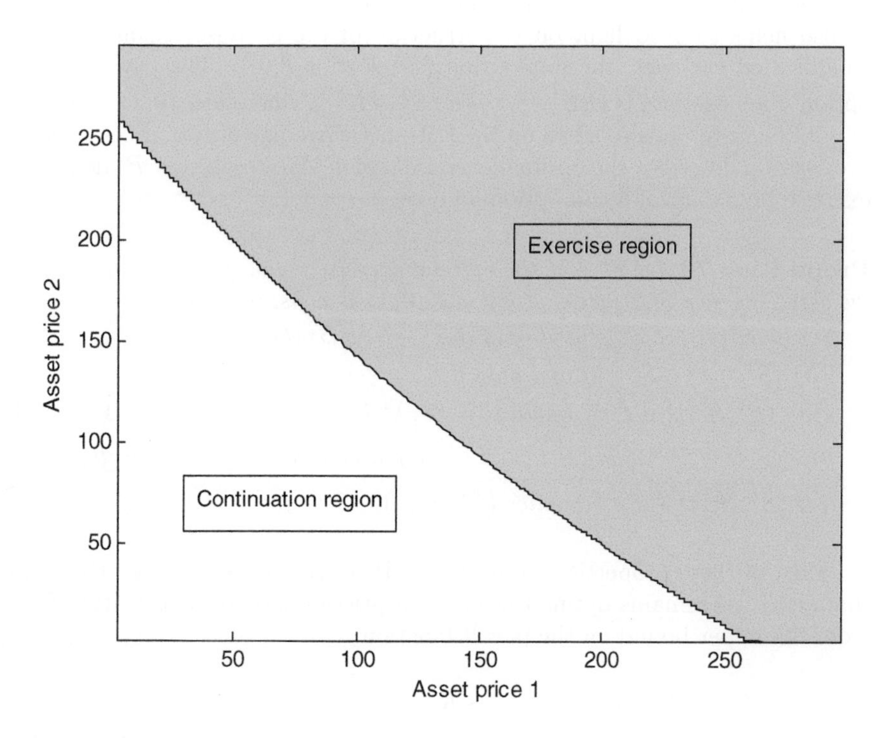

Figure 6.6: This figure displays the exercise region (the shaded region) of an equally-weighted average call option 6 months before maturity. Parameter values are $\delta_1 = \delta_2 = 6\%$, $\sigma_1 = \sigma_2 = 0.20$, $\rho = 0$, $r = 6\%$ and $K = 100$. Computations use a bivariate version of the binomial model with 400 time steps.

of two assets and $\pi_t^w \equiv \pi^w \left(S^1, S^2, t; B^{c,w}\right)$ *is the early exercise premium,*

$$\begin{aligned}
\pi_t^w &= \int_t^T w\delta_1 S_t^1 e^{-\delta_1(v-t)} \phi_1(v)\, dv \\
&\quad + \int_t^T (1-w)\, \delta_2 S_t^2 e^{-\delta_2(v-t)} \phi_2(v)\, dv \\
&\quad - \int_t^T rKe^{-r(v-t)} \phi_3(v)\, dv
\end{aligned}$$

where

$$\phi_1(v) = \Phi\left(S^2, B^{c,w}(\cdot, v), v-t, \rho, 0, \sigma_1\sqrt{v-t}\right)$$

$$\phi_2(v) = \Phi\left(S^2, B^{c,w}(\cdot, v), v-t, \rho, \sigma_2\sqrt{(1-\rho^2)(v-t)}, \sigma_2\rho\sqrt{v-t}\right)$$

$$\phi_3(v) = \Phi\left(S^2, B^{c,w}(\cdot, v), v-t, \rho, 0, 0\right)$$

and

$$\Phi\left(S^2, B^{c,w}(\cdot, v), h, \rho, x, y\right)$$
$$\equiv \int_{-\infty}^{+\infty} n(z-y) N\left(-d\left(S^2, B^{c,w}\left(S^1(z), v\right), h, \rho, z\right) + x\right) dz$$

$$d\left(S^2, B^{c,w}\left(S^1(z), v\right), h\right) \equiv \frac{1}{\sigma_2 \sqrt{h}}\left[\log\left(\frac{B^{c,w}\left(S^1(z), v\right)}{S^2}\right) - \alpha_2 h\right],$$

with $S^1(z) = S^1 \exp\left(\alpha_1 (v-t) + \sigma_1 z \sqrt{v-t}\right)$ *and* $\alpha_i = r - \delta_i - (1/2)\sigma_i^2, i = 1, 2$. *The optimal exercise boundary solves*

$$wS^1 + (1-w) B^{c,w}\left(S^1, t\right) - K = C^w\left(S^1, B^{c,w}\left(S^1, t\right), t; B^{c,w}\right), \quad \textit{for } t \in [0, T]$$

subject to the boundary condition $w\delta_1 S^1 + (1-w)\delta_2 B^{c,w}\left(S^1, T_-\right) = rK \vee \left(\delta_2 K + w(\delta_1 - \delta_2) S^1\right)$.

6.6 Call Options on the Minimum of 2 Prices

An American call option on the minimum of 2 asset prices has an exercise payoff given by $\left(S^1 \wedge S^2 - K\right)^+$. Let $C^m\left(S^1, S^2, t\right)$ be its price. The immediate exercise region is the set

$$\mathcal{E}^{c,m} \equiv \left\{\left(S^1, S^2, t\right) \in \mathbb{R}_+^2 \times [0, T] : C^m\left(S^1, S^2, t\right) = \left(S^1 \wedge S^2 - K\right)^+\right\}.$$

Its complement is the continuation region $\mathcal{C}^{c,m}$.

6.6.1 Exercise Region of a Min-Call Option

As for the max-option, it is of interest to examine the behavior when one asset, or the other one, is more expensive. This partition corresponds to the subregions $\mathcal{G}_i = \left\{\left(S^1, S^2, t\right) \in \mathbb{R}_+^2 \times [0, T] : S^i = \max\left(S^1, S^2\right)\right\}, i = 1, 2$. Taking the intersection with the exercise region leads to the subsets $\mathcal{E}_i^{c,m} = \mathcal{E}^{c,m} \cap \mathcal{G}_i$, for $i = 1, 2$. Here $\mathcal{E}_i^{c,m}$ is the subset of the exercise region in which asset i is more expensive. With these definitions, it is clear that the intersection $\mathcal{E}_1^{c,m} \cap \mathcal{E}_2^{c,m}$, if non-empty, belongs to the diagonal. Given that some of the results presented involve options on individual assets we also need the exercise region $\mathcal{E}_i$ for a call option on asset i along with the boundary B^i, $i = 1, 2$. Similarly, we recall that $\mathcal{E}^{c,w}$ represents the exercise region of a call option on a weighted average of two assets. The corresponding boundary is $B^{c,w}$.

The next proposition identifies key properties of $\mathcal{E}^{c,m}, \mathcal{E}_1^{c,m}$ and $\mathcal{E}_2^{c,m}$.

Proposition 80 *The immediate exercise region of an American min-call option,* $\mathcal{E}^{c,m}$ *has the following properties*

(i) $(S, t) \in \mathcal{E}^{c,m}$ *implies* $(S, s) \in \mathcal{E}^{c,m}$ *for all* $t \le s \le T$.

(ii) $(S^2,t) \in \mathcal{E}_2$ *and* $(S^1,S^2,t) \in \mathcal{G}_1$ *implies* $(S^1,S^2,t) \in \mathcal{E}_1^{c,m}$.
(iii) $(S^1,S^2,t) \in \mathcal{E}_1^{c,m}$ *implies* $(\lambda S^1,\lambda S^2) \in \mathcal{E}_1^{c,m}$ *for all* $\lambda \geq 1$.
(iv) $(S^1,S^2,t) \in \mathcal{E}_1^{c,m}$ *implies* $(S^1,\lambda S^2,t) \in \mathcal{E}_1^{c,m}$ *for all* $1 \leq \lambda \leq S^1/S^2$.
(v) $(S^1,S^2,t) \in \mathcal{E}_1^{c,m}$ *implies* $(\lambda S^1,S^2,t) \in \mathcal{E}_1^{c,m}$ *for all* $S^2/S^1 \leq \lambda \leq 1$.
(vi) $(S^1,S^2,t) \in \mathcal{E}^{c,w}$ *for* $w \in (0,1)$ *and* $S^1 = S^2$, *implies* $(S^1,S^2,t) \in \mathcal{E}^{c,m}$.
(vii) $(0,S^2,t) \notin \mathcal{E}^{c,m}$; $(S^1,0,t) \notin \mathcal{E}^{c,m}$.
Results (ii)-(v) also hold for the subregion $\mathcal{E}_2^{c,m}$.

Property (i) is standard. Result (ii) states that immediate exercise is optimal for the min-option if it is optimal for a standard option on asset 2 and asset 2 is cheapest. This is true because the min-option payoff can attain, but never exceed, the payoff of a call option on asset i, $i = 1$ or 2. The min-option price is then bounded above by the prices of the two call options written on the individual asset prices. If immediate exercise is optimal for one of these two calls and the min-option attains the same payoff, it will also be optimal to exercise the min-option. Property (iii) shows that the subregion $\mathcal{E}_1^{c,m}$ is ray-connected: increasing both prices by a common factor preserves the optimality of exercise. The subregion $\mathcal{E}_1^{c,m}$ is also up-connected (see (iv)) and left-connected (see (v)). Of course, by the definition of $\mathcal{E}_1^{c,m}$, these connectedness properties only extend up to the diagonal.

Property (vi) highlights an interesting relation between the weighted average option and the min-option. It is easy to verify that the payoff of the min-option is bounded above by the payoff of the average option

$$\left(S^1 \wedge S^2 - K\right)^+ \leq \left(wS^1 + (1-w)S^2 - K\right)^+$$

with equality along the diagonal (where $S^1 = S^2$). As a result, if immediate exercise is optimal along the diagonal for the holder of the average option, it will also be optimal for the min-option.

The last statement, (vii), expresses the simple intuition that immediate exercise cannot be optimal along the two axes because the payoff is null.

Proposition 80 suggests an exercise region whose boundary can be described by two surfaces $B_2^{c,m}(S^1,t)$ and $B_1^{c,m}(S^2,t)$ corresponding, respectively, to the subregions $\mathcal{E}_1^{c,m}$ and $\mathcal{E}_2^{c,m}$. These surfaces meet and merge along the diagonal. Their common lower extremity, along the diagonal, is a curve $B_d^{c,m}(t)$ parametrized by time. Thus, $(S^1,S^2,t) \in \mathcal{E}_1^{c,m}$ if and only if $S^2 = \min(S^1,S^2) \geq B_2^{c,m}(S^1,t)$. Similarly, $(S^1,S^2,t) \in \mathcal{E}_2^{c,m}$ if and only if $S^1 = \min(S^1,S^2) \geq B_1^{c,m}(S^2,t)$. Finally, $(S,S,t) \in \mathcal{E}^{c,m}$ if and only if $S \geq B_d^{c,m}(t)$. Figure 6.7 illustrates this structure.

Associated with the properties of the exercise region are the following properties of the boundaries.

Proposition 81 *The boundary* $B_1^{c,m}(S^2,t)$ *has the following properties over its domain,*

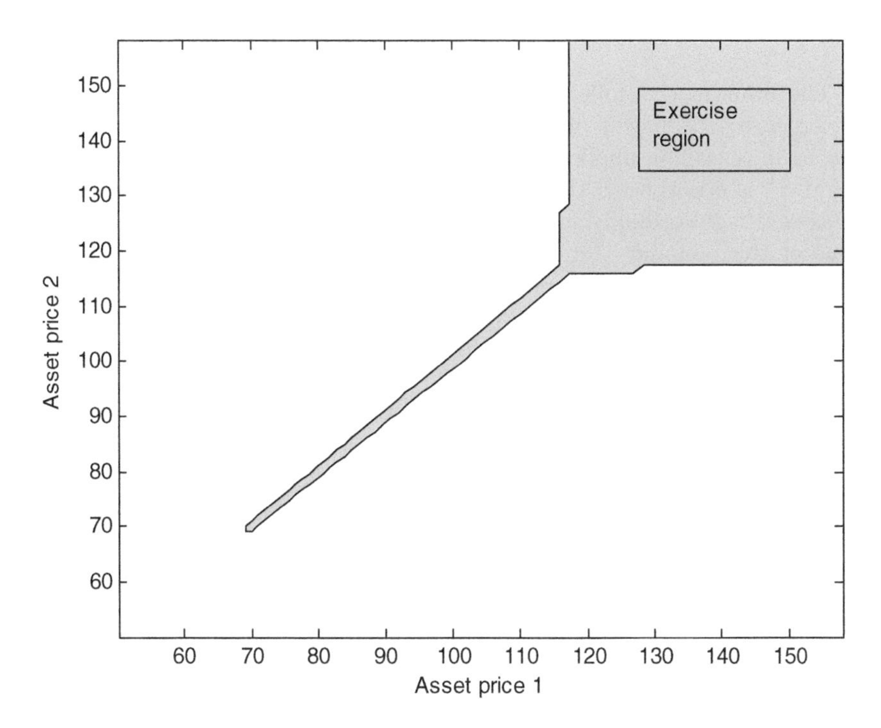

Figure 6.7: This figure displays the exercise region of a min-call option 6 months before maturity. Parameter values are $\delta_1 = \delta_2 = 6\%$, $\sigma_1 = \sigma_2 = 0.50$, $\rho = 0$, $r = 6\%$ and $K = 60$. Computations use a bivariate version of the binomial model with 1500 time steps.

(i) $B_1^{c,m}(S^2, t)$ *is a non-increasing function of time.*

(ii) $B_1^{c,m}(S^2, t)$ *is non-decreasing in* S^2 *and bounded above by* $B^1(t)$.

(iii) If $B_1^{c,m}(S_i^2, t) = S_i^2$ *for* $i = 1, 2$ *and* $S_1^2 < S_2^2$, *then* $B_1^{c,m}(x, t) = x$ *for all* $x \in [S_1^2, S_2^2]$.

(iv) $\lim_{S^2 \uparrow \infty} B_1^{c,m}(S^2, t) = B^1(t)$

(v) $\lim_{t \uparrow T} B_1^{c,m}(S^2, t) = K(1 \vee r/\delta_1)$ *if* $S^2 > K(1 \vee r/\delta_1)$; $\lim_{t \uparrow T} B_d^{c,m}(t) = K$.

(vi) Let $B_d^{c,m}(t) = \inf\{x : x = B_1^{c,m}(x, t)\}$. *Then* $B_d^{c,m}(t) \leq B^{c,w}(x^*(t), t)$ *where* $x^*(t)$ *uniquely solves* $B^{c,w}(x, t) = x$.

Analogous properties hold for the boundary $B_2^{c,m}(S^1, t)$ *of the region* $\mathcal{E}_1^{c,m}$.

The first 2 properties are direct consequences of (i), (ii) and (v) of Proposition 80. Property (iii) states that the diagonal boundary component is connected in the price direction (i.e., if two pairs of asset prices lie on the diagonal and belong to the exercise boundary then every intermediate diagonal pair also belongs to the exercise boundary). This parallels the behavior of the exercise region, along the cap, for a capped call option (see chapter 5).

The limit in (iv) follows from the fact that the payoff of the min-option converges to the payoff of an option on asset 1 alone as S^2 becomes large. The same reasoning applies for the first limit in (v). As maturity approaches, and if S^1 is strictly less than S^2 then the min-option converges to an option on asset 1. Accordingly, the min-option boundary converges to the limiting value of the boundary for that single asset option. For the second limit in (v), along the diagonal, convergence is toward the strike, even if the dividend yields on both assets are low. This may appear puzzling at first sight. The explanation is similar to the one for the suboptimality of exercise along the diagonal for max-options. To simplify matters consider the case where asset prices are independent, have identical volatility coefficients and dividend yields equal to the interest rate. The probability of an improvement in the minimum of two prices, along the diagonal, is only about 1/4 (the likelihood of a decrease is about 3/4). Thus, as maturity approaches, it will pay to cash in any positive payoff no matter how small. In the limit, the policy of waiting to exercise is dominated as long as the option is in the money and asset prices are equal.

The results above show that the stopping time problem reduces to the identification of the two surfaces $B_1^{c,m}\left(S^2,t\right)$ and $B_2^{c,m}\left(S^1,t\right)$ and of the curve $B_d^{c,m}(t)$ along the diagonal (compare this to the (t_e, t^*, t_f)-policies that are optimal for capped options with growing caps). Integral equations for these curves are provided below.

6.6.2 The EEP Representation

As usual the derivation of integral equations for the boundary components relies on the EEP representation of the option price. To establish this formula in the case of a min-call one has to deal with a non-smooth payoff function with discontinuous derivatives across the diagonal. The existence of these discontinuities implies that the standard version of Ito's lemma cannot be used to describe the behavior of the minimum price process $S^1 \wedge S^2$. Instead we must rely on an extension of the Ito rule, the Tanaka-Meyer formula (see Karatzas and Shreve [1988, Chapter 3, Proposition 6.8]), that applies to convex functions.

Define $X_s = S_s^1 \wedge S_s^2$ to represent the minimum price. For any fixed t, and $s \in [t,T]$, let $F_{t,s} \equiv e^{-r(s-t)}\left(X_s - K\right)^+$ be the discounted payoff of the min-option. An application of the Tanaka-Meyer formula establishes

$$\begin{aligned}
F_{t,s} &= F_{t,t} + \int_t^s (-r) F_{t,u} du + \int_t^s r F_{t,u} du \\
&+ \int_t^s 1_{\{X_u > K\}} e^{-r(u-t)} rK du + \frac{1}{2} \int_t^s e^{-r(u-t)} dL(u, 0; X - K) \\
&- \int_t^s 1_{\{X_u > K\}} e^{-r(u-t)} \left(\delta_1 S_u^1 1_{\{S_u^1 < S_u^2\}} + \delta_2 S_u^2 1_{\{S_u^1 > S_u^2\}} \right) du \\
&- \frac{1}{2} \sigma_m^2 \int_t^s 1_{\{X_u > K\}} e^{-r(u-t)} X_u dL\left(u, 0; \ln\left(S^1\right) - \ln\left(S^2\right)\right) \\
&+ \int_t^s 1_{\{X_u > K\}} e^{-r(u-t)} \left(\sigma_1 S_u^1 1_{\{S_u^1 < S_u^2\}} d\widetilde{z}_u^1 + \sigma_2 S_u^2 1_{\{S_u^1 > S_u^2\}} d\widetilde{z}_u^2 \right),
\end{aligned}$$

where, for a process x, $L(u, 0; x)$ is the local time of x at the point zero and $\sigma_m^2 \equiv \sigma_1^2 - 2\sigma_1 \rho \sigma_2 + \sigma_2^2$ is the local variance of the difference in returns (or, equivalently, in log-prices).[1] This gives a decomposition of the form (3.8) for the payoff process Y. From Theorem 21 we now obtain the following EEP representation of the option price (note that $dL(u, 0; X - K) = 0$ in the exercise region of the min-option),

Theorem 82 *The price of the min-call option has the early exercise premium representation*

$$C_t^m\left(S_t^1, S_t^2\right) = c_t^m\left(S_t^1, S_t^2\right) + \pi_t^m\left(S_t^1, S_t^2; B_1^{c,m}, B_2^{c,m}, B_d^{c,m}\right),$$

where $c_t^m\left(S_t^1, S_t^2\right)$ *is the price of the European min-call option and the function* $\pi_t^m\left(S_t^1, S_t^2; B_1^{c,m}, B_2^{c,m}, B_d^{c,m}\right) = \pi_1^m + \pi_2^m + \pi_3^m$ *is the early exercise premium with components*

$$\pi_1^m = \widetilde{E}_t\left[\int_t^T e^{-r(v-t)} \left(\delta_1 S_v^1 - rK\right) 1_{\left\{B_1^{c,m}(S_v^2, v) \le S_v^1 < S_v^2\right\}} dv\right]$$

$$\pi_2^m = \widetilde{E}_t\left[\int_t^T e^{-r(v-t)} \left(\delta_2 S_v^2 - rK\right) 1_{\left\{B_2^{c,m}(S_v^1, v) \le S_v^2 < S_v^1\right\}} dv\right]$$

$$\pi_3^m = \frac{1}{2}\sigma_m^2 \widetilde{E}_t\left[\int_t^T e^{-r(v-t)} 1_{\left\{S_v^1 = S_v^2 \ge B_d^{c,m}(v)\right\}} X_v dL\left(v, 0; \ln\left(S^1\right) - \ln\left(S^2\right)\right)\right]$$

and where $\sigma_m^2 \equiv \sigma_1^2 - 2\sigma_1 \rho \sigma_2 + \sigma_2^2$.

As usual the EEP formula decomposes the option value into a European component, capturing the value of waiting to maturity, and an early exercise premium, representing the benefits associated with early exercise. Unlike the

[1] When x is a Brownian motion the process $L(u, 0; x)$ is the Brownian local time defined as $L(u, 0; x) \equiv \lim_{n \to \infty} \frac{n}{2} \int_0^u 1_{\{|x_s| \le 1/n\}} ds$.

standard case, the exercise premium is the sum of two parts. The first part is the present value of the net flow gains (i.e., dividends net of interest costs) from early exercise, $\pi_1^m + \pi_2^m$. The dividend benefits, for this contract, are those associated with asset 1 (resp. asset 2) when S^1 (resp. S^2) is the minimum of the two prices. The second part, π_3^m, is due to the lack of smoothness in the payoff function, in the exercise region, along the diagonal. Every time the minimum of the two prices crosses the diagonal the derivatives, with respect to prices, of the payoff function jump. The local time process $L\left(t, 0; \ln\left(S^1\right) - \ln\left(S^2\right)\right)$ compensates for these jumps: the increment dL is non-negative at all times, and strictly positive along the diagonal. Thus, π_3^m is non-negative. Intuitively, exercising the option along the diagonal saves value given the higher likelihood of a decrease in the minimum price, and the option payoff, over the next instant of time.

The formula for π_3^m provides additional perspective about the behavior of the contract along the diagonal. Note, for example, that the variance of the return differential is positive (i.e. $\sigma_m^2 > 0$) except when returns are perfectly correlated and carry the same risk (i.e., $\sigma_1 = \sigma_2$ and $\rho = 1$). When these conditions prevail underlying prices satisfy the deterministic relation $S_t^2 = S_t^1 H(t)$, with $H(t) \equiv \left(S_0^2/S_0^1\right) \exp\left(\left(\delta_1 - \delta_2\right) t\right)$. The min-option is then equivalent to a call option written on a single asset, with a quantity adjustment that depends on time: the option's payoff takes the form $\left(Q(t) S_t^1 - K\right)^+$ where $Q(t) = 1 \wedge H(t)$. The smoothness of this contractual structure, with respect to the underlying Brownian motion, implies that the local time component disappears from the EEP representation.

Note also that the min-option is identical to a capped call option with time-dependent cap when either $\sigma_1 = 0$ or $\sigma_2 = 0$ (see chapter 5 and section 6.4.3 above). Our previous analysis of this contract led to a pricing formula emphasizing the gains from exercising before hitting the cap (see Theorem 53). The price decomposition in Theorem 82 provides an additional representation of capped option prices, highlighting the benefits of exercising before the maturity date of the contract. This expression, in the spirit of traditional EEP formulas, prices the early exercise premium relative to a European capped call option.

When $\sigma_m^2 > 0$, the premium component π_3^m is positive and the exercise boundary consists of the three parts $B_1^{c,m}\left(S^2, t\right), B_2^{c,m}\left(S^1, t\right)$ and $B_d^{c,m}(t)$ identified above. If the underlying assets do not pay dividends ($\delta_1 = \delta_2 = 0$). there are no incentives to exercise off the diagonal. In this situation the premia π_1^m, π_2^m vanish and the boundary curves $B_1^{c,m}\left(S^2, t\right), B_2^{c,m}\left(S^1, t\right)$ collapse to the diagonal: the exercise region is the subset of the diagonal above $B_d^{c,m}(t)$.

Before moving on to the next section let us also point out that some of the EEP components can be expressed in more explicit form. A formula for the European min-option price can be found in Johnson [1987]. More explicit formulas can also be derived for the components π_1^m and π_2^m of the early exercise premium.

6.6.3 Integral Equations for the Boundary Components

Given that the EEP formula is valid in the exercise region, it can be used in the traditional manner to characterize the exercise boundary components. This leads to the following integral equations.

Theorem 83 *The components* $B_1^{c,m}\left(S^2,t\right), B_2^{c,m}\left(S^1,t\right), B_d^{c,m}(t)$ *of the immediate exercise boundary solve the integral equations*

$$\begin{aligned} B_2^{c,m}\left(S^1,t\right)-K &= c_t^m\left(S^1,B_2^{c,m}\left(S^1,t\right)\right) \\ &\quad +\pi_t^m\left(S^1,B_2^{c,m}\left(S^1,t\right);B_1^{c,m},B_2^{c,m},B_d^{c,m}\right) \end{aligned}$$

$$\begin{aligned} B_1^{c,m}\left(S^2,t\right)-K &= c_t^m\left(B_1^{c,m}\left(S^2,t\right),S^2\right) \\ &\quad +\pi_t^m\left(B_1^{c,m}\left(S^2,t\right),S^2;B_1^{c,m},B_2^{c,m},B_d^{c,m}\right) \end{aligned}$$

$$\begin{aligned} B_d^{c,m}(t)-K &= c_t^m\left(B_d^{c,m}(t),B_d^{c,m}(t)\right) \\ &\quad +\pi_t^m\left(B_d^{c,m}(t),B_d^{c,m}(t);B_1^{c,m},B_2^{c,m},B_d^{c,m}\right) \end{aligned}$$

subject to the boundary conditions $\lim_{t\to T} B_2^{c,m}\left(S^1,t\right)=K\vee(r/\delta_2)K$ *for* $S^1>K\vee(r/\delta_2)K$, $\lim_{t\to T} B_1^{c,m}\left(S^2,t\right)=K\vee(r/\delta_1)K$ *for* $S^2>K\vee(r/\delta_1)K$ *and* $\lim_{t\to T} B_d^{c,m}(t)=K$.

This theorem shows that the boundary surfaces $B_1^{c,m}\left(S^2,t\right)$ and $B_2^{c,m}\left(S^1,t\right)$ satisfy a system of coupled integral equations. As in the case of max-options the linkage between these integral equations is natural. It follows from the fact that the exercise premium depends on both boundary components. As $B_d^{c,m}(t)$ coincides with the lower extremities of $B_1^{c,m}\left(S^2,t\right)$ and $B_2^{c,m}\left(S^1,t\right)$ along the diagonal the third equation is, in fact, redundant.

The system of integral equations in Theorem 83 can, in principle, serve as a starting point for a numerical scheme designed to implement the model. The presence of the local time component is an aspect that sets this contract apart from the other multiasset claims studied in this chapter. This element, along with the price dependence of the boundary and the coupling of the integral equations involved, are likely to pose interesting challenges for implementation.

6.7 Appendix A: Derivatives on Multiple Assets

In this appendix we provide general results about the structure of exercise regions for derivatives written on n underlying assets. We suppose that the price of asset i satisfies (under the risk neutral measure)

$$dS_t^i = S_t^i\left[(r-\delta_i)\,dt+\sigma_i d\widetilde{z}_t^i\right] \tag{6.16}$$

where $\widetilde{z}^i, i=1,...,n$ are standard Brownian motion processes and the correlation between $\widetilde{z}^i$ and $\widetilde{z}^j$ is ρ_{ij}. As before, r is the constant rate of interest, $\delta_i \geq 0$ is the dividend rate of asset i and the price processes in (6.16) are represented in their risk neutral form. We use this setting with constant coefficients for ease of exposition. However, many of the results presented below hold in more general settings.

Consider an American contingent claim written on the n assets with prices (6.16) and that matures at time T. Suppose that its payoff, if exercised at time t, is $F\left(S_t^1, S_t^2, ..., S_t^n\right) \geq 0$. For convenience, let S_t represent the vector $\left(S_t^1, S_t^2, ..., S_t^n\right)$. Denote the value of this claim at time t by $V^F\left(S_t, t\right)$. From chapter 3 we have

$$V^F\left(S_t, t\right) = \sup_{\tau \in \mathcal{S}_{t,\tau}} \widetilde{E}_t\left[e^{-r(\tau - t)} F\left(S_\tau\right)\right],$$

where $\mathcal{S}_{t,T}$ is the set of stopping times of the filtration with values in $[t, T]$. The immediate exercise region is $\mathcal{E}^F \equiv \left\{(S, t) \in \mathbb{R}^n \times [0, T] : V_t^F(S) = F(S)\right\}$.

We can now build on the transformation used in the proof of Proposition 31. Let $\alpha_i \equiv r - \delta_i - (1/2)\sigma_i^2$, $i = 1, ..., n$ and consider a stopping time $\tau \in \mathcal{S}_{t,T}$. Given that the filtration generated by the increment $\left\{\widetilde{z}_s^i - \widetilde{z}_t^i : s \in [t, T]\right\}$ is identical to the filtration generated by the new Brownian motion

$$\widehat{z}^i = \left\{\widetilde{z}_{s(T-t)}^i : s \in [0, 1]\right\}$$

we can write

$$S_\tau^i = S^i \exp\left(\alpha_i \widehat{\tau}\left(T - t\right) + \sigma_i \widehat{z}_{\widehat{\tau}(T-t)}\right) \equiv S^i N_{\widehat{\tau}(T-t)}^i$$

where $\tau = \widehat{\tau}\left(T - t\right)$ and $\widehat{\tau} \in \widehat{\mathcal{S}}_{0,1}$ is a stopping time, taking values in $[0, 1]$, of the filtration generated by $\widehat{z}^i$. Let $N_t \equiv \left(N_t^1, ..., N_t^n\right)$. In what follows, we write $SN = \left(S^1 N^1, ..., S^n N^n\right)$, to indicate the product, term by term, of two vectors. With these definitions we see that

$$V^F(S, t) = \sup_{\widehat{\tau} \in \widehat{\mathcal{S}}_{0,1}} \widehat{E}\left[e^{-r\widehat{\tau}(T-t)} F\left(SN_{\widehat{\tau}(T-t)}\right)\right].$$

where the expectation is taken relative to the process $\widehat{z}^i, i = 1, ..., n$. To simplify notation we will simply write, throughout this section, E for this expectation and $\mathcal{S}_{0,1}$ for the set of relevant stopping times.

Our first proposition provides information about the structure of the exercise region as maturity approaches.

Proposition 84 *Suppose that immediate exercise is optimal at time t with asset prices S, i.e., $(S, t) \in \mathcal{E}^F$. Then immediate exercise is optimal at all later times at the same asset prices. That is, $(S, s) \in \mathcal{E}^F$ for all s such that $t \leq s \leq T$.*

Proof of Proposition 84: Consider the new stopping time $\tau' \equiv \tau k$ where $k = \frac{T-t}{T-s} > 1$ for $s > t$. As $\tau \in \mathcal{S}_{0,1}$, we have $\tau' \in \mathcal{S}_{0,k}$. Using this relation and $\tau'\left(T - s\right) = \tau\left(T - t\right)$ enables us to write

$$\begin{aligned}
V^F(S,t) &= \sup_{\tau \in \mathcal{S}_{0,1}} E\left[e^{-r\tau(T-t)} F\left(SN_{\tau(T-t)}\right)\right] \\
&= \sup_{\tau' \in \mathcal{S}_{0,k}} E\left[e^{-r\tau'(T-s)} F\left(SN_{\tau'(T-s)}\right)\right] \\
&\geq \sup_{\tau' \in \mathcal{S}_{0,1}} E\left[e^{-r\tau'(T-s)} F\left(SN_{\tau'(T-s)}\right)\right] \\
&= V^F(S,s).
\end{aligned}$$

The inequality in the next to last line follows from $\mathcal{S}_{0,1} \subset \mathcal{S}_{0,k}$ for $k > 1$. Suppose now that $(S,s) \notin \mathcal{E}^F$. Then $V^F(S,s) > F(S)$ and the inequality above implies $V^F(S,t) > F(S)$. This contradicts $(S,t) \in \mathcal{E}^F$. ■

Define $\lambda \circ_i S$ by

$$\lambda \circ_i S \equiv \left(S^1, S^2, ..., S^{i-1}, \lambda S^i, S^{i+1}, ..., S^n\right),$$

i.e., the operator $\circ_i$ scales the ith component of the vector S by the factor λ. Proposition 85 gives a sufficient condition for immediate exercise to be optimal at time t with asset prices $\lambda \circ_i S_t$ and $\lambda \geq 1$ if immediate exercise is optimal at time t with asset prices S_t.

Proposition 85 *(Right/up-connectedness): Consider an American claim with maturity T and exercise payoff $F(S)$. Suppose that immediate exercise is optimal at time t with asset prices S_t, i.e., $(S_t, t) \in \mathcal{E}^F$. Fix an index i and let $\lambda \geq 1$. Suppose that the payoff function F satisfies*

$$F(\lambda \circ_i S_t) = F(S_t) + cS_t^i \tag{6.17}$$

where $c \geq 0$ is a constant that is independent of S_t^i, but may depend on λ. Also suppose that

$$F(\lambda \circ_i S) \leq F(S) + cS^i \tag{6.18}$$

for all $S \in \mathbb{R}_+^n$ (with the same c as in (6.17)). Then $(\lambda \circ_i S_t, t) \in \mathcal{E}^F$.

Proof of Proposition 85: Suppose that immediate exercise is not optimal at $(\lambda \circ_i S_t, t)$, i.e., suppose $V^F(\lambda \circ_i S_t, t) > F(\lambda \circ_i S_t)$ for some fixed i and $\lambda \geq 1$. We have

$$\begin{aligned}
V^F(\lambda \circ_i S_t, t) &= \sup_{\tau \in \mathcal{S}_{0,1}} E\left[e^{-r\tau(T-t)} F\left((\lambda \circ_i S_t) N_{\tau(T-t)}\right)\right] \\
&\leq \sup_{\tau \in \mathcal{S}_{0,1}} E\left[e^{-r\tau(T-t)} \left(F\left(S_t N_{\tau(T-t)}\right) + cS_t^i N_{\tau(T-t)}^i\right)\right] \\
&\leq V^F(S_t, t) + cS_t^i \\
&= F(S_t) + cS_t^i \qquad \text{(because } (S,t) \in \mathcal{E}^F\text{)} \\
&= F(\lambda \circ_i S_t) \qquad \text{(by assumption (6.17)).}
\end{aligned}$$

This contradicts the assumption $V^F(\lambda \circ_i S_t, t) > F(\lambda \circ_i S_t)$. ■

Conditions (6.17) and (6.18) are satisfied by the following option payoff functions (for the values of i indicated):

Option payoff function	Valid i and λ
(a) $F(S_t) = \left(\max(S_t^1, ..., S_t^n) - K\right)^+$	$i : S_t^i = \max\left(S_t^1, ..., S_t^n\right)$
(b) $F\left(S_t^1, S_t^2\right) = \left(S_t^2 - S_t^1 - K\right)^+$	$i = 2$

First consider payoff function (a). We prove that conditions (6.17) and (6.18) hold for all i such that $S_t^i = \max\left(S_t^1, ..., S_t^n\right)$. Note that if (S_t, t) belongs to $\mathcal{E}^F$ then $F(S_t) = \left(S_t^i - K\right)^+ = S_t^i - K > 0$. For $\lambda > 1$, we have

$$\begin{aligned} F(\lambda \circ_i S_t) &= \lambda S_t^i - K \\ &= S_t^i - K + (\lambda - 1) S_t^i \\ &= F(S_t) + cS_t^i. \end{aligned}$$

So (6.17) holds for $c = \lambda - 1$. To prove (6.18), define $l = \arg\max_{j=1,...,n} \lambda \circ_i S_\tau^j$ and note that if $l \neq i$,

$$\begin{aligned} F(\lambda \circ_i S_\tau) &= \left(S_\tau^l - K\right)^+ \\ &\leq \left(S_\tau^l - K\right)^+ + (\lambda - 1) S_\tau^i \\ &= F(S_\tau) + cS_\tau^i. \end{aligned}$$

If $l = i$, then

$$\begin{aligned} F(\lambda \circ_i S_\tau) &= \left(\lambda S_\tau^i - K\right)^+ \\ &= \left(\left(S_\tau^i - K\right) + (\lambda - 1) S_\tau^i\right)^+ \\ &\leq \left(S_\tau^i - K\right)^+ + (\lambda - 1) S_\tau^i \\ &\leq F(S_\tau) + cS_\tau^i. \end{aligned}$$

The first inequality follows because $(a+b)^+ \leq a^+ + b^+$ for any $a, b \in \mathbb{R}_+$.

For payoff function (b), conditions (6.17) and (6.18) hold for $i = 2$. To prove this, note that $(S_t, t) \in \mathcal{E}^F$ implies $F(S_t) = S_t^2 - S_t^1 - K > 0$. Thus, for $\lambda > 1$, we have

$$\begin{aligned} F(\lambda \circ_i S_t) &= \lambda S_t^2 - S_t^1 - K \\ &= S_t^2 - S_t^1 - K + (\lambda - 1) S_t^2 \\ &= F(S_t) + cS_t^2, \end{aligned}$$

so (6.17) holds for $c = \lambda - 1$. To prove (6.18), note that

$$\begin{aligned} F(\lambda \circ_i S_\tau) &= \left(\lambda S^2_\tau - S^1_\tau - K\right)^+ \\ &= \left(\left(S^2_\tau - S^1_\tau - K\right) + (\lambda - 1) S^2_\tau\right)^+ \\ &\leq \left(S^2_\tau - S^1_\tau - K\right)^+ + (\lambda - 1) S^2_\tau \\ &= F(S_\tau) + cS^2_\tau. \end{aligned}$$

The next proposition, Proposition 86, gives a sufficient condition for the optimality of immediate exercise at time t with asset prices $\lambda \circ_i S_t$ and $0 \leq \lambda \leq 1$ if immediate exercise is optimal at time t with asset prices S_t.

Proposition 86 *Consider an American claim with maturity T and exercise payoff $F(S)$. Suppose that immediate exercise is optimal at time t with asset prices S_t, i.e., $(S_t, t) \in \mathcal{E}^F$. Fix an index i and a constant λ such that $0 \leq \lambda \leq 1$. Suppose that the payoff function F satisfies*

$$F(\lambda \circ_i S_t) = F(S_t). \tag{6.19}$$

Also suppose that

$$F(\lambda \circ_i S) \leq F(S) \tag{6.20}$$

for all $S \in \mathbb{R}^n_+$. Then $(\lambda \circ_i S_t, t) \in \mathcal{E}^F$.

Proof of Proposition 86: The proof is similar to the proof of Proposition 85. Suppose that immediate exercise is not optimal at $(\lambda \circ_i S_t, t)$, i.e., suppose $V^F(\lambda \circ_i S_t, t) > F(\lambda \circ_i S_t)$. For this claim we have

$$\begin{aligned} V^F(\lambda \circ_i S_t, t) &= \sup_{\tau \in \mathcal{S}_{0,1}} E\left[e^{-r\tau(T-t)} F\left((\lambda \circ_i S_t) N_{\tau(T-t)}\right)\right] \\ &\leq \sup_{\tau \in \mathcal{S}_{0,1}} E\left[e^{-r\tau(T-t)} F\left(S_t N_{\tau(T-t)}\right)\right] \quad \text{(by assumption (6.20))} \\ &= V^F(S_t, t) \\ &= F(S_t) \quad \text{(as } (S_t, t) \in \mathcal{E}^f\text{)}. \end{aligned}$$

We conclude that $V^F(\lambda \circ_i S_t, t) \leq F(S_t) = F(\lambda \circ_i S_t)$ by (6.19). This contradicts $V^F(\lambda \circ_i S_t, t) > F(\lambda \circ_i S_t)$. ■

Conditions (6.19) and (6.20) are satisfied by the following option payoff functions (for the values of i that are indicated),

Option payoff function	Valid i
(a) $F(S_t) = \left(\max\left(S^1_t, ..., S^n_t\right) - K\right)^+$	$i : S^i_t < \max\left(S^i_t, ..., S^n_t\right)$
(b) $F\left(S^1_t, S^2_t\right) = \left(S^2_t - S^1_t - K\right)^+$	$i = 1.$

Verification is straightforward.

For a constant α define αS by the usual scalar multiplication

$$\alpha S = \left(\alpha S^1, \alpha S^2, ..., \alpha S^n\right).$$

Proposition 87 gives a sufficient condition for immediate exercise to be optimal for time t with asset prices αS_t and $\alpha \geq 1$ if immediate exercise is optimal at time t with asset prices S_t.

Proposition 87 *(Ray-connectedness): Consider an American claim with maturity T and exercise payoff $F(S)$. Suppose that immediate exercise is optimal at time t with asset prices S_t, i.e., $(S_t, t) \in \mathcal{E}^F$. Also suppose that for all $\alpha \geq 1$, the payoff function F satisfies*

$$F(\alpha S_t) = \alpha F(S_t) + c \tag{6.21}$$

where $c \geq 0$ is a constant that is independent of S_t, but may depend on α. Also suppose that

$$F(\alpha S) \leq \alpha F(S) + c \tag{6.22}$$

for all $S \in \mathbb{R}^n_+$. Then for all $\alpha \geq 1$, we have $(\alpha S_t, t) \in \mathcal{E}^F$.

Proof of Proposition 87: Suppose not, i.e., suppose $V^F(\alpha S_t, t) > F(\alpha S_t)$ for some $\alpha > 1$. A contradiction follows from the following string of inequalities

$$\begin{aligned}
V^F(\alpha S_t, t) &= \sup_{\tau \in \mathcal{S}_{0,1}} E\left[e^{-r\tau(T-t)} F\left(\alpha S_t N_{\tau(T-t)}\right)\right] \\
&\leq \sup_{\tau \in \mathcal{S}_{0,1}} E\left[e^{-r\tau(T-t)}\left(\alpha F\left(S_t N_{\tau(T-t)}\right) + c\right)\right] \quad \text{(by (6.22))} \\
&\leq \alpha V^F(S_t, t) + c \quad \text{(as } (S_t, t) \in \mathcal{E}^F) \\
&= F(\alpha S_t) \quad \text{(by (6.21))}.
\end{aligned}$$

This completes the proof of the proposition. ■

Conditions (6.21) and (6.22) are satisfied by the option payoff functions

Option payoff function
(a) $F(S_t) = \left(\max\left(S_t^1, ..., S_t^n\right) - K\right)^+$
(b) $F\left(S_t^1, S_t^2\right) = \left(S_t^2 - S_t^1 - K\right)^+$
(c) $F(S_t) = \left(\min\left(S_t^1, ..., S_t^n\right) - K\right)^+$

To prove that the conditions hold for (a) note that $(S_t, t) \in \mathcal{E}^F$ implies $F(S_t) > 0$. We then have

$$\begin{aligned}
F(\alpha S_t) &= \max_{j=1,...,n} \alpha S_t^j - K \\
&= \alpha\left(\max_{j=1,...,n} S_t^j - K\right) + (\alpha - 1) K \\
&= \alpha F(S_t) + c,
\end{aligned}$$

so (6.21) holds for $c = (\alpha - 1) K$. To prove (6.22), define $l = \arg\max_{j=1,\ldots,n} S^j$ and note that

$$\begin{aligned} F(\alpha S) &= \left(\alpha S^l - K\right)^+ \\ &= \left(\alpha\left(S^l - K\right) + (\alpha - 1) K\right)^+ \\ &\leq \alpha\left(S^l - K\right)^+ + (\alpha - 1) K \\ &= \alpha F(S) + c. \end{aligned}$$

To prove that the conditions hold for (b) note that $(S_t, t) \in \mathcal{E}^F$ implies $F(S_t) = S_t^2 - S_t^1 - K > 0$. Then

$$\begin{aligned} F(\alpha S_t) &= \alpha S_t^2 - \alpha S_t^1 - K \\ &= \alpha\left(S_t^2 - S_t^1 - K\right) + (\alpha - 1) K \\ &= \alpha F(S_t) + c, \end{aligned}$$

so (6.21) holds for $c = (\alpha - 1) K$. To prove (6.22),

$$\begin{aligned} F(\alpha S) &= \left(\alpha S^2 - \alpha S^1 - K\right)^+ \\ &= \left(\alpha\left(S^2 - S^1 - K\right) + (\alpha - 1) K\right)^+ \\ &\leq \alpha\left(S^2 - S^1 - K\right)^+ + (\alpha - 1) K \\ &= \alpha F(S) + c. \end{aligned}$$

To prove the conditions for (c) note that $(S_t, t) \in \mathcal{E}^F$ implies $F(S_t) > 0$. We then have

$$\begin{aligned} F(\alpha S_t) &= \min_{j=1,\ldots,n} \alpha S_t^j - K \\ &= \alpha\left(\min_{j=1,\ldots,n} S_t^j - K\right) + (\alpha - 1) K \\ &= \alpha F(S_t) + c, \end{aligned}$$

so (6.21) holds for $c = (\alpha - 1) K$. To prove (6.22) note that

$$\begin{aligned} F(\alpha S) &= \left(\min_{j=1,\ldots,n} \alpha S^j - K\right)^+ \\ &= \left(\alpha\left(\min_{j=1,\ldots,n} S^j - K\right) + (\alpha - 1) K\right)^+ \\ &\leq \alpha\left(\min_{j=1,\ldots,n} S^j - K\right)^+ + (\alpha - 1) K \\ &= \alpha F(S) + c. \end{aligned}$$

Proposition 88 shows that claims with convex payoff function, also have convex pricing functions.

Proposition 88 *(Convexity): Consider an American claim with maturity T and exercise payoff $F(S)$. Suppose that F is a (strictly) convex function. Then $V^F(S,t)$ is (strictly) convex with respect to S.*

Proof of Proposition 88: Using the convexity of the payoff function, we can write

$$\begin{aligned}
& V^F(S(\lambda), t) \\
= \; & \sup_{\tau \in \mathcal{S}_{0,1}} E\left[e^{-r\tau(T-t)} F\left(\lambda S N_{\tau(T-t)} + (1-\lambda)\widetilde{S} N_{\tau(T-t)}\right)\right] \\
\leq \; & \sup_{\tau \in \mathcal{S}_{0,1}} E\left[e^{-r\tau(T-t)} \left(\lambda F\left(S N_{\tau(T-t)}\right) + (1-\lambda) F\left(\widetilde{S} N_{\tau(T-t)}\right)\right)\right] \\
\leq \; & \sup_{\tau \in \mathcal{S}_{0,1}} E\left[e^{-r\tau(T-t)} \lambda F\left(S N_{\tau(T-t)}\right)\right] \\
& \quad + \sup_{\tau \in \mathcal{S}_{0,1}} E\left[e^{-r\tau(T-t)} (1-\lambda) F\left(\widetilde{S} N_{\tau(T-t)}\right)\right] \\
\leq \; & \lambda V^F(S,t) + (1-\lambda) V^F\left(\widetilde{S}, t\right). \blacksquare
\end{aligned}$$

Important variations of Propositions 85 and 86 appear in our next two results. These propositions cover the case of min-option payoffs.

Proposition 89 *(Up-segment-connectedness): Consider an American claim with maturity T and exercise payoff $F(S)$. Suppose that immediate exercise is optimal at time t with asset prices S_t, i.e., $(S_t,t) \in \mathcal{E}^F$. Fix an index i and suppose that there exists a function $h(\cdot): \mathbb{R}^n_+ \to \mathbb{R}$ such that, for $h(S_t)/S^i_t \geq \lambda \geq 1$, the payoff function F satisfies*

$$F(\lambda \circ_i S_t) = \lambda F(S_t) + c \tag{6.23}$$

where $c \geq 0$ is a constant that is independent of S_t, but may depend on λ. Also suppose that

$$F(\lambda \circ_i S) \leq \lambda F(S) + c \tag{6.24}$$

for all $S \in \mathbb{R}^n_+$ (with the same c as in (6.23)). Then $(\lambda \circ_i S_t, t) \in \mathcal{E}^F$ for all $h(S_t)/S^i_t \geq \lambda \geq 1$.

Proof of Proposition 89: Suppose that this is not the case, i.e., suppose $V^F(\lambda \circ_i S_t, t) > F(\lambda \circ_i S_t)$ for some fixed i and some λ such that $h(S_t)/S^i_t \geq \lambda \geq 1$. We have

$$\begin{aligned}
V^F(\lambda \circ_i S_t, t) &= \sup_{\tau \in \mathcal{S}_{0,1}} E\left[e^{-r\tau(T-t)} F\left((\lambda \circ_i S_t) N_{\tau(T-t)}\right)\right] \\
&\leq \sup_{\tau \in \mathcal{S}_{0,1}} E\left[e^{-r\tau(T-t)} \left(\lambda F\left(S_t N_{\tau(T-t)}\right) + c\right)\right] \\
&\leq \lambda V^F(S_t, t) + c \\
&= \lambda F(S_t) + c \qquad \text{(given that } (S,t) \in \mathcal{E}^F\text{)} \\
&= F(\lambda \circ_i S_t) \qquad \text{(by assumption (6.23)).}
\end{aligned}$$

This contradicts our assumption $V^F(\lambda \circ_i S_t, t) > F(\lambda \circ_i S_t)$. ■

Conditions (6.23) and (6.24) are satisfied by the min-option payoff $F(S_t) = \left(\min(S_t^1, ..., S_t^n) - K\right)^+$ for the index $i : S_t^i = \min\left(S_t^1, ..., S_t^n\right)$ and the function $h(S_t) = \min\left(S_t^1, ..., S_t^{i-1}, S_t^{i+1}, ..., S_t^n\right)$. The result stated holds for all $\lambda \geq 1$ such that $\lambda S_t^i \leq \min\left(S_t^1, ..., S_t^{i-1}, S_t^{i+1}, ..., S_t^n\right)$. The constant $c = (\lambda - 1) K$. Note in particular that $(\min \lambda \circ_i S - K)^+ \leq (\min \lambda S - K)^+ = (\lambda \min S - K)^+ \leq \lambda (\min S - K)^+ + (\lambda - 1) K = \lambda F(S) + c$.

Proposition 90 *(Left-segment-connectedness): Consider an American claim with maturity T and exercise payoff $F(S)$. Suppose that immediate exercise is optimal at time t with asset prices S_t, i.e., $(S_t, t) \in \mathcal{E}^F$. Fix an index j and suppose that there exists a function $h(\cdot) : \mathbb{R}_+^n \to \mathbb{R}$ such that, for $h(S_t) \leq \lambda \leq 1$, the payoff function F satisfies*

$$F(\lambda \circ_j S_t) = F(S_t). \tag{6.25}$$

Also suppose that

$$F(\lambda \circ_j S) \leq F(S) \tag{6.26}$$

for all $S \in \mathbb{R}_+^n$. Then $(\lambda \circ_j S_t, t) \in \mathcal{E}^F$ for $h(S_t) \leq \lambda \leq 1$.

Proof of Proposition 90: Suppose that immediate exercise is not optimal at $(\lambda \circ_j S_t, t)$, i.e., $V^F(\lambda \circ_j S_t, t) > F(\lambda \circ_j S_t)$. We have

$$\begin{aligned} V^F(\lambda \circ_j S_t, t) &= \sup_{\tau \in \mathcal{S}_{0,1}} E\left[e^{-r\tau(T-t)} F\left((\lambda \circ_j S_t) N_{\tau(T-t)}\right)\right] \\ &\leq \sup_{\tau \in \mathcal{S}_{0,1}} E\left[e^{-r\tau(T-t)} F\left(S_t N_{\tau(T-t)}\right)\right] \qquad \text{(by (6.26))} \\ &= V^F(S_t, t) \\ &= F(S_t) \qquad \text{(as } (S_t, t) \in \mathcal{E}^F\text{)}. \end{aligned}$$

Hence $V^F(\lambda \circ_j S_t, t) \leq F(S_t) = F(\lambda \circ_j S_t)$ by (6.25). This contradicts the assumption that $V^F(\lambda \circ_j S_t, t) > F(\lambda \circ_j S_t)$. ■

For min-options the proposition holds for any index $j \neq i$ where $i : S_t^i = \min(S_t^1, ..., S_t^n)$. The function $h(\cdot)$ is $h(S_t) = S_t^i / S_t^j$.

6.8 Appendix B: Proofs

Proof of Proposition 66: The argument following the proposition shows that the American-style max-option price is bounded below by the value of a package consisting of a forward contract and a European exchange option. That is,

$$C^x\left(S_t^1, S_t^2, t\right) \geq S^1 e^{-\delta_1(s-t)} - K e^{-r(s-t)} + \widetilde{E}_t\left[e^{-r(s-t)}\left(S_s^2 - S_s^1\right)^+\right]$$

for $s \in [t, T]$. Let $V(t, s)$ denote the value of the package at date t, when the constituent securities have maturity date s. Suppose equal underlying prices, $S_t^1 = S_t^2$ and assume that $S_t^1 = S_t^2 > K$ (so that immediate exercise has positive value). As $s \to t$ the limits

$$S_t^1 e^{-\delta_1(s-t)} - K e^{-r(s-t)} \to S_t^1 - K$$

$$e^{-r(s-t)} \widetilde{E}_t \left[\left(S_s^2 - S_s^1 \right)^+ \right] \to 0$$

hold. Lemma 91, below, shows that convergence is from above. As a result there exists a time $w \in [t, T]$ such that $V(t, w) > S_t^1 - K$. It follows that $C^x\left(S_t^1, S_t^1, t\right) > S_t^1 - K = (S_t^1 - K)^+$. This argument holds for all $t < T$. ■

Lemma 91 *Assume that $S_t^1 = S_t^2 > K$ and $t < T$. Then there exists a time $w \in (t, T)$, such that*

$$S_t^1 \left(e^{-\delta_1(w-t)} - 1 \right) - K \left(e^{-r(w-t)} - 1 \right) + e^{-r(w-t)} \widetilde{E}_t \left[\left(S_w^2 - S_w^1 \right)^+ \right] > 0.$$

Proof of Lemma 91: Let $w = t + h$ and $\Psi(h) \equiv e^{-rh} \widetilde{E}_t \left[\left(S_w^2 - S_w^1 \right)^+ \right]$. By Theorem 75 this European exchange option is worth

$$\begin{aligned} \Psi(h) &= S_t^1 e^{-\delta_2 h} N\left(\left(\delta_1 - \delta_2 + \frac{1}{2}\sigma_e^2 \right) \sqrt{h}/\sigma_e \right) \\ &\quad - S_t^1 e^{-\delta_1 h} N\left(\left(\delta_1 - \delta_2 - \frac{1}{2}\sigma_e^2 \right) \sqrt{h}/\sigma_e \right) \end{aligned}$$

with $\sigma_e = \sqrt{\sigma_1^2 - 2\sigma_1 \rho \sigma_2 + \sigma_2^2}$. Clearly $\Psi(0) = 0$. Also, given that

$$\begin{aligned} \Psi'(h) &= -\delta_2 S_t^1 e^{-\delta_2 h} N\left(\left(\delta_1 - \delta_2 + \frac{1}{2}\sigma_e^2 \right) \sqrt{h}/\sigma_e \right) \\ &\quad + \delta_1 S_t^1 e^{-\delta_1 h} N\left(\left(\delta_1 - \delta_2 - \frac{1}{2}\sigma_e^2 \right) \sqrt{h}/\sigma_e \right) \\ &\quad + S_t^1 e^{-\delta_2 h} n\left(\left(\delta_1 - \delta_2 + \frac{1}{2}\sigma_e^2 \right) \sqrt{h}/\sigma_e \right) \left(\delta_1 - \delta_2 + \frac{1}{2}\sigma_e^2 \right) \frac{1}{2} \frac{1}{\sqrt{h}\sigma_e} \\ &\quad - S_t^1 e^{-\delta_1 h} n\left(\left(\delta_1 - \delta_2 - \frac{1}{2}\sigma_e^2 \right) \sqrt{h}/\sigma_e \right) \left(\delta_1 - \delta_2 - \frac{1}{2}\sigma_e^2 \right) \frac{1}{2} \frac{1}{\sqrt{h}\sigma_e} \end{aligned}$$

it is easy to verify that $\Psi'(0) = +\infty$. Let $\Gamma(h) \equiv S_t^1\left(e^{-\delta_1 h} - 1\right) - K\left(e^{-rh} - 1\right)$. Simple computations show that $\Gamma(0) = 0$ and $\Gamma'(0) = rK - \delta_1 S_t^1$. We conclude that there exists an $h > 0$ such that the lemma holds (see Figure 6.1 for an illustration). ■

Proof of Proposition 67: Property (i) is an application of Proposition 84 in Appendix A. For (ii) use Proposition 85 and the remarks concerning the payoff function (a) following that proposition. For (iii) use Proposition 86 and the subsequent remarks for payoff function (a). For (iv) note that $S_t^2 = 0$ implies $S_s^2 = 0$ for all $s \geq t$. In this event the max-option reduces to a single asset option written on the price S^1, whose exercise boundary is B^1. To prove (v) note that the convexity of the payoff function and Proposition 88 enable us to write

$$C^x(S(\lambda), t) \leq \lambda C^x(S, t) + (1-\lambda) C^x\left(\widetilde{S}, t\right).$$

By assumption $(S,t) \in \mathcal{E}_i^{c,x}$ and $(\widetilde{S}, t) \in \mathcal{E}_i^{c,x}$. It then follows that $C^x(S(\lambda), t) \leq \lambda (S^i - K) + (1-\lambda)\left(\widetilde{S}^i - K\right) = S^i(\lambda) - K$. Suppose now that immediate exercise is suboptimal at $(S(\lambda), t)$, that is $C^x(S(\lambda), t) > \left(S^1(\lambda) \vee S^2(\lambda) - K\right)^+ = S^i(\lambda) - K$. This contradicts the previous inequality. We conclude that $(S(\lambda), t) \in \mathcal{E}_i^{c,x}$. ■

Proof of Proposition 68: Let $\mathcal{E}^{c,x}(t, K)$ be the exercise region at date t, parametrized by the strike K and let $C^x\left(S^1, S^2, t; K\right)$ be the corresponding max-option price. The proof uses the next three lemmas documenting properties of the exercise region.

Lemma 92 *$K_1 > K_2 \geq 0$ implies $\mathcal{E}^{c,x}(t, K_1) \subset \mathcal{E}^{c,x}(t, K_2)$. In particular, $K_1 > K_2 = 0$ implies $\mathcal{E}^{c,x}(t, K_1) \subset \mathcal{E}^{c,x}(t, 0)$.*

Proof of Lemma 92: The proof proceeds by contradiction. Let $K_1 > K_2 \geq 0$ and suppose that $(S,t) \in \mathcal{E}^{c,x}(t, K_1)$ but $(S,t) \notin \mathcal{E}^{c,x}(t, K_2)$. We show that this combination of assumptions is not possible.

The following string of relations holds

$$\begin{aligned}
\left(S^1 \vee S^2 - K_2\right)^+ &< C^x\left(S^1, S^2, t; K_2\right) \\
&= \sup_{\tau \in \mathcal{S}_{t,T}} \widetilde{E}_t\left[e^{-r(\tau - t)}\left(S_\tau^1 \vee S_\tau^2 - K_2\right)^+\right] \\
&= \widetilde{E}_t\left[e^{-r(\tau_2 - t)}\left(S_{\tau_2}^1 \vee S_{\tau_2}^2 - K_1 + K_1 - K_2\right)^+\right] \\
&\leq \widetilde{E}_t\left[e^{-r(\tau_2 - t)}\left(S_{\tau_2}^1 \vee S_{\tau_2}^2 - K_1\right)^+\right] \\
&\quad + \widetilde{E}_t\left[e^{-r(\tau_2 - t)}\left(K_1 - K_2\right)^+\right] \\
&\leq C^x\left(S^1, S^2, t; K_1\right) + K_1 - K_2 \\
&= S^1 \vee S^2 - K_1 + K_1 - K_2 = S^1 \vee S^2 - K_2.
\end{aligned}$$

The inequality on the first line expresses the suboptimality of exercise at (S,t) for the max-call with strike K_2 (i.e. $(S,t) \notin \mathcal{E}^{c,x}(t, K_2)$). The second line is the pricing formula for the American-style max-option. The third line is a straightforward decomposition of the option payoff ($\tau_2 > t$ is the optimal exercise policy for the option with strike K_2). The inequality on the fourth line results for the bound $(a+b)^+ \leq a^+ + b^+$. The next inequality follows from

$r \geq 0$. Finally, the last line uses the optimality of exercise at (S, t) for the max-call with strike K_1 (i.e. $(S, t) \in \mathcal{E}^{c,x}(t, K_1)$). The resulting contradiction shows that immediate exercise must be optimal at (S, t) for the strike K_2 if it is optimal for the strike K_1.

The case $K_2 = 0$ is an application of the general result. ■

Lemma 93 *(Ray connectedness)* $(S^1, S^2, t) \in \mathcal{E}^{c,x}(t, 0)$ *implies* $(\lambda S^1, \lambda S^2, t) \in \mathcal{E}^{c,x}(t, 0)$, *for all* $\lambda > 0$.

Proof of Lemma 93: The proof is by contradiction. Let $(S^1, S^2, t) \in \mathcal{E}^{c,x}(t, 0)$ and assume that there exists $\lambda > 0$ such that $(\lambda S^1, \lambda S^2, t) \notin \mathcal{E}^{c,x}(t, 0)$. Letting $\tau_\lambda \in \mathcal{S}_{t,T}$ denote the optimal policy at $(\lambda S^1, \lambda S^2, t)$ (for an option with null strike) we can write

$$\begin{aligned}
\lambda S^1 \vee \lambda S^2 &< C^x\left(\lambda S^1, \lambda S^2, t; 0\right) \\
&= \widetilde{E}_t\left[e^{-r(\tau_\lambda - t)}\left(\lambda S^1_{\tau_\lambda} \vee \lambda S^2_{\tau_\lambda}\right)\right] \\
&= \lambda \widetilde{E}_t\left[e^{-r(\tau_\lambda - t)}\left(S^1_{\tau_\lambda} \vee S^2_{\tau_\lambda}\right)\right] \\
&\leq \lambda C^x\left(S^1, S^2, t; 0\right).
\end{aligned}$$

(the last inequality follows from the feasibility of τ_λ at (S^1, S^2, t)). We conclude that $S^1 \vee S^2 < C^x(S^1, S^2, t; 0)$, in contradiction with the initial assumption that $(S^1, S^2, t) \in \mathcal{E}^{c,x}(t, 0)$. ■

Lemma 94 *(Diagonal behavior)* $(S, S, t) \notin \mathcal{E}^{c,x}(t, 0)$ *for* $t < T$.

Proof of Lemma 94: The lemma is a particular case of Proposition 66 with $K = 0$. ■

Let us now return to the proof of Proposition 68. By Lemma 94 $(S, S, t) \notin \mathcal{E}^{c,x}(t, 0)$. Given that the continuation region is an open set there exists an open neighborhood $\mathcal{N}$ of the point (S, S) such that $(S^1, S^2, t) \notin \mathcal{E}^{c,x}(t, 0)$ for all $(S^1, S^2) \in \mathcal{N}$. Lemma 93 then shows the existence of an open cone $R(\lambda_1, \lambda_2)$, containing the diagonal, such that $R(\lambda_1, \lambda_2) \cap \mathcal{E}^{c,x}(t, 0) = \emptyset$. Lemma 92 enables us to conclude that $R(\lambda_1, \lambda_2) \cap \mathcal{E}^{c,x}(t, K) = \emptyset$. ■

Proof of Proposition 69: Property (i) can be proved by using arguments in Broadie and Detemple [1997] and Jaillet, Lamberton and Lapeyre [1990]. The property rests, in particular, on the continuity of the discounted payoff function and the continuity of the flow of the stochastic differential equation for $S^i, i = 1, 2$. Property (ii) can be proved by using the Comparison Theorem for solutions of SDEs (see Karatzas and Shreve [1988]) and the fact that the payoff function is non-decreasing in prices. Property (iii) is the counterpart of property (i) in Proposition 67. Convexity, in property (iv), results from the convexity of the payoff function (see Proposition 88). ■

Proof of Theorem 70: The EEP formula follows from Theorem 21 extended to the case of multiple underlying assets. The particular structure of the exercise premium follows from the geometry of the exercise region delimited by the two boundary surfaces $B_1^{c,x}$ and $B_2^{c,x}$. The recursive integral equations for the boundaries are obtained by evaluating the contract along each of the boundaries. ■

Proof of Proposition 71: Property (i) follows from Proposition 84 in Appendix A. For (ii) and (iii) apply Propositions 85 and 86. Property (iv) holds because the dual strike max-option collapses to an option on asset one alone when $S^2 = 0$. The case $S^2 = S^1 + K_2 - K_1$, in property (v), corresponds to the case of the diagonal for max-options. When $S^2 - K_2 = S^1 - K_1$ a movement away from the line $S^2 = S^1 + K_2 - K_1$ provides good odds of improving the payoff. This can be formally demonstrated using the arguments in the proof of Proposition 66. The convexity property (vi) follows from Proposition 88. ■

Proof of Proposition 72: Properties (i), (ii) and (iii) are straightforward. Property (iv) follows because the payoff of a min-put option is bounded below by the payoff of a plain vanilla put on asset 1. Under the assumption of property (v) the min-put option has maximal payoff K, which implies optimality of immediate exercise. For (vi), suppose that there is no threshold $S^{2*} > S^1$ such that $(S^1, S^2, t) \in \mathcal{E}_1^{p,m}$ for all $S^2 \geq S^{2*}$. It must then be that $\lim_{S^2 \uparrow \infty} P^m(S^1, S^2, t) > K - S^1$. But the limit on the left hand side is the value of a put on asset 1. This contradicts the assumption $(S^1, t) \in \mathcal{E}_1^p$. For (vii) note that a deviation from the diagonal has high probability of increasing the payoff. Dominance of waiting can be formally proved as in the proof of Proposition 66. Convexity in (viii) follows from Proposition 88. ■

Proof of Proposition 73: For assertion (i) note that immediate exercise when $S_t^2 \leq S_t^1 + K$ would generate a non-positive payoff. Property (ii) is a special case of Proposition 84 in Appendix A. Assertion (iii) follows from Proposition 85 and the subsequent remarks for payoff function (b). For (iv) use Proposition 86 and the remarks that follow for payoff function (b). To prove (v) note that $S_t^1 = 0$ implies $S_v^1 = 0$ for all $v \geq t$. In this situation the spread option reduces to an option on on asset S^2, with exercise boundary B^2. The proof of (vi) follows from Proposition 88. ■

Proof of Theorem 74:: The EEP formula for the spread option is an application of Theorem 21 generalized to multiple assets. The particular structure of the exercise premium follows from the geometry of the exercise region detailed in Proposition 73. Standard arguments lead to the recursive integral equation. ■

Proof of Theorem 75: The proof applies the change of numeraire outlined in section 2.7. To establish the proposition we need to verify the dynamics of the process S^e under the new pricing measure.

Let $S^e = S^2/S^1$. An application of Ito's lemma gives

$$dS_t^e = S_t^e \left[(r - \delta_e)\, dt + \sigma_2 d\widetilde{z}_t^2 - \sigma_1 d\widetilde{z}_t^1 \right]$$

where $\delta_e \equiv r + \delta_2 - \delta_1 - \sigma_1^2 + \rho\sigma_1\sigma_2$. Under the pricing measure Q^* such that

$$dQ^* = \exp\left(-\frac{1}{2}\sigma_1^2 T + \sigma_1 \widetilde{z}_T^1 \right) dQ$$

the translated process $z_t^* = -\widetilde{z}_t^1 + \sigma_1 t$ is a Brownian motion. Using $d\widetilde{z}_t^2 = \rho d\widetilde{z}_t^1 + \sqrt{1-\rho^2} d\widetilde{z}_t^3$ for some orthogonal Q-Brownian motion $\widetilde{z}^3$ yields

$$\begin{aligned} dS_t^e &= S_t^e \left[(r - \delta_e)\, dt + \sigma_2 d\widetilde{z}_t^2 - \sigma_1 d\widetilde{z}_t^1 \right] \\ &= S_t^e \left[(r - \delta_e)\, dt + (\sigma_2\rho - \sigma_1)\, d\widetilde{z}_t^1 + \sigma_2\sqrt{1-\rho^2} d\widetilde{z}_t^3 \right] \\ &= S_t^e \left[(\delta_1 - \delta_2)\, dt - (\sigma_2\rho - \sigma_1)\, dz_t^* + \sigma_2\sqrt{1-\rho^2} d\widetilde{z}_t^3 \right] \\ &= S_t^e \left[(\delta_1 - \delta_2)\, dt + \sigma_e dz_t^e \right] \end{aligned}$$

with the definitions $\sigma_e = \sqrt{\sigma_1^2 - 2\sigma_1\rho\sigma_2 + \sigma_2^2}$ and $\sigma_e dz_e = -(\sigma_2\rho - \sigma_1) dz_t^* + \sigma_2\sqrt{1-\rho^2} d\widetilde{z}_t^3$.

Theorem 8 can now be applied, with numeraire $S_t^1 \exp(\delta_1 t)$, to establish the symmetry result stated. The EEP representation and the integral equation for the boundary follow immediately. ■

Proof of Theorem 76: The option value can be transformed in the following manner

$$\begin{aligned} C^{ec}\left(S^1, S^2, t\right) &= \sup_{\tau \in \mathcal{S}_{t,T}} \widetilde{E}_t \left[e^{-r(\tau - t)} \left(S_\tau^2 \wedge L S_\tau^1 - S_\tau^1 \right)^+ \right] \\ &= \sup_{\tau \in \mathcal{S}_{t,T}} \widetilde{E}_t \left[e^{-r(\tau - t)} S_\tau^1 \left(S_\tau^e \wedge L - 1 \right)^+ \right] \\ &= S_t^1 \sup_{\tau \in \mathcal{S}_{t,T}} E_t^1 \left[e^{-\delta_1(\tau - t)} \left(S_\tau^e \wedge L - 1 \right)^+ \right] \end{aligned}$$

where the last equality uses the passage to the new measure in which asset 1 is the numeraire (see proof of Theorem 75). The results in the proposition follow from this transformation. ■

Proof of Theorem 77: The option with geometric sum payoff is identical to an option on a single asset with price dynamics

$$dY_t = Y_t \left[(r - \delta_Y)\, dt + \sigma_Y dz_t \right]$$

where

$$\delta_Y = (1 - \gamma)\, r + \gamma_1\delta_1 + \gamma_2\delta_2 - \frac{1}{2}\left(\sigma_Y^2 - \gamma_1\sigma_1^2 - \gamma_2\sigma_2^2 \right)$$

$$\sigma_Y = \sqrt{(\gamma_1\sigma_1)^2 + 2\gamma_1\gamma_2\sigma_1\rho\sigma_2 + (\gamma_2\sigma_2)^2}$$

and

$$dz_t = \frac{\gamma_1 \sigma_1 d\widetilde{z}_t^1 + \gamma_2 \sigma_2 d\widetilde{z}_t^2}{\sigma_Y}.$$

Let B denote the exercise boundary of the option written on Y. The boundary of the average option is obtained by solving the equation $\left(S^1\right)^{\gamma_1} B^{gs}\left(S^1,t\right)^{\gamma_2} = B(t)$. The value of the option is the value of the option written on asset Y. ■

Proof of Proposition 78: The claims (i) and (ii) are straightforward. The proof of (iii) is by contradiction. Suppose that immediate exercise is optimal at $\left(S^1, S^2, t\right)$ but not at $\left(\lambda_1 S^1, \lambda_2 S^2, t\right)$ where $\lambda_1, \lambda_2 \geq 1$. Let $\tau_\lambda > t$ be the optimal policy at $\left(\lambda_1 S^1, \lambda_2 S^2, t\right)$. Standard arguments can be applied to show that

$$\begin{aligned}
C^w\left(\lambda_1 S^1, \lambda_2 S^2, t\right) &= \widetilde{E}_t\Big[e^{-r(\tau_\lambda - t)}\left(wS^1_{\tau_\lambda} + (1-w)\, S^2_{\tau_\lambda} - K\right. \\
&\qquad \left.+ w\left(\lambda_1 - 1\right) S^1_{\tau_\lambda} + (1-w)\left(\lambda_2 - 1\right) S^2_{\tau_\lambda}\right)^+\Big] \\
&\leq \widetilde{E}_t\left[e^{-r(\tau_\lambda - t)}\left(wS^1_{\tau_\lambda} + (1-w)\, S^2_{\tau_\lambda} - K\right)^+\right] \\
&\qquad + w\left(\lambda_1 - 1\right)\widetilde{E}_t\left[e^{-r(\tau_\lambda - t)} S^1_{\tau_\lambda}\right] \\
&\qquad + (1-w)\left(\lambda_2 - 1\right)\widetilde{E}_t\left[e^{-r(\tau_\lambda - t)} S^2_{\tau_\lambda}\right] \\
&\leq C^w\left(S^1, S^2, t\right) + w\left(\lambda_1 - 1\right) S^1 + (1-w)\left(\lambda_2 - 1\right) S^2 \\
&= wS^1 + (1-w)\, S^2 - K \\
&\qquad + w\left(\lambda_1 - 1\right) S^1 + (1-w)\left(\lambda_2 - 1\right) S^2 \\
&= w\lambda_1 S^1 + (1-w)\,\lambda_2 S^2 - K.
\end{aligned}$$

The resulting contradiction proves (iii).

The convexity result in (iv) is a special case of Proposition 88 in Appendix A. The proof of (v) is standard (see Proposition 84). ■

Proof of Theorem 79: The EEP representation gives

$$C^w\left(S^1, S^2, t\right) = c^w\left(S^1, S^2, t\right) + \pi^w\left(S^1, S^2, t\right)$$

where

$$c^w\left(S^1, S^2, t\right) = \widetilde{E}_t\left[e^{-r(T-t)}\left(wS^1_T + (1-w)\, S^2_T - K\right)^+\right].$$

$$\pi^w\left(S^1, S^2, t\right) = \int_t^T e^{-r(v-t)} \phi(t,v)\, dv$$

$$\phi(t,v) \equiv \widetilde{E}_t\left[\left(w\delta_1 S^1_v + (1-w)\,\delta_2 S^2_v - rK\right) 1_{\{S^2_v \geq B^{c,w}(S^1_v, v)\}}\right].$$

The event of exercise can be written as $\left\{S^2_v \geq B^{c,w}\left(S^1_v, v\right)\right\}$ and is equivalent to $\left\{z^2 \geq d\left(S^2, B^{c,w}\left(S^1_v, v\right), v-t\right)\right\}$ with

$$d\left(S^2, B^{c,w}\left(S^1_v, v\right), v-t\right) = \frac{1}{\sigma_2\sqrt{v-t}}\left[\log\left(\frac{B^{c,w}\left(S^1_v, v\right)}{S^2}\right) - \alpha_2\left(v-t\right)\right]$$

and $\alpha_2 \equiv r - \delta_2 - \frac{1}{2}\sigma_2^2$. Using $z^2 = \rho z^1 + \sqrt{1-\rho^2}u$ we obtain

$$\left\{S_v^2 \geq B^{c,w}\left(S_v^1, v\right)\right\} = \left\{u \geq d\left(S^2, B^{c,w}\left(S_v^1, v\right), v-t, \rho, z^1\right)\right\}$$

with

$$d\left(S^2, B^{c,w}\left(S_v^1, v\right), v-t, \rho, z^1\right) \equiv \frac{d\left(S^2, B^{c,w}\left(S_v^1, v\right), v-t\right)}{\sqrt{1-\rho^2}} - \frac{\rho z_1}{\sqrt{1-\rho^2}}.$$

With $h \equiv v-t$, the conditional expectation inside the early exercise premium π^w is

$$\phi(t,v) = w\delta_1 S_t^1 e^{(r-\delta_1)h}\phi_1(v) + (1-w)\delta_2 S_t^2 e^{(r-\delta_2)h}\phi_2(v) - rK\phi_3(v)$$

where

$$\phi_1(v) = \int_{-\infty}^{+\infty}\int_d^{+\infty} \frac{1}{\sqrt{2\pi}}\exp\left(-\frac{1}{2}\left(z^1 - \sigma_1\sqrt{h}\right)^2\right) n(u)\,dudz^1$$

$$\begin{aligned}\phi_2(v) &= \int_{-\infty}^{+\infty}\int_d^{+\infty} \exp\left(-\frac{1}{2}\sigma_2^2 h + \sigma_2\left(\rho z^1 + \sqrt{1-\rho^2}u\right)\sqrt{h}\right) \\ &\quad \times n\left(z^1\right) n(u)\,dudz^1\end{aligned}$$

$$\phi_3(v) = \int_{-\infty}^{+\infty}\int_d^{+\infty} n\left(z^1\right) n(u)\,dudz^1.$$

To simplify the notation let $z \equiv z^1$. Simple transformations give

$$\phi_1(v) = \int_{-\infty}^{+\infty} n\left(z - \sigma_1\sqrt{h}\right) N\left(-d\left(S^2, B^{c,w}\left(S_v^1(z), v\right), h, \rho, z\right)\right) dz$$

$$\begin{aligned}\phi_2(v) &= \int_{-\infty}^{+\infty}\int_d^{+\infty} \frac{1}{2\pi}\exp\left(-\frac{1}{2}\left(z - \sigma_2\rho\sqrt{h}\right)^2\right. \\ &\quad \left. -\frac{1}{2}\left(u - \sigma_2\sqrt{1-\rho^2}\sqrt{h}\right)^2\right) dudz\end{aligned}$$

$$\phi_3(v) = \int_{-\infty}^{+\infty} n(z) N\left(-d\left(S^2, B^{c,w}\left(S_v^1(z), v\right), h, \rho, z\right)\right) dz$$

and it is straightforward to verify that the double integral in $\phi_2(v)$ equals

$$\int_{-\infty}^{+\infty} n\left(z - \sigma_2\rho\sqrt{h}\right) N\left(-d\left(S^2, B^{c,w}\left(S_v^1(z), v\right), h, \rho, z\right) + \sigma_2\sqrt{1-\rho^2}\sqrt{h}\right) dz.$$

Substituting the function

$$\begin{aligned}\Phi\left(S^2, B^{c,w}(\cdot, v), h, \rho, x, y\right) &\equiv \int_{-\infty}^{+\infty} n(z-y) \\ &\quad \times N\left(-d\left(S^2, B^{c,w}(\cdot, v), h, \rho, z\right) + x\right) dz\end{aligned}$$

in the formulas above gives the result stated. ■

Proof of Proposition 80: Property (i) follows from Proposition 84. For (ii) it suffices to note that the min-option payoff is bounded above by the payoff of the option on asset $i, i = 1, 2$. Properties (iii)-(v) are special cases of Propositions 87, 89 and 90.

For (vi) note that the min-option payoff function satisfies the bound

$$\left(S^1 \wedge S^2 - K\right)^+ \leq \left(wS^1 + (1-w)\, S^2 - K\right)^+,$$

for $w \in (0, 1)$. The price upper bound $C^m\left(S^1, S^2, t\right) \leq C_t^w\left(S^1, S^2, t\right)$ follows. Suppose now that immediate exercise is optimal at a point along the diagonal, for the weighted average option: $\left(S^1, S^2, t\right) \in \mathcal{E}^{c,w}$ and $S^1 = S^2$. At this point the previous inequality becomes $C_t^m\left(S^1, S^2\right) \leq \left(wS^1 + (1-w)\, S^2 - K\right)^+ = (S^i - K)^+$, $i = 1, 2$. Given that immediate exercise is feasible for the min-option we also have the lower bound $C_t^m\left(S^1, S^2\right) \geq \left(S^1 \wedge S^2 - K\right)^+ = \left(S^i - K\right)^+$, $i = 1, 2$. The result follows.

For (vii) it suffices to note that the payoff is null along the axes. ■

Proof of Proposition 81: Statements (i) and (ii) are direct consequences of (i), (ii) and (v) in Proposition 80.

To prove (iii) let $B_1^{c,m}\left(S_i^2, t\right) = S_i^2, i = 1, 2$ with $S_1^2 < S_2^2$. Consider a point $S \in \left(S_1^2, S_2^2\right)$. The ray connectedness property of the exercise region (property (iii) in Proposition 80) implies that immediate exercise is optimal along the diagonal above $\left(S_1^2, S_1^2\right)$, hence at the point (S, S). Thus, $B_1^{c,m}(S, t) \leq S$. Suppose that the inequality is strict: $B_1^{c,m}(S, t) < S$. Given that the exercise region is closed, a property that follows from the continuity of the price function, we have $(B_1^{c,m}(S, t), S, t) \in \mathcal{E}_2^{c,m}$. Ray-connectedness then shows that $(\lambda B_1^{c,m}(S, t), \lambda S, t) \in \mathcal{E}_2^{c,m}$ for any $\lambda \geq 1$. Selecting $\lambda = S_2^2/S > 1$ establishes that $\lambda B_1^{c,m}(S, t) \geq B_1^{c,m}(\lambda S, t) = B_1^{c,m}\left(S_2^2, t\right)$. But, by assumption, $B_1^{c,m}(S, t) < S$. Multiplying each side of this inequality by $\lambda = S_2^2/S$ produces $\lambda B_1^{c,m}(S, t) < \lambda S = S_2^2 = B_1^{c,m}\left(S_2^2, t\right)$, a contradiction. It follows that $B_1^{c,m}(S, t) = S$.

Let us now establish the limit in (iv). By (ii) we know that $B_1^{c,m}\left(S^2, t\right)$ is non-decreasing in S^2 and $B_1^{c,m}\left(S^2, t\right) \leq B^1(t)$ for all $S^2 > S^1$. Suppose that $\lim_{S^2 \to \infty} B_1^{c,m}\left(S^2, t\right) < B^1(t)$ and consider a price S^1 such that

$$S^1 \in \left(\lim_{S^2 \to \infty} B_1^{c,m}\left(S^2, t\right), B^1(t)\right).$$

For all $S^2 \geq S^1$ we have $S^1 - K = C^m\left(S^1, S^2, t\right)$. Taking the limit as $S^2 \to \infty$, and using the continuity in $\left(S^1, S^2\right)$ of the price function gives

$$S^1 - K = \lim_{S^2 \to \infty} C^m\left(S^1, S^2, t\right) = C^m\left(S^1, \infty, t\right).$$

Given that $C^m(S^1,\infty,t) = C(S^1,t)$ (i.e., when S^2 goes to infinity any exercise policy for the option on asset 1 becomes feasible for the min-option) we conclude $C(S^1,t) = S^1 - K$. This contradicts our assumption $S^1 < B^1(t)$ (i.e., immediate exercise is suboptimal for $C(S^1,t)$ at S^1, so that $C(S^1,t) > S^1 - K$). This proves the limit stated.

For the first limit in (v), note that (ii) also implies $\lim_{t\to T} B_1^{c,m}(S^2,t) \leq \lim_{t\to T} B^1(t) = K \vee rK/\delta_1$. To establish the reverse inequality we need to show that $(S^1,S^2,t) \in \mathcal{E}_2^{c,m}$ implies $S^1 \geq K \vee rK/\delta_1$. The case $K \geq rK/\delta_1$ is trivial because $(S^1,S^2,t) \in \mathcal{E}_2^{c,m}$ implies $S^1 \geq K$. We consider the case $rK/\delta_1 > K$ next.

The proof is by contradiction. Suppose that $(S^1,S^2,t) \in \mathcal{E}_2^{c,m}$, while $S^2 > S^1$ and $S^1 < rK/\delta_1$. We show the existence of an arbitrage opportunity.

The optimality of immediate exercise implies $C^m(S^1,S^2,t) = S^1 - K \geq 0$. Consider the portfolio comprised of a long position in one call option, a short position in one share of the asset 1 and an investment of K dollars at the riskless rate. The initial cost of this portfolio is null. Suppose that we adopt the portfolio liquidation policy

$$\tau = \inf\left\{u \in [t,T) : S_v^1 = S_v^2 \text{ or } S_v^1 = rK/\delta_1\right\} \wedge T.$$

Clearly $\tau > t$ because $S^2 > S_t^1$ and $S^1 < rK/\delta_1$. The portfolio payoffs generated by this policy are given by

	time t	time τ
Buy Call	$-(S_t^1 - K)$	$(S_\tau^1 - K)^+ \vee C^m(S^1,S^2,t)$
Sell Stock	$+S_t^1$	$-S_\tau^1 - \int_t^\tau e^{r(\tau-v)}\delta_1 S_v^1 dv$
Invest K	$-K$	$Ke^{r(\tau-t)} = K + \int_t^\tau e^{r(\tau-v)} rK dv$
Total	0	$(S_\tau^1 - K)^+ \vee C^m(S^1,S^2,t) - (S_\tau^1 - K)$ $- \int_t^\tau e^{r(\tau-v)}(\delta_1 S_v^1 - rK) dv.$

As $rK - \delta_1 S_v^1 > 0$ for all $v \in [t,\tau)$ the cash flow at τ is strictly positive (the portfolio represents an arbitrage opportunity). We conclude that $S^1 \geq rK/\delta_1$. The limit stated follows.

In order to establish the limit $\lim_{t\to T} B_d^{c,m}(t) = K$ we use the behavior of the option with respect to dividends. Let $C^m(S^1,S^2,t;\delta)$ be the price function and $\mathcal{E}^{c,m}(\delta)$, $\mathcal{E}_1^{c,m}(\delta)$, $\mathcal{E}_2^{c,m}(\delta)$ the exercise region components for the model with dividends $\delta = (\delta_1,\delta_2)$. Elementary arguments show that $C^m(S^1,S^2,t;\delta) \leq C^m(S^1,S^2,t;\mathbf{0})$ for all $(S^1,S^2) \in \mathbb{R}_+^2$, where $\mathbf{0} = (0,0)$. Consider a diagonal point (S,S) and suppose that immediate exercise is optimal at time t in the absence of dividends, i.e., $(S,S,t) \in \mathcal{E}^{c,m}(\mathbf{0}) = \mathcal{E}_1^{c,m}(\mathbf{0}) \cap \mathcal{E}_2^{c,m}(\mathbf{0})$. Combining these elements shows that

$$S \wedge S - K \leq C^m(S,S,t;\delta) \leq C^m(S,S,t;\mathbf{0}) = S \wedge S - K,$$

where the inequality of the left hand side follows from the feasibility of exercise at (S, S, t) in the model with dividends. Thus, $(S, S, t) \in \mathcal{E}^{c,m}(\delta)$ and therefore $\mathcal{E}_1^{c,m}(\delta) \cap \mathcal{E}_2^{c,m}(\delta) \supseteq \mathcal{E}_1^{c,m}(\mathbf{0}) \cap \mathcal{E}_2^{c,m}(\mathbf{0})$. If follows immediately that $K \leq B_d^{c,m}(t;\delta) \leq B_d^{c,m}(t;\mathbf{0})$. By Proposition 2.4 in Villeneuve (1999) the limit result, $B_d^{c,m}(t;\mathbf{0}) \to K$ as $t \to T$, holds. This, combined with the previous inequality, proves the claim.

Finally, property (vi) follows from (vi) in Proposition 80. ■

Proof of Theorem 82: From the Tanaka-Meyer formula for the discounted payoff $F_{0,t} \equiv e^{-rt}(X_t - K)^+$ (see the beginning of section 6.6.2) we obtain $F_{0,t} = F_{0,0} + M_t^F + A_t^F$ with

$$
\begin{aligned}
dA_t^F &= 1_{\{X_t>K\}}e^{-rt}rKdt + \frac{1}{2}e^{-rt}dL(t,0;X-K) \\
&\quad -1_{\{X_t>K\}}e^{-rt}\left[\delta_1 S_t^1 1_{\{S_t^1<S_t^2\}} + \delta_2 S_t^2 1_{\{S_t^1>S_t^2\}}\right]dt \\
&\quad -\frac{1}{2}\sigma_m^2 1_{\{X_t>K\}}e^{-rt}X_t dL\left(t,0;\ln\left(S^1\right)-\ln\left(S^2\right)\right)
\end{aligned}
$$

and where $\sigma_m^2 \equiv \sigma_1^2 - 2\sigma_1\rho\sigma_2 + \sigma_2^2$. In the exercise region the local time increment $dL(t,0;X-K) = 0$ (because immediate exercise at $X \leq K$ is suboptimal for $t < T$). It follows that $R_t\left(rY_t dt - dA_t^Y\right) = -dA_t^F$ is given by

$$
\begin{aligned}
&1_{\{X_t>K\}}e^{-rt}\left[\left(\delta_1 S_t^1 - rK\right)1_{\{S_t^1<S_t^2\}} + \left(\delta_2 S_t^2 - rK\right)1_{\{S_t^1>S_t^2\}}\right]dt \\
&+\frac{1}{2}\sigma_m^2 1_{\{X_t>K\}}e^{-rt}X_t dL\left(t,0;\ln\left(S^1\right)-\ln\left(S^2\right)\right)
\end{aligned}
$$

in the exercise region. The EEP representation follows from Theorem 21. ■

Proof of Theorem 83: The proof follows directly from the EEP representation in Theorem 82. The limiting conditions are those from Proposition 81. ■

[illegible] inequality of the left-hand side follows [illegible] in the model with [illegible] and the [illegible]. It follows immediately that [illegible]. By Theorem 2.1 in [illegible] holds. This, combined with the [illegible] inequality, proves the claim.

[illegible] follows from [illegible] and Proposition 3.6. ∎

Proof of Theorem 8.2. Using [illegible] bounds for the [illegible] of [illegible] (Section 6.2) we obtain

$$[illegible]$$

and where [illegible] is the [illegible] of the [illegible] for [illegible] is a [illegible] for [illegible]

$$[illegible]$$

in the [illegible]. The [illegible] representation follows from Theorem 7.1. ∎

Proof of Theorem 8.3. The proof follows directly from [illegible] representation of [illegible] in Theorem 8.2 [illegible] Proposition 6.1. ∎

Chapter 7

Occupation Time Derivatives

This chapter provides an introduction to the valuation of occupation time derivatives (OTDs), both European- and American-style. Following a brief review of the literature associated with occupation time derivatives (section 7.1), the general class of OTDs is presented (section 7.2) and a general symmetry property for contracts of this sort is established (section 7.3). The next sections are devoted to specific contracts, such as quantile options (section 7.4), Parisian options (section 7.5), cumulative Parisian options (section 7.6) and step options (section 7.7). Most of the presentation in these sections focuses on the case of European contracts that pay off at a fixed maturity date, provided they are alive at that date. The valuation of American-style OTDs is addressed next (section 7.8). The chapter concludes with a discussion of the multivariate case (7.9).

7.1 Background and Literature

Occupation Time Derivatives (OTDs) are recent innovations that have appeared in derivatives markets. Broadly speaking, an *occupation time derivative* is a security whose payoff depends on the time spent, by the underlying asset price, in certain pre-specified regions of the state space. Various contractual forms embedding features of this sort have been quoted in the OTC market. Examples include step options, whose payoffs are discounted at random rates depending on the time spent above (or below) a barrier, Parisian options that expire, or come into existence, if the asset price spends more than a prespecified amount of time above (or below) a barrier and quantile options that are contingent on order statistics computed from the path of the underlying price.

The literature dealing with OTDs is still nascent but grows rapidly. Parisian options and cumulative Parisian options are priced by Chesney, Jeanblanc-Picque and Yor [1997]. Cumulative Parisian claims are valued by Hugonnier [1999]. Step options are considered by Linetsky [1999] and quantile options by

Miura [1992], Akahori [1995] and Dassios [1995], among others.

Occupation time derivatives have applications for default risk modelling. Corporate finance has long relied on option pricing theory to examine default events and value defaultable securities. In legal systems with limited liability, firms financed with combinations of equity and zero-coupon debt can be analyzed as portfolios of vanilla options. This well known feature of standard corporate securities, originally discussed in the seminal papers of Black and Scholes [1973] and Merton [1973], has given rise to an abundant literature dealing with incentives and conflicts of interest between different classes of claimholders. Debts with protective covenants can often be viewed as barrier options. These contracts typically include early liquidation provisions stipulating a transfer of assets, or even ownership, to debtholders when certain liquidity conditions fail to be satisfied. Black and Cox [1976] recognized this connection and developed valuation formulas accounting for early liquidation provisions. In their setting the firm is liquidated (assets are transferred to debtholders) at the first time at which the firm value reaches a prespecified barrier. In practice, liquidation covenants often provide grace periods allowing the firm to take corrective actions, before mandating settlement of the claims. Even in the absence of explicit provisions to that effect, grace periods are sometimes implicit as it takes time to determine the best course of action once a violation occurs. Liquidity thresholds may then be violated for some time before legal procedures are engaged. Actions leading to dissolution of the firm or restructuring of the claims only take place if the time spent below the prespecified threshold exceeds a given amount of time. As a result debt and equity claims can be understood as occupation time derivatives where the liquidation trigger is the length of an excursion below a barrier. This insight is the basis for some recent studies using Parisian option theory in order to value corporate securities and assess the risk of default. Debt valuation models in this class are developed by Francois and Morellec [2002], Moraux [2004] and Galai, Raviv and Wiener [2005], among others.

7.2 Definitions

Suppose that the underlying asset price S satisfies the Ito process (2.2) of section 2.2 and that the interest rate is a progressively measurable process. An occupation time contingent claim has exercise payoff

$$Y = F(S, O(S, A))$$

for some measurable function $F(\cdot,\cdot) : \mathbb{R}_+ \times [0,T] \to \mathbb{R}$, where $O(S, A)$ is an occupation time process defined by

$$O_{0,t}(S, A) = \int_0^t 1_{\{S_v \in A(v,\omega)\}} dv, \quad t \in [0, T]$$

for some random, closed set $A(\cdot,\cdot) : ([0,T] \times \Omega, \mathcal{B}([0,T]) \otimes \mathcal{F}) \to (\mathbb{R}_+, \mathcal{B}(\mathbb{R}_+))$, that is progressively measurable. The process $O_{s,t}(S, A)$ measures the amount of time spent by S in the set A, during the time interval $[s,t]$. When $s=0$ the abbreviated notation $O_t(S,A) \equiv O_{0,t}(S,A)$ will also be used.

The formulation adopted here is general. Note, in particular, that the set $A(\cdot,\cdot)$ can be a random set, as long as it remains progressively measurable. Simple examples that have been extensively studied in the literature involve constant sets of the form $A^+(L) = \{x \in \mathbb{R}^+ : x \geq L\}$ or $A^-(L) = \{x \in \mathbb{R}^+ : x \leq L\}$ with L constant. In these cases the occupation time of A is the amount of time spent above or below a constant barrier L during the period of reference. Immediate generalizations of these examples are obtained by letting the barrier be a function of time or of a progressively measurable stochastic process that lies in the asset span.

The pricing principles reviewed in chapters 2 and 3 apply to occupation time derivatives. The price of a European-style OTD, issued at date 0, is given by the present value formula

$$v_t(Y) = \widetilde{E}_t\left[R_{t,T}F\left(S_T, O_{0,T}(S,A)\right)\right]. \tag{7.1}$$

Prior to exercise, the value of an American-style OTD, issued at date 0, satisfies

$$V_t(Y) = \sup_{\tau \in \mathcal{S}_{t,T}} \widetilde{E}_t\left[R_{t,\tau}F\left(S_\tau, O_{0,\tau}(S,A)\right)\right] \tag{7.2}$$

where $\mathcal{S}_{t,T}$ is the set of stopping times of the filtration with values in $[t,T]$.

Before focusing on the valuation of specific OTDs it is useful to establish a general symmetry property.

7.3 Symmetry Properties

As for standard options and "one-touch" barrier options, a symmetry property links different types of occupation time derivatives. Write the claim's payoff as $Y_T = F(S_T, K, O_{0,T}(S,A))$ to emphasize the dependence on a constant parameter K. This parameter can be thought of as a strike price or as a cap, depending on the context. Suppose also that the payoff function is homogeneous of degree one in the pair (S,K). The change of measure described in section 2.7 can be applied to obtain,

Theorem 95 *Consider an American-style occupation time derivative with maturity date T and exercise payoff $Y_\tau = F(S_\tau, K, O_\tau(S,A))$ at time τ, where $O_\tau(S,A)$ is the occupation time of the set A during the period $[0,\tau]$. Suppose that the payoff function $F(S,K,O(S,A))$ is homogeneous of degree one with respect to (S,K). Let $V(S,K,\ O(S,A), r, \delta; \mathcal{F}_t)$ be the value of the claim in the financial market with filtration $\mathcal{F}_{(\cdot)}$, asset price S satisfying (2.2) and progressively measurable interest rate r. Prior to exercise*

$$V(S_t, K, O_t(S,A), r, \delta; \mathcal{F}_t) = V^*(S_t^*, S, O_t(S^*, A^*), \delta, r; \mathcal{F}_t) \tag{7.3}$$

where $A^* = \{A^*(v,\omega), v \in [t,T]\}$ with

$$A^*(v,\omega) = \{x \in \mathbb{R}_+ : x = KS/y \text{ and } y \in A(v,\omega)\} \tag{7.4}$$

and $O_{0,t}(S^*, A^*) \equiv O_{0,t}(S, A)$. *Here* $O_{t,v}(S^*, A^*)$ *is the occupation time of the set* A^* *by the price* S^* *during* $[t, v]$. *The quantity* $V^*(S_t^*, S, O_t(S^*, A^*), \delta, r; \mathcal{F}_t)$ *is the value at* t *of the symmetric claim with payoff* $F^*(S_\tau^*, S, O_\tau(S^*, A^*)) \equiv F(S, S_\tau^*, O_\tau(S^*, A^*))$ *at time* τ, *parameter* $S \equiv S_t$, *occupation time* $O_\tau(S^*, A^*)$ *and maturity date* T. *The symmetric claim is valued in an auxiliary financial market with interest rate* δ *and in which the underlying asset price follows the Ito process*

$$dS_v^* = S_v^*[(\delta_v - r_v)\,dt + \sigma_v dz_v^*], \quad \text{for } v \in [t, T] \tag{7.5}$$

with initial condition $S_t^* = K$. *The process* z^* *is defined by* $dz_v^* = -d\widetilde{z}_v + \sigma_v dv, v \in [0, T], z_0^* = 0$. *The optimal exercise time of the symmetric claim in the auxiliary financial market is the same as the optimal exercise time of the initial claim in the original market.*

This symmetry result is reminiscent of the earlier property stated in Theorem 26 for American options. It is somewhat more involved because an adjustment in the occupation time is now necessary to obtain the symmetric claim. Note, in particular, that the occupation time of the new claim, $O_{0,t}(S^*, A^*)$, is based on the set (7.4). This follows from the equality of events

$$\{S_v \in A(v, \omega)\} = \{S_v^* = KS/S_v \in A^*(v, \omega)\}$$

implying that the occupation times

$$O_{0,\tau}(S, A) = \int_0^\tau 1_{\{S_v \in A(v,\omega)\}} dv = \int_0^\tau 1_{\{S_v^* \in A^*(v,\omega)\}} dv = O_{0,\tau}(S^*, A^*)$$

coincide for any stopping time τ of the filtration.

The symmetry property for occupation time derivatives in Theorem 95 helps to price a number of contracts by just using formulas derived for the symmetric claims. Several applications of this principle will appear in the next sections devoted to specific contractual forms.

7.4 Quantile Options

Quantile options were introduced by Miura [1992] and studied by Akahori [1995] and Dassios [1995]. Complementary results about the α-quantile of a Brownian motion with drift can be found in Embrecht, Rogers and Yor [1995], Yor [1995], Takács [1996] and Fusai [2000].

7.4.1 Contractual Specification

A European-style α-quantile call option pays off $(M(\alpha, T) - K)^+$ upon exercise where

$$M(\alpha, t) = \inf\left\{x : \int_0^t 1_{\{S_v \leq x\}} dv > \alpha t\right\} = \inf\{x : O_t(S, A^-(x)) > \alpha t\}$$

and $A^{-}(x) = \{y \in \mathbb{R}_{+} : y - x \leq 0\}$. The process $M(\alpha, t)$ is the smallest threshold such that the fraction of time spent by the underlying asset price at or below $M(\alpha, t)$, during the period $[0, t]$, exceeds α. The quantile call payoff can be written in the equivalent form $\left(S_0 \exp\left(\sigma M^Z(\alpha, T)\right) - K\right)^{+}$ where $M^Z(\alpha, T)$ is the α-quantile of $Z_t \equiv \log\left(S_t/S_0\right)/\sigma$, the cumulative continuously compounded asset return per unit volatility. This follows because α-quantiles are preserved by monotone increasing transformations (the α-quantile of a transformed process is the transform of the α-quantile of the underlying process), so that $M(\alpha, T) = S_0 \exp\left(\sigma M^Z(\alpha, T)\right)$ (see Lemma 104 in the Appendix).

7.4.2 The Distribution of an α-Quantile

In order to price α-quantile options it is useful to have a closed form expression for the density function of the threshold $M^Z(\alpha, T)$. For tractability, assume the standard financial market with constant coefficients (the Black-Scholes setting). When the cumulative continuously compounded return per unit volatility is a Brownian motion with drift m this density is (see Dassios [1995], Theorem 1, with $\sigma = 1$)

$$P\left(M^Z(\alpha, t) \in dx\right) = g(x; \alpha, t)\, dx$$

where

$$g(x; \alpha, m, t) = \int_{-\infty}^{\infty} g^{-}(x - y; m, \alpha t)\, g^{+}(y; m, (1-\alpha)t)\, dy \tag{7.6}$$

with

$$g^{-}(x; m, t) = \left(\left(\frac{2}{\pi t}\right)^{1/2} \exp\left(-\frac{(x - mt)^2}{2t}\right) - 2me^{2mx} N\left(-\frac{x + mt}{\sqrt{t}}\right)\right) 1_{\{x>0\}} \tag{7.7}$$

$$g^{+}(x; m, t) = \left(\left(\frac{2}{\pi t}\right)^{1/2} \exp\left(-\frac{(x - mt)^2}{2t}\right) + 2me^{2mx} N\left(\frac{x + mt}{\sqrt{t}}\right)\right) 1_{\{x<0\}} \tag{7.8}$$

and $N(x)$ the cumulative standard normal distribution function. The complementary distribution function, $G(z; \alpha, m, t) \equiv \int_z^{\infty} g(x; \alpha, m, t)\, dx$, is the probability of the event $\left\{M^Z(\alpha, t) > z\right\}$. The representation (7.6)-(7.8) is the consequence of a deep result, showing that the law of the quantile $M^Z(\alpha, t)$ is identical to the law of $\sup_{0 \leq s \leq \alpha t} Z_s + \inf_{0 \leq s \leq (1-\alpha)t} Z'_s$, where Z' is an independent copy of Z (see Dassios [1995], Theorem 2 and Embrecht, Rogers and Yor [1995], Theorem 1). From this identity it follows that the density of the quantile is the convolution of the densities of $\sup_{0 \leq s \leq \alpha t} Z_s$ and $\inf_{0 \leq s \leq (1-\alpha)t} Z'_s$. The latter correspond to the functions $g^{-}(\cdot; m, \alpha t)$ and $g^{+}(\cdot; m, (1-\alpha)t)$ appearing in (7.6) and defined in (7.7)-(7.8).

7.4.3 Pricing Quantile Options

Given the explicit nature of the probability law of an α-quantile, the European quantile option becomes easy to price.

Theorem 96 *(Dassios [1995]) Suppose that the underlying price S follows the geometric Brownian motion process (4.1) with dividend yield δ. The value of a European-style α-quantile call with strike K and maturity date T is then given by*

$$\begin{aligned} c^{\alpha}(S_t, t) &= R_{t,T} S_t \int_{K/S_t}^{\infty} G\left(\frac{\log(z)}{\sigma}; \frac{\alpha T - O_t(S, A^-(S_t z))}{T-t}, m, T-t\right) \\ &\quad \times 1_{\{t-(1-\alpha)T \le O_t(S, A^-(S_t z)) < \alpha T\}} dz \\ &\quad + R_{t,T} S_t \int_{K/S_t}^{\infty} 1_{\{O_t(S, A^-(S_t z)) < t-(1-\alpha)T\}} dz \end{aligned} \tag{7.9}$$

where $O_t(S, A^-(S_t z))$ is the occupation time at t of the set

$$A^-(S_t z) = \{x \in \mathbb{R}^+ : x \le S_t z\}$$

and $m = \left(r - \delta - \frac{1}{2}\sigma^2\right)/\sigma$.

The α-quantile pricing formula (7.9) is an adaptation of Dassios [1995] to the case of dividend-paying underlying asset. In order to gain some intuition it is useful to consider first the valuation formula at the issue date of the contract, date 0. Using standard arguments and the properties of α-quantiles, it is clear that the option price can be written as

$$\begin{aligned} c^{\alpha}(S_0, 0) &= \widetilde{E}\left[R_{0,T}\left(S_0 \exp\left(\sigma M^Z(\alpha, T)\right) - K\right)^+\right] \\ &= R_{0,T} S_0 \widetilde{E}\left[\left(\exp\left(\sigma M^Z(\alpha, T)\right) - \frac{K}{S_0}\right)^+\right] \\ &= R_{0,T} S_0 \int_{\log(K/S_0)/\sigma}^{\infty} \left(\exp(\sigma x) - \frac{K}{S_0}\right) g(x; \alpha, m, T)\, dx \end{aligned}$$

where $g(x; \alpha, m, T)$ is the density of $M^Z(\alpha, T)$ under the risk neutral measure. An integration by parts argument (using $-dG(x; \alpha, m, T) = g(x; \alpha, m, T)\, dx$) gives

$$\begin{aligned} c^{\alpha}(S_0, 0) &= R_{0,T} S_0 \int_{K/S_0}^{\infty} \int_{\log(z)/\sigma}^{\infty} g(x; \alpha, m, T)\, dx dz \\ &= R_{0,T} S_0 \int_{K/S_0}^{\infty} G\left(\frac{\log(z)}{\sigma}; \alpha, m, T\right) dz \end{aligned} \tag{7.10}$$

where $G(\log(z)/\sigma; \alpha, m, T)$ is the probability of the set $\left\{M^Z(\alpha, T) > \frac{\log(z)}{\sigma}\right\}$. This expression corresponds to the pricing formula (7.9) indicated in the theorem because $O_0(S, A^-(S_0, z)) = 0$ for all values of z at the initial date $t = 0$. The alternative pricing formula (7.10) reveals a two-stage decomposition of the option value. The first stage consists in conditioning on a level z of the α-quantile of the normalized asset price S_T/S_0 (i.e., conditioning on $z = \exp\left(\sigma M^Z(\alpha, T)\right)$) and in computing the cumulative probability that the quantile exceeds $\log(z)/\sigma$. This gives the function $G(\log(z)/\sigma; \alpha, m, T)$, parametrized by z. The second stage integrates over all levels z for which the option is in the money at maturity.

A similar decomposition applies if the option is valued at a later date $t > 0$. In this case, however, a correction is necessary to account for the passage of time and the fact that the occupation time $O_t(S, A^-(S_t, z))$ becomes positive. Conditional on information at date t

$$
\begin{aligned}
c^\alpha(S_t, t) &= \widetilde{E}_t\left[R_{t,T}\left(S_0 \exp\left(\sigma M^Z(\alpha, T)\right) - K\right)^+\right] \\
&= R_{t,T} S_t \widetilde{E}_t\left[\left(\frac{S_0}{S_t}\exp\left(\sigma M^Z(\alpha, T)\right) - \frac{K}{S_t}\right)^+\right] \\
&= R_{t,T} S_t \widetilde{E}_t\left[\left(\exp\left(\sigma\left(M^Z(\alpha, T) - Z_t\right)\right) - \frac{K}{S_t}\right)^+\right] \\
&= R_{t,T} S_t \int_{\log(K/S_t)/\sigma}^{\infty}\left(\exp(\sigma x) - \frac{K}{S_t}\right)\widehat{g}(x, t; \alpha, m, T)\,dx \\
&= R_{t,T} S_t \int_{K/S_t}^{\infty}\int_{\log(z)/\sigma}^{\infty}\widehat{g}(x, t; \alpha, m, T)\,dxdz
\end{aligned}
$$

where $\widehat{g}(x, t; \alpha, m, T)$ is the conditional density at t of $M^Z(\alpha, T) - Z_t$ under the risk neutral measure. Given that

$$
\begin{aligned}
\left\{M^Z(\alpha, T) - Z_t > w\right\} &= \left\{\int_0^T 1_{\{Z_v \le w + Z_t\}}dv \le \alpha T\right\} \\
&= \left\{\int_t^T 1_{\{Z_v - Z_t \le w\}}dv \le \alpha T - \int_0^t 1_{\{Z_v - Z_t \le w\}}dv\right\} \\
&= \left\{\int_t^T 1_{\{\widehat{Z}_v \le w\}}dv \le \widehat{\alpha}(t, w)(T - t)\right\} \\
&= \left\{M^{\widehat{Z}}\left(\widehat{\alpha}(t, w), T - t\right) > w\right\}
\end{aligned}
$$

with $\widehat{Z}_v = Z_v - Z_t$ and

$$
\widehat{\alpha}(t, w) \equiv \frac{\alpha T - \int_0^t 1_{\{Z_v - Z_t \le w\}}dv}{T - t}
$$

it becomes apparent that the tail probabilities (setting $w = \log(z)/\sigma$)

$$
\int_{\log(z)/\sigma}^{\infty}\widehat{g}(x, t; \alpha, m, T)\,dx = \int_{\log(z)/\sigma}^{\infty} g\left(x; \widehat{\alpha}(t, \log(z)/\sigma), m, T - t\right)dx
$$

coincide. The identification

$$
\int_0^t 1_{\{Z_v - Z_t \le w\}}dv = \int_0^t 1_{\{S_v \le S_t \exp(\sigma w)\}}dv = O_t\left(S, A^-\left(S_t \exp(\sigma w)\right)\right)
$$

combined with the change of variables $w = \log(z)/\sigma$ then gives, with $c_t^\alpha \equiv c^\alpha(S_t, t)$,

$$\begin{aligned} c_t^\alpha &= R_{t,T} S_t \int_{K/S_t}^\infty \int_{\log(z)/\sigma}^\infty g\left(x; \frac{\alpha T - O_t(S, A^-(S_t z))}{T-t}, m, T-t\right) dx dz \\ &= R_{t,T} S_t \int_{K/S_t}^\infty G\left(\frac{\log(z)}{\sigma}; \frac{\alpha T - O_t(S, A^-(S_t z))}{T-t}, m, T-t\right) dz. \quad (7.11) \end{aligned}$$

To reconcile this last expression with the result in the theorem it is useful to consider two cases. The first case is for $t < \min\{\alpha T, (1-\alpha)T\}$. As $t-(1-\alpha)T < 0$ the event $\{O_t(S, A^-(S_t z)) < t - (1-\alpha)T\}$ is empty. On the contrary, as $O_t(S, A^-(S_t z)) \leq t < \alpha T$, the event $\{t - (1-\alpha)T \leq O_t(S, A^-(S_t z)) < \alpha T\}$ is the full set. Formula (7.9) then reduces precisely to (7.11). In this instance the conditional distribution of the $\widehat{\alpha}(t, \log(z)/\sigma)$-quantile is well defined for all values of z in the range where the option pays off at maturity. The option formula is then identical to the formula at date 0, except for the correction parameter $\widehat{\alpha}(t, \cdot)$ in the density function of the quantile induced by the realized occupation time at the valuation date t. The second case is when $t \geq \min\{\alpha T, (1-\alpha)T\}$. In this instance the event $\{O_t(S, A^-(S_t z)) < t - (1-\alpha)T\} = \{\widehat{\alpha}(t, \log(z)/\sigma) > 1\}$ may become non-empty for some values of z and, in those cases, the probability of the event $\left\{M^{\widehat{Z}}(\widehat{\alpha}(t, w), T-t) > w\right\} = \left\{\int_t^T 1_{\{\widehat{Z}_v \leq w\}} dv \leq \widehat{\alpha}(t, w)(T-t)\right\}$ is one.

7.4.4 A Reduction in Dimensionality

The formula (7.9), provided in the theorem, is difficult to implement because the occupation times $O_t(S, A^-(S_t z))$ must be recorded for all levels $z \geq K/S_t$. As revealed by the discussion above, computations appear much simpler for newly issued contracts. But even in this case, implementation requires the computation of a triple integral, given that the complementary probability function $G(z; \alpha, m, T)$ is a double integral.

A simpler expression is derived by Takács [1996] (see also Fusai [2000], for additional results). For the case $z \geq 0$ he shows that the probability of the event $\{M^Z(\alpha, T) > z\}$ can be written in the alternative form

$$\begin{aligned} G(z; \alpha, m, T) &\equiv \int_z^\infty g(x; \alpha, m, T) dx \\ &= 2\int_0^\alpha g_1\left(u; m\sqrt{T}\right) g_2\left(u; \frac{z}{\sqrt{T}}, m\sqrt{T}\right) du \end{aligned}$$

for $\alpha \in [0, 1]$, where

$$g_1(u; m) = \frac{1}{\sqrt{2\pi}} \frac{1}{\sqrt{1-u}} \exp\left(-\frac{m^2(1-u)}{2}\right) + mN\left(m\sqrt{1-u}\right)$$

$$g_2(u; z, m) = \frac{1}{\sqrt{2\pi}} \frac{1}{\sqrt{u}} \exp\left(-\frac{(z-mu)^2}{2u}\right) - me^{2mz} N\left(-\frac{z+mu}{\sqrt{u}}\right).$$

The density function can be obtained by differentiating $G(z;\alpha,m,T)$ with respect to z and taking the negative of the resulting expression. In this formula the product $2g_1\left(u;m\sqrt{T}\right)g_2\left(u;z/\sqrt{T},m\sqrt{T}\right)$ represents the density of the average occupation time

$$\frac{O_T\left(Z,A^-(z)\right)}{T}=\frac{1}{T}\int_0^T 1_{\{Z_v\leq z\}}dv.$$

Given that the events $\left\{M^Z(\alpha,T)>z\right\}$ and $\left\{O_T\left(Z,A^-(z)\right)/T\leq\alpha\right\}$ are the same, the expression above follows.

The case $z<0$ can be treated by symmetry. It is easy to verify the the random variables $O_T\left(Z,A^-(z)\right)/T$ and $1-O_T\left(-Z,A^-(-z)\right)/T$ have the same distributions. The variate $O_T\left(-Z,A^-(-z)\right)/T$ is the average occupation time by the Brownian motion $-Z$ with drift $-m$ of the set $A^-(-z)$, where $-z\geq 0$. Its distribution is obtained from the expressions above substituting $(-z,-m)$ in place of (z,m).

7.4.5 Quantile Contingent Claims

A European α-quantile contingent claim pays off $Y_T\equiv F\left(M\left(\alpha,T\right),K,T\right)$ at the maturity date, for some measurable function F. Here K represents a parameter or a vector of parameters affecting the payment. Examples of claims of this sort include quantile forward contracts, quantile vanilla options, capped quantile options, quantile straddles or quantile digital options. An adaptation of Theorem 96 provides the following valuation formula,

Theorem 97 *Suppose that the underlying price S follows the geometric Brownian motion process (4.1) with dividend yield δ. The value, at initiation, of a European-style α-quantile contingent claim with parameter K and maturity date T is given by*

$$cc^{\alpha}\left(S_0,0\right)=R_{0,T}\int_{-\infty}^{\infty}F\left(S_0\exp\left(\sigma x\right),K,T\right)g\left(x;\alpha,m,T\right)dx$$

with $m=\left(r-\delta-\frac{1}{2}\sigma^2\right)/\sigma$.

For differentiable claims, an expression involving the cumulative distribution function of the quantile, as in (7.10), can also be retrieved by using an integration by parts argument. For non-linear contracts this representation will involve the derivative of the claim in addition to the cumulative distribution function. Pricing formulas for subsequent times $t>0$ can also be obtained using the arguments underlying the derivation of (7.9), with suitable modifications to account for the general payoff structure.

7.5 Parisian Options

The Parisian option is another example of an occupation time derivative. This class of derivative security was studied by Chesney, Jeanblanc-Picque and Yor

[1997]. The rationale for considering a contract of this sort is to remedy some of the problems (e.g. vulnerability to price manipulations) associated with standard "one-touch" barrier options that lose all proceeds (knock-out options) or come alive (knock-in options), suddenly, when the underlying asset price reaches a contractually specified level. Parisian options, in contrast, lose value or come into existence progressively if the underlying asset price stays above, or below, a barrier for a sustained period of time.

7.5.1 Contractual Specification

A Parisian knock-out option with window D, barrier L and maturity date T will lose all value if the underlying asset price has an excursion of duration D above or below the barrier L during the option's life. If the loss of value is contingent upon an excursion above (below) the barrier, the option is an up-and-out (down-and-out) Parisian option. A Parisian knock-in option with window D, barrier L and maturity date T comes alive if there is an excursion of duration D before maturity. If birth is contingent upon an excursion above (below) the barrier the option is an up-and-in (down-and-in) Parisian option.

The stopping time representing an excursion of duration D is needed to describe the contract payoff. Fix a time $t \in [0,T]$ (the current time), a barrier L and suppose that no excursion of length D has taken place strictly prior to t. Also suppose that the initial asset price lies below (above) the barrier in the case of a knock-in or knock-out up (down) option.[1] Let $g(L,t) = \sup\{s \leq t : S_s = L\}$ be the last time before t at which the process S has reached the barrier L in the interval $[0,T]$ (if such a time does not exist set $g(L,t) = t$). The random time

$$O_t\left(S, A^+(t,L)\right) = \int_{g(L,t)}^{t} 1_{\{S_v > L\}} dv = \int_0^t 1_{\{(v,S_v) \in A^+(t,L)\}} dv, \tag{7.12}$$

where $A^+(t,L) = \{(v,x) : t \geq v \geq g(L,t), x > L\}$, measures the age of a current excursion above the level L. The random time

$$H^+(t,L,D) = \inf\left\{v \in [t,T] : O_v\left(S, A^+(v,L)\right) \geq D\right\} \tag{7.13}$$

is the first time in $[t,T]$ at which an excursion of length D above L is recorded; if no such time exists set $H^+(t,L,D) = \infty$. Note that this excursion starts at $g(L,t)$, a time that may be strictly prior to t. The payoff of a Parisian up-and-out call with window D is $1_{\{H^+(t,L,D)>T\}}(S_T - K)^+$.

Likewise, the payoff of a down-and-out Parisian call is $1_{\{H^-(t,L,D)>T\}}(S_T - K)^+$. Here

$$H^-(t,L,D) = \inf\left\{v \in [t,T] : O_v\left(S, A^-(v,L)\right) \geq D\right\}$$

with $A^-(t,L) = \{(v,x) : t \geq v \geq g(L,t), x < L\}$ represents the first time in $[t,T]$ at which an excursion of length D below L is recorded (as before set

[1] This assumption only serves to simplify the description of the contracts. If it is violated the definition of the random time $g(L,t)$ must be adjusted.

$H^{-}(t,L,D)=\infty$ is no such time exists). Parisian knock-in calls come into existence if an excursion of length D occurs before maturity. The payoffs are $1_{\{H^{+}(t,L,D)\leq T\}}(S_T-K)^{+}$ for an up-and-in call and $1_{\{H^{-}(t,L,D)\leq T\}}(S_T-K)^{+}$ for a down-and-in call. Parisian put option payoffs are obtained by substituting $(K-S_T)^{+}$ in place of $(S_T-K)^{+}$ in the expressions above.

7.5.2 Parity and Symmetry Relations

Before proceeding to value Parisian options it is useful to identify certain relationships tying the prices of various contracts together. It is assumed, throughout this section, that the primitives (S,r) satisfy the conditions of section 2.2.

Parity formulas relate the prices of European-style Parisian knock-in and knock-out options. Let us focus on call options; the same relations apply for puts. Let $c^{pdo}(S_t,t,K,T,D,L)$ be the price of a down-and-out call with strike K, maturity date T, window D and barrier L. Let $c^{pdi}(S_t,t,K,T,D,L)$ be the price of the down-and-in call with the same contractual characteristics. The following no-arbitrage relation ties the two prices together

$$\begin{aligned}
& c^{pdo}(S_t,t,K,T,D,L) \\
= \; & \widetilde{E}_t\left[R_{t,T}1_{\{H^{-}(t,L,D)>T\}}(S_T-K)^{+}\right] \\
= \; & \widetilde{E}_t\left[R_{t,T}(S_T-K)^{+}\right]-\widetilde{E}_t\left[R_{t,T}1_{\{H^{-}(t,L,D)\leq T\}}(S_T-K)^{+}\right] \\
= \; & c(S_t,t,K,T)-c^{pdi}(S_t,t,K,T,D,L)
\end{aligned}$$

where $c(S_t,t,K,T)$ is the European vanilla call price. A portfolio composed of a down-and-out and a down-and-in call, both written on the same underlying asset and with the same strike, maturity date, window and barrier, is therefore identical to a standard call option. It is also straightforward to verify that Parisian up-and-out and up-and-in call options are related in a similar manner. These parity relations mirror identical relations linking "one-touch" barrier knock-in and knock-out calls and standard call options.

An application of Theorem 95 also shows that American-style Parisian call and put options are symmetric to each other. Let $t\in[0,T]$ and suppose that no excursion of age D has occurred before t. This symmetry relation can be stated as

$$C^{puo}(S_t,t,K,T,D,L,r,\delta;\mathcal{F}_t)=P^{pdo}(S_t^{*},t,S,T,D,KS/L,\delta,r;\mathcal{F}_t)$$

where $P^{pdo}(S_t^{*},t,S,T,D,KS/L,\delta,r;\mathcal{F}_t)$ is the price of an American Parisian down-and-out put option with strike S, maturity T, window D and barrier KS/L in an auxiliary market with interest rate δ and where the underlying asset S^{*} pays dividends at the rate r.

To get insights about the nature of this symmetry property recall that $S_v^{*}=KS/S_v$ and note that $g(L,t)=\sup\{s\leq t:S_s=L\}=\sup\{s\leq t:S_s^{*}=KS/L\}$

$= g^*(KS/L, t)$. It follows that the occupation times

$$\begin{aligned} O_t\left(S, A^+(t,L)\right) &= \int_{g(L,t)}^{t} 1_{\{S_v > L\}} dv \\ &= \int_{g^*(KS/L,t)}^{t} 1_{\{KS/L > S_v^*\}} dv = O_t\left(S^*, A^-(t, KS/L)\right), \end{aligned}$$

and therefore the stopping times

$$H^+(t, L, D) = \inf\left\{v \in [t,T] : O_v\left(S, A^+(v,L)\right) \geq D\right\}$$

$$H^{*-}(t, KS/L, D) = \inf\left\{v \in [t,T] : O_v\left(S^*, A^-(v, KS/L)\right) \geq D\right\}$$

at which the call and put options lose all value also coincide. One concludes that

$$\begin{aligned} & C^{puo}(S_t, t, K, T, D, L, r, \delta; \mathcal{F}_t) \\ = & \sup_{\tau \in \mathcal{S}_{t,T}} \widetilde{E}_t\left[R_{t,\tau} 1_{\{H^+(t,L,D) > \tau\}} (S_\tau - K)^+\right] \\ = & \sup_{\tau \in \mathcal{S}_{t,T}} E_t^*\left[\exp\left(-\int_t^\tau \delta_v dv\right) 1_{\{H^{*-}(t,KS/L,D) > \tau\}} (S - S_\tau^*)^+\right] \\ = & P^{pdo}(S_t^*, t, S, T, D, KS/L, \delta, r; \mathcal{F}_t) \end{aligned}$$

where E_t^* is the conditional expectation at t under the measure Q^*. An American-style Parisian up-and-out call with strike K, maturity T, window D and barrier L has therefore the same value as an American-style Parisian down-and-out put with strike $S = S_t$, maturity T, window D and barrier KS/L, in an auxiliary financial market with interest rate δ and in which the underlying asset price follows the Ito process described in Theorem 95. Chesney, Jeanblanc-Picque and Yor [1997] derive this symmetry property for European Parisian options in a financial market with constant coefficients.

7.5.3 Pricing Parisian Options

Let us now focus on the valuation of European-style Parisian knock-in call options. The parity and symmetry relations discussed above show that the formulas presented in the next theorem also apply to knock-out call options and knock-in and knock-out put options, modulo simple transformations.

Assume the standard model with constant interest rate and geometric Brownian motion price process (Black-Scholes setting). For a constant γ let $R_T^\gamma = e^{-\gamma T}$ be the discount factor at the rate γ. A pricing formula for the option, in this environment, is derived by Chesney, Jeanblanc-Picque and Yor [1997].

Theorem 98 *(Chesney, Jeanblanc-Picque and Yor [1997]) Let $S_0 = S$. The value of a Parisian down-and-in call option with window D, barrier L, strike K and maturity date T is given by*

$$c^{pdi}(S, 0, K, T, D, L) = R_T^\gamma \int_{d_o(S,K)}^{\infty} e^{my}\left(Se^{\sigma y} - K\right) h_l(T, y) dy \qquad (7.14)$$

where $\gamma = r + \frac{1}{2}m^2$, $m \equiv \left(r - \delta - \frac{1}{2}\sigma^2\right)/\sigma$, $d_o(S,K) = \log(K/S)/\sigma$ *and* $l \equiv d_o(S,L) = \log(L/S)/\sigma$. *The function* $h_l(T,y)$ *is characterized by its Laplace transform* $\widehat{h}_l(\lambda,y) \equiv \int_0^\infty e^{-\lambda t} h_l(t,y)\,dt$ *given by the formulas in Lemma 107 in the Appendix.*

To gain insights into the structure of this formula note that a change of measure can be invoked to write the price as (for parsimony the notation c^{pdi} is used for the price without specifying the arguments)

$$\begin{aligned} c^{pdi} &= R_T \widetilde{E}\left[1_{\{H^-(0,L,D)\leq T\}}\left(S_T - K\right)^+\right] \\ &= R_T E^s\left[1_{\{H_z^-(0,l,D)\leq T\}} e^{mZ_T - \frac{1}{2}m^2T}\left(Se^{\sigma Z_T} - K\right)^+\right] \\ &= e^{-(r+\frac{1}{2}m^2)T} E^s\left[1_{\{H_z^-(0,l,D)\leq T\}} e^{mZ_T}\left(Se^{\sigma Z_T} - K\right)^+\right] \end{aligned} \qquad (7.15)$$

where $Z_v = W_v + mv$ for $v \in [0,T]$ is a Brownian motion process under the measure Q^s such that $dQ^s = \exp\left(-mW_T - \frac{1}{2}m^2T\right)dQ = \exp\left(-mZ_T + \frac{1}{2}m^2T\right)dQ$. The expectation $E^s[\cdot]$ is under Q^s and the random time

$$H_z^-(0,l,D) = \inf\left\{v \in [0,T] : O_v\left(Z, A_z^-(v,l)\right) \geq D\right\}$$

is defined with

$$A_z^-(v,l) = \{(s,x) : v \geq s \geq g_z(l,v), x < l\}$$

$$g_z(l,v) = \sup\{s \leq v : Z_s = l\}.$$

Note that the equality in the second line of (7.15) uses $g_z(l,v) = g(L,v)$ and $\{H^-(0,L,D) \leq T\} = \{H_z^-(0,l,D) \leq T\}$. Inspection of the expression on the right hand side of (7.15) shows that the Parisian down-and-in call in the original market has the same value as a random number e^{mZ_T} of Parisian down-and-in calls in a financial market with interest rate $r^s \equiv r + \frac{1}{2}m^2$ and in which the underlying price $S_t^s \equiv Se^{\sigma Z_t}$ evolves according to

$$dS_t^s/S_t^s = (r^s - \delta^s)\,dt + \sigma dZ_t = \frac{1}{2}m^2dt + \sigma dZ_t$$

and thus pays dividends at the rate $\delta^s = r$.

Next, use the law of iterated expectations to condition on the time of birth $H_z^-(0,l,D)$ and write

$$c^{pdi} = R_T^\gamma E^s\left[1_{\{H_z^-(0,l,D)\leq T\}} E^s\left[e^{mZ_T}\left(Se^{\sigma Z_T} - K\right)^+\middle| \mathcal{F}_{H_z^-(0,l,D)}\right]\right] \qquad (7.16)$$

where

$$\begin{aligned} & E^s\left[e^{mZ_T}\left(Se^{\sigma Z_T} - K\right)^+\middle| \mathcal{F}_{H_z^-(0,l,D)}\right] \\ &= \int_{-\infty}^{+\infty} e^{my}\left(Se^{\sigma y} - K\right)^+ n\left(y; Z_{H_z^-(0,l,D)}, T - H_z^-(0,l,D)\right)dy \\ &\equiv v^{pdi}\left(Z_{H_z^-(0,l,D)}, T - H_z^-(0,l,D); m\right) \end{aligned} \qquad (7.17)$$

and where $n(y;\mu,\sigma^2)$ is the normal density function with mean μ and variance σ^2. Modulo a discount factor correction (equal to $R^{\gamma}_{H_z^-(0,l,D),T}$) the quantity $v^{pdi}\left(Z_{H_z^-(0,l,D)},T-H_z^-(0,l,D);m\right)$ in (7.17) is nothing else but the price of the option at the time of birth $H_z^-(0,l,D)$. Formula (7.16) is the present value at initiation of the option value at birth. This present value is calculated by taking the expectation $E^s[\cdot]$ over the random variables $Z_{H_z^-(0,l,D)}$ and $H_z^-(0,l,D)$ appearing in the components of the formula.

The valuation formula (7.16)-(7.17) may seem difficult to exploit, at first sight, as it requires integration over a non-trivial bivariate distribution. The usefulness of the representation becomes apparent once it is found out that the random variables $Z_{H_z^-(0,l,D)}$ and $H_z^-(0,l,D)$ are in fact independent of each other. This remarkable property relates to the information content of $H_z^-(0,l,D)$. In essence knowing that $H_z^-(0,l,D)$ has materialized reveals the last hitting time $g_z(l,v)=\sup\{s\leq v: Z_s=l\}$ as well as the occurrence of an excursion of length D below the barrier. But it does not convey further information about the trajectory followed from l to $Z_{H_z^-(0,l,D)}$. The set of possible trajectories from l to $Z_{H_z^-(0,l,D)}$ is therefore unrelated to the particular values of $H_z^-(0,l,D)$ drawn. This intuitively suggests that the two variables are independent (see the Appendix in Chesney, Jeanblanc-Picque and Yor [1997] for additional insights and mathematical notions associated with this result). From this independence property it now follows that the expectation in (7.15) can be written as

$$\begin{aligned}
& E^s\left[1_{\{H_z^-(0,l,D)\leq T\}}v^{pdi}\left(Z_{H_z^-(0,l,D)},T-H_z^-(0,l,D);m\right)\right] \\
= {} & E^s\left[1_{\{H_z^-(0,l,D)\leq T\}}\int_{-\infty}^{+\infty}v^{pdi}\left(z,T-H_z^-(0,l,D);m\right)\phi(dz)\right] \\
= {} & \int_{-\infty}^{+\infty}E^s\left[1_{\{H_z^-(0,l,D)\leq T\}}v^{pdi}\left(z,T-H_z^-(0,l,D);m\right)\right]\phi(dz) \\
= {} & \int_{-\infty}^{+\infty}e^{my}\left(Se^{\sigma y}-K\right)^+ \\
& \times\left(\int_{-\infty}^{+\infty}E^s\left[1_{\{H_z^-(0,l,D)\leq T\}}n\left(y;z,T-H_z^-(0,l,D)\right)\right]\phi(dz)\right)dy
\end{aligned}$$

where $\phi(dz)$ is the probability law of the random variable $Z_{H_z^-(0,l,D)}$ and the inside expectation is over the random variable $H_z^-(0,l,D)$. Setting

$$h_l(T,y)\equiv\int_{-\infty}^{+\infty}E^s\left[1_{\{H_z^-(0,l,D)\leq T\}}n\left(y;z,T-H_z^-(0,l,D)\right)\right]\phi(dz)$$

yields the form (7.14) in the theorem. The derivation of the Laplace transform of $h_l(T,y)$ is carried out in Chesney, Jeanblanc-Picque and Yor [1997].

7.5.4 Parisian Contingent Claims

A Parisian contingent claim is a generalization of the Parisian option obtained by replacing the option payoff with a more general function. In its most general form a Parisian claim (European-style) with window D, barrier L and maturity date T pays off $Y_T \equiv F(S_T, K, T, D, O_T(S, A^{\pm}(\cdot, L)))$ at maturity. In this contractual form, parametrized by the occupation time (7.12), the final payment depends on the age of an excursion above or below the barrier. Knock-in and knock-out Parisian claims are special cases where the claim comes into existence or expires if the underlying asset price has an excursion of duration D above or below the barrier L during the life of the contract. Up and down versions of these claims are defined as usual, contingent upon an excursion above or below the barrier. For instance, a Parisian up-and-out claim with window D pays off $1_{\{H^+(t,L,D)>T\}}F(S_T, K, T)$ at the maturity date, where $H^+(t, L, D)$, defined in (7.13), measures the first time at which an excursion of length D above L is recorded.

Knock-in and knock-out Parisian claims satisfy parity relations that are similar to those tying pairs of knock-in and knock-out options together. For example, it is immediate to verify that the combination of a down-and-out and a down-and-in Parisian claims, both with exercise payoffs $F(S_T, K, T)$, gives a European claim with payoff $F(S_T, K, T)$. Likewise symmetry relations hold for knock-in and knock-out claims that are homogeneous of degree one in the pair (S_T, K). For valuation purposes, note that the following variation of Theorem 98 holds,

Theorem 99 *Let $S_0 = S$ and suppose that the standard model with constant interest rate and geometric Brownian motion price process applies. The value of a European-style Parisian down-and-in contingent claim with exercise payoff $F(S_T, K, T)$, window D, barrier L, parameter K and maturity date T is given by*

$$cc^{pdi}(S, 0, K, T, D, L) = R_T^{\gamma} \int_{-\infty}^{\infty} e^{my} F(Se^{\sigma y}, K, T)\, h_l(T, y) dy \tag{7.18}$$

where $\gamma = r + \frac{1}{2}m^2$, $m \equiv \left(r - \delta - \frac{1}{2}\sigma^2\right)/\sigma$ and $l = \log(L/S)/\sigma$. The function $h_l(T, y)$ is the same as in Theorem 98.

The proof of (7.18) is straightforward and parallels (7.15). Parisian down-and-out claims can be priced from (7.18) by using the parity relation tying in and out claims to plain vanilla claims.

More general contractual specifications can be priced by using the joint distribution of the occupation time and the underlying asset price at the maturity date. This approach is especially useful when the joint distribution, or its Laplace transform, has an analytical representation. Cumulative Parisian claims, examined next, provide an example where this is the case.

7.6 Cumulative Parisian Contingent Claims

Cumulative Parisian options, like Parisian options, were analyzed by Chesney, Jeanblanc-Picque and Yor [1997]. Cumulative Parisian contingent claims are discussed and priced by Hugonnier [1999].

7.6.1 Definitions and Parity/Symmetry Relations

A *cumulative Parisian option*, also called a *cumulative barrier option* or a *delayed barrier option*, pays off contingent on the cumulative amount of time spent above or below a prespecified barrier L. For European-style contracts the occupation time of relevance is

$$O_T\left(S, A^{\pm}(L)\right) = \int_0^T 1_{\{S_v \in A^{\pm}(L)\}} dv$$

with $A^{\pm}(L) \equiv \{x \in \mathbb{R}_+ : \pm(x - L) \geq 0\}$. Cumulative knock-in options pay off in the event $\{O_T(S, A^{\pm}(L)) \geq D\}$, i.e., if the underlying price spends an amount of time above or below the barrier in excess of D during the option's life $[0, T]$. For cumulative up-and-in and down-and-in calls the payoffs are

$$1_{\{O_T(S,A^+(L)) \geq D\}} (S_T - K)^+ \quad \text{and} \quad 1_{\{O_T(S,A^-(L)) \geq D\}} (S_T - K)^+ .$$

Cumulative knock-in puts are defined by substituting $(K - S_T)^+$ for $(S_T - K)^+$ in these expressions. Knock-out options are obtained by performing the usual modification.

More generally, a European-style cumulative (Parisian) barrier contingent claim has a payoff of the form

$$Y_T \equiv F\left(S_T, K, T, D, O_T\left(S, A^{\pm}(L)\right)\right)$$

for some measurable function F. Particular cases include cumulative Parisian options, cumulative binary and digital options, and cumulative asset-or-nothing options. Cumulative corridor claims, such as corridor options, can also be defined by writing the payoff as a function of the occupation time of a band $[L_1, L_2]$ (or letting O be a two dimensional vector of occupation times). Further generalizations are obtained by using multiple bands (letting O be n-dimensional) to define payment contingencies.

The general symmetry result in Theorem 95 applies to cumulative barrier claims. If the function $F(S, K, T, D, O(S, A^{\pm}))$ is homogeneous of degree one with respect to the pair (S, K) the value of the cumulative barrier claim is identical to the value of a symmetric claim with payoff

$$Y_T^* \equiv F\left(S, S_T^*, T, D, O_T\left(S^*, A^{\mp}(KS/L)\right)\right)$$

in an auxiliary economy with interest rate δ and in which the underlying price follows the Ito process described in the Theorem. The occupation time of relevance is based on the set $A^{\mp}(KS/L) \equiv \{y \in \mathbb{R}_+ : \pm(KS/L - y) \geq 0\}$ because

$x \geq L$ is equivalent to $KS/L \geq KS/x \equiv y$, for positive reals K, S, L and x. Applied to option payoffs this result establishes that a cumulative up-and-in call is symmetric to a cumulative down-and-in put.

Note also that cumulative barrier payoffs satisfy the relation

$$1_{\{O_T(S,A^+(L))\geq D\}}F\left(S_T,K,T\right)+1_{\{O_T(S,A^+(L))<D\}}F\left(S_T,K,T\right)=F\left(S_T,K,T\right)$$

where $F(S_T, K, T)$ is any measurable function. Parity formulas tying knock-in and knock-out cumulative barrier contingent claim prices to the corresponding "plain vanilla" claim price follow.

7.6.2 Pricing Cumulative Barrier Claims

For pricing purposes let us consider the model with constant coefficients (Black-Scholes setting). In this context, Hugonnier [1999] is able to solve for the joint law of the occupation time and the asset price and to provide explicit valuation formulas for a variety of claims, such as those described in section 7.6.3. His derivation of the joint law relies on a characterization of the distribution of certain functionals of Brownian motion established by Kac [1949], [1951].

A cumulative barrier claim with payoff $Y_T \equiv F\left(S_T, K, T, D, O_T\left(S, A^{\pm}\left(L\right)\right)\right)$ can be priced by using the change of measure in (7.15) to write

$$\begin{aligned} v_0^{cb,\pm}\left(S,K,T,D,L\right) &= R_T\widetilde{E}\left[F\left(S_T,K,T,O_T\left(S,A^{\pm}\left(L\right)\right)\right)\right] \\ &= e^{-(r+\frac{1}{2}m^2)T}E^s\left[e^{mZ_T}F\left(Se^{\sigma Z_T},K,T,O_T\left(Z,A^{\pm}\left(l\right)\right)\right)\right] \end{aligned}$$

where $O_T\left(Z, A^{\pm}\left(l\right)\right)$ is the occupation time of $A^{\pm}\left(l\right) \equiv \left\{x \in \mathbb{R}_+ : \pm\left(x-l\right) \geq 0\right\}$ by Z, with $l \equiv \log\left(L/S\right)/\sigma$. If $k_l^{\pm}\left(t, ds, dx\right) \equiv P\left(O_t\left(Z, A^{\pm}\left(l\right)\right) \in ds, Z_t \in dx\right)$ represents the joint law of Z_t and $O_t\left(Z, A^{\pm}\left(l\right)\right)$ one can write

$$v_0^{cb,\pm}=R_T^{\gamma}\int_{-\infty}^{+\infty}\int_0^{\infty}e^{mx}F\left(Se^{\sigma x},K,T,s\right)k_l^{\pm}\left(T,ds,dx\right)dsdx.$$

Assume that the asset price is below the barrier (i.e., $S < L$ or $l > 0$). Using Kac's characterization Hugonnier shows that the joint law of Z_t and $O_t\left(Z, A^+\left(L\right)\right)$ is given by

$$\begin{aligned} k_l^+\left(t,ds,dx\right) &= 1_{\{x\geq l\}}\Gamma\left(l,x-l,s,t-s\right)dsdx \\ &\quad +1_{\{x<l\}}\left(\Gamma\left(2l-x,0,s,t-s\right)ds-\delta_0\left(ds\right)\Lambda_l\left(t,x\right)\right)dx \end{aligned}$$

where $\delta_0(\cdot)$ is the Dirac delta function at zero and

$$\Gamma\left(a,b,u,v\right)\equiv\int_0^{\infty}\frac{\left(z+a\right)\left(z+b\right)}{\pi\left(uv\right)^{3/2}}\exp\left(-\frac{\left(z+a\right)^2}{2v}\right)\exp\left(-\frac{\left(z+b\right)^2}{2u}\right)dz \tag{7.19}$$

$$\Lambda_l\left(t,x\right)\equiv\frac{1}{\sqrt{2\pi t}}\exp\left(-\frac{x^2}{2t}\right)-\frac{1}{\sqrt{2\pi t}}\exp\left(-\frac{\left(2l-x\right)^2}{2t}\right). \tag{7.20}$$

When the asset price is above the barrier (i.e., $S > L$ or $l < 0$) the joint law can be recovered from the relation $k_l^+(t, ds, dx) = k_{|l|}^+(t, t - ds, -dx)$, for $t > 0$ and $0 \leq s < t$. Finally, for excursions below the barrier it is given by $k_l^-(t, ds, dx) = k_l^+(t, t - ds, dx)$, for all $l \in \mathbb{R}$ and $t \geq 0$.

For intermediate times $t \in [0, T]$ straightforward arguments show that

$$v_t^{cb,\pm} = R_{t,T}^{\gamma} \int_{-\infty}^{+\infty} \int_0^{\infty} e^{mx} F\left(Se^{\sigma x}, K, T, O_t + s\right) k_l^{\pm}(T - t, ds, dx)\, dsdx$$

where $S = S_t$ is the underlying asset price at t, $O_t = O_t\left(S, A^{\pm}(L)\right)$ is the occupation time at t and $R_{t,T}^{\gamma} = R_T^{\gamma}/R_t^{\gamma}$.

7.6.3 Standard and Exotic Cumulative Barrier Options

In order to obtain the values of specific contracts it suffices to specialize the payoff functions in the formulas above.

For a cumulative Parisian knock-in call option with strike K, maturity T, window D and barrier L this gives

$$\begin{aligned}
& c_0^{cb,\pm}(S, K, T, D, L) \\
= \; & R_T^{\gamma} \int_{-\infty}^{+\infty} \int_0^{T} e^{mx} \left(Se^{\sigma x} - K\right)^+ 1_{\{s \geq D\}} k_l^{\pm}(T, ds, dx)\, dsdx \\
= \; & R_T^{\gamma} \int_{d_o(S,K)}^{+\infty} e^{mx} \left(Se^{\sigma x} - K\right) \int_D^{T} k_l^{\pm}(T, ds, dx)\, dsdx \\
= \; & R_T^{\gamma} \left(S\Psi_{m+\sigma}^{\pm}\left(T, d_o(S, K), D, l\right)\right. \\
& \left. - K\Psi_m^{\pm}\left(T, d_o(S, K), D, l\right)\right)
\end{aligned} \tag{7.21}$$

where $d_o(S, K) = \log(K/S)/\sigma$, $l \equiv d_o(S, L) = \log(L/S)/\sigma$ and the function $\Psi_{\nu}^{\pm}$ is defined as

$$\Psi_{\nu}^{\pm}(T, d, D, l) \equiv \int_d^{+\infty} e^{\nu x} \int_D^{T} k_l^{\pm}(T, ds, dx)\, dsdx.$$

A cumulative digital knock-in call with strike K, maturity T, window D and barrier L pays off a dollar in the event $\{S_T \geq K\} \cap \{O_T\left(S, A^{\pm}(L)\right) \geq D\}$ and zero in the complementary event. In this instance the formula simplifies to

$$\begin{aligned}
& dc_0^{cb,\pm}(S, K, T, D, L) \\
= \; & R_T^{\gamma} \int_{-\infty}^{+\infty} \int_0^{\infty} e^{mx} 1_{\{Se^{\sigma x} \geq K\}} 1_{\{s \geq D\}} k_l^{\pm}(T, ds, dx)\, dsdx \\
= \; & R_T^{\gamma} \int_{d_o(S,K)}^{+\infty} e^{mx} \int_D^{\infty} k_l^{\pm}(T, ds, dx)\, dsdx \\
= \; & R_T^{\gamma} \Psi_m^{\pm}\left(T, d_o(S, K), D, l\right).
\end{aligned}$$

Multiplying this expression by H also gives the value of a cash-or-nothing cumulative binary call with strike K, maturity T, window D, barrier L and payment H in the event $\{S_T \geq K\} \cap \{O_T(S, A^{\pm}(L)) \geq D\}$ (and zero otherwise).

For an asset-or-nothing cumulative binary knock-in call with strike K, maturity T, window D and barrier L, that pays the underlying asset price in the event $\{S_T \geq K\} \cap \{O_T(S, A^{\pm}(L)) \geq D\}$, one gets

$$\begin{aligned}
& ab_0^{cb,\pm}(S, K, T, D, L) \\
= \; & R_T^{\gamma} \int_{-\infty}^{+\infty} \int_0^{\infty} e^{mx} S e^{\sigma x} 1_{\{Se^{\sigma x} \geq K\}} 1_{\{s \geq D\}} k_l^{\pm}(T, ds, dx)\, ds dx \\
= \; & R_T^{\gamma} S \int_{d_o(S,K)}^{+\infty} e^{(m+\sigma)x} \int_D^{\infty} k_l^{\pm}(T, ds, dx)\, ds dx \\
= \; & R_T^{\gamma} S \Psi_{m+\sigma}^{\pm}(T, d_o(S, K), D, l).
\end{aligned}$$

These formulas show that a cumulative knock-in call is the difference between an asset-or-nothing cumulative knock-in call and a cash-or-nothing cumulative knock-in call with payment $H = K$. This parallels the relation that ties together the prices of standard and binary (non-cumulative) call options.

Computation of prices can be performed by numerical integration. Some of the expressions presented above can be simplified, as shown by Hugonnier [1999] and Moraux [2002] (see Lemma 108 in the Appendix).

7.7 Step Options

Step options were also designed to bypass some of the difficulties associated with "one-touch" barrier options. Like Parisian options, they lose value more gradually than the classic barrier options. Detailed motivation for the contractual design and pricing formulas can be found in Linetsky [1999].

7.7.1 Contractual Specification

A *step option* is an option whose payoff is discounted at a rate that depends on the occupation time of a set above or below a barrier. A proportional step call, with strike K and barrier L, pays off

$$Y_t = \exp\left(-\rho O_t\left(S, A^{\pm}(L)\right)\right)(S_t - K)^+$$

at exercise, for some $\rho > 0$ where $A^{\pm}(L)$ is as defined above. The depreciation rate ρ is called the *knock-out* rate and $\exp\left(-\rho O_t\left(S, A^{\pm}(L)\right)\right)$ is the *knock-out* factor. Proportional step options are also known as *geometric* or *exponential* step options.

Simple step options use a piecewise linear amortization scheme instead of an exponential one. A *down-and-out simple step call*, with strike K and barrier L, has exercise payoff

$$Y_t = \max\left(1 - \rho O_t\left(S, A^{\pm}(L)\right), 0\right)(S_t - K)^+.$$

This option is knocked out at the random time

$$\zeta \equiv \inf\left\{v \in [0,T] : O_v\left(S, A^{\pm}(L)\right) \geq 1/\rho\right\}$$

(if such a time does not exist in $[0,T]$ set $\zeta = \infty$), called the *knock-out-time*. This is the smallest occupation time of the set above, or below the barrier, at which the payoff vanishes.

Step put options are defined by substituting the put payoff function for the call payoff function in the expressions above. Step options are European- or American-style depending on whether they can be exercised only at maturity T or at any time at or before maturity.

7.7.2 Pricing European-style Step Options

Step options are particular examples of cumulative barrier contingent claims. Assuming the standard Black-Scholes framework they can be priced by specializing the previous formulas.

Consider, for instance, a proportional step call with strike K, maturity T and barrier L. Its price satisfies

$$\begin{aligned}
& c_0^{ps,\pm}(S,K,T,L) \\
= \; & R_T^{\gamma} \int_{-\infty}^{+\infty} \int_0^T e^{-\rho s} e^{mx} \left(S e^{\sigma x} - K\right)^{+} k_l^{\pm}(T, ds, dx)\, ds dx \\
= \; & R_T^{\gamma} \int_{d_o(S,K)}^{+\infty} e^{mx} \left(S e^{\sigma x} - K\right) \int_0^T e^{-\rho s} k_l^{\pm}(T, ds, dx)\, ds dx \\
= \; & R_T^{\gamma} \left(S \Phi_{m+\sigma}^{\pm}(\rho, T, d_o(S,K), 0, T, l) - K \Phi_m^{\pm}(\rho, T, d_o(S,K), 0, T, l)\right)
\end{aligned}$$

where

$$\Phi_{\nu}^{\pm}(\rho, t, d, D, T, l) \equiv \int_d^{+\infty} \int_D^T e^{\nu x} e^{-\rho s} k_l^{\pm}(t, ds, dx)\, ds dx.$$

For a simple step call the price is

$$\begin{aligned}
& c_0^{ss,\pm}(S,K,T,L) \\
= \; & R_T^{\gamma} \int_{-\infty}^{+\infty} \int_0^T (1-\rho s)^{+} e^{mx} \left(S e^{\sigma x} - K\right)^{+} k_l^{\pm}(T, ds, dx)\, ds dx \\
= \; & R_T^{\gamma} \int_{d_o(S,K)}^{+\infty} e^{mx} \left(S e^{\sigma x} - K\right) \int_0^{T \wedge 1/\rho} (1-\rho s)\, k_l^{\pm}(T, ds, dx)\, ds dx \\
= \; & R_T^{\gamma} \left(S \Phi_{m+\sigma}^{\pm}(0, T, d_o(S,K), 0, T_\rho, l) - K \Phi_m^{\pm}(0, T, d_o(S,K), 0, T_\rho, l)\right) \\
& + \rho R_T^{\gamma} S \partial_\rho \Phi_{m+\sigma}^{\pm}(0, T, d_o(S,K), 0, T_\rho, l) \\
& - \rho R_T^{\gamma} K \partial_\rho \Phi_m^{\pm}(0, T, d_o(S,K), 0, T_\rho, l)
\end{aligned}$$

where $T_\rho \equiv T \wedge 1/\rho$ and $\partial_\rho \Phi_{\nu}^{\pm}$ is the derivative with respect to the first argument, ρ, of the function $\Phi_{\nu}^{\pm}$ (and those derivatives are evaluated at the point $\rho = 0$).

Note also that the contracts in 7.6.3 could have been priced using the function $\Phi_{\nu}^{\pm}$. Indeed, simple inspection of the formulas shows that $\Psi_{\nu}^{\pm}(t, d, D, l) \equiv \Phi_{\nu}^{\pm}(0, t, d, D, t, l)$.

7.8 American Occupation Time Derivatives

The valuation of American-style OTDs can be approached from several angles. This section provides results pertaining to the early exercise premium associated with such a contract.

7.8.1 Early Exercise Premium Representation

Consider an American-style OTD with exercise payoff $Y = F(S, O(S, A))$ where $O(S, A)$ is the occupation time process

$$O_{0,t}(S, A) = \int_0^t 1_{\{S_v \in A(v,\omega)\}} dv, \quad t \in [0, T] \tag{7.22}$$

defined for a progressively measurable, closed set $A(\cdot, \cdot) : ([0, T] \times \Omega, \mathcal{B}([0, T]) \otimes \mathcal{F}) \to (\mathbb{R}_+, \mathcal{B}(\mathbb{R}_+))$. In the general market with Ito price process and progressively measurable interest rate the claim's price, $V_t(Y)$, is given by (7.2). Recall the Snell envelope $Z_t \equiv \sup_{\tau \in \mathcal{S}_{t,T}} E_t[D_\tau]$ of the deflated process $D_t \equiv \xi_{0,t} Y_t$ and the stopping time

$$\tau_t \equiv \inf\{s \in [t, T] : Z_s = D_s\}, \tag{7.23}$$

which exists, if D is continuous. Invoking Theorem 21 and specializing to the payoff function under consideration gives the early exercise premium representation.

Theorem 100 *Consider an American-style OTD with exercise payoff* $Y = F(S, O(S, A))$ *where* $O(S, A)$ *is given in (7.22). Suppose that* $Y \in \mathcal{I}_+^*$ *and that the function* $F(x, y) \in \mathcal{C}^{2,1}(\mathbb{R}_+ \times [0, T])$. *The value of the claim at* $t \in [0, \tau_0]$ *has the early exercise premium representation*

$$V_t(f, Y) = \widetilde{E}_t[R_{t,T} Y_T] + \widetilde{E}_t\left[\int_{\tau_0}^T R_{t,s} 1_{\{\tau_s = s\}} (r_s Y_s ds - dA_s^Y)\right] \tag{7.24}$$

where $\tau_t = \inf\{v \in [t, T] : Z_v = D_v\}$ *and*

$$dA_s^Y = \frac{\partial F}{\partial S} S_s (r_s - \delta_s) + \frac{1}{2}\frac{\partial^2 F}{\partial S^2} S_s^2 \sigma_s^2 + \frac{\partial F}{\partial O} 1_{\{S_s \in A(s,\omega)\}}. \tag{7.25}$$

If the claim expires at a random time τ_e *the representation (7.24) holds with the substitution of* $\tau_e \wedge T$ *in place of* T *and the convention that* $Y_{\tau_e \wedge T}$ *is the rebate (which may be null) in the event of expiration prior to maturity.*

As usual, the EEP formula (7.24) splits the value of the American claim into two parts, the value of a European claim and an early exercise premium. The benefits from early exercise, $r_s Y_s ds - dA_s^Y$, depend on the appreciation of the payoff described in (7.25). Two non-standard components appear in this expression. The first one corresponds to the quadratic variation resulting from the non-linear payoff structure. The second one captures the impact of the occupation time. These components could add or subtract value depending on the curvature of the payoff and the nature of its dependence on the occupation time. The expressions obtained show that immediate exercise is optimal only if

$$r_s F - \frac{\partial F}{\partial S} S_s (r_s - \delta_s) - \frac{1}{2}\frac{\partial^2 F}{\partial S^2} S_s^2 \sigma_s^2 - \frac{\partial F}{\partial O} 1_{\{S_s \in A(s,\omega)\}} > 0.$$

Ceteris paribus, contracts that are convex in the exercise region, entail reduced local benefits. Likewise contracts that are positively related to the occupation time lose value if exercised when the underlying price lies in the set $A(s,\omega)$.

Contracts, such as knock-out cumulative Parisian claims or knock-out step claims, that are not continuous with respect to the occupation time, are still valued according to (7.24) provided the maturity date is replaced by $\tau_e \wedge T$ with τ_e the random expiration date.

Finally, it is worth noting that the EEP formula (7.24) can apply even if the payoff function is not $\mathcal{C}^{2,1}$ over the whole domain. Smoothness in the exercise region, i.e., over the event $\{\tau_s = s\}$, is enough for its validity.

7.8.2 Valuation in the Standard Model

More explicit expressions can be written in the standard model with constant coefficients r, δ and σ (Black-Scholes setting). Let us focus on the valuation of a cumulative down-and-out call with strike K, maturity T, window D, barrier L and null rebate in the event of a knock-out. Exercise, at the stopping time τ, yields the payoff $1_{\{O_\tau(S,A^-(L))<D\}}(S_\tau - K)^+$ where

$$O_\tau(S, A^-(L)) = \int_0^\tau 1_{\{S_v \in A^-(L)\}} dv$$

is the occupation time of the set $A^-(L) \equiv \{x \in \mathbb{R}_+ : -(x-L) \geq 0\}$ during the period $[0,\tau]$. Given that the pair of underlying variables, (S,O), follows a diffusion process the price

$$C^{cdo}(S,O,t) = \sup_{\tau \in \mathcal{S}_{t,T}} \widetilde{E}_t \left[R_{t,\tau} 1_{\{O_\tau(S,A^-(L))<D\}} (S_\tau - K)^+ \right]$$

is a function of (S,O) and time t. The notation $C^{cdo}(S,O,t;K,T,D,L)$ is also used below to emphasize the dependence on the contractual characteristics K,T,D,L. The call price can be viewed as a function defined over the domain $(S,O,t) \in \mathbf{D} \equiv \mathbb{R}_+ \times [0,D] \times [0,T]$, given that the option is worthless if $O \geq D$. Moreover, for t fixed, the domain of the occupation time is $[0, t \wedge D]$, because $O_t \leq t$. The immediate exercise region is

$$\mathcal{E}^{cdoc} \equiv \{(S,O,t) \in \mathbf{D} : C^{cdo}(S,O,t) = (S-K)^+\}.$$

Its (t,O)-section is $\mathcal{E}^{cdoc}(t,O) \equiv \{S \in \mathbb{R}_+ : (S,O,t) \in \mathcal{E}\}$.

Basic properties of the immediate exercise region are recorded in the next proposition. Using $\mathcal{E}$ to denote the exercise region of a standard call option and $\mathcal{E}^{doc}$ that of a one-touch down-and-out call, the following holds

Proposition 101 *The immediate exercise region has the properties*

(i) $(S, O, t) \in \mathcal{E}^{cdoc}$ *implies* $(S, O, s) \in \mathcal{E}^{cdoc}$ *for all* $t \in [0, T]$ *and* $s \in [t, T]$.

(ii) $(S, O, t) \in \mathcal{E}^{cdoc}$ *implies* $(\lambda_1 S, \lambda_2 O, t) \in \mathcal{E}^{cdoc}$ *for all* $\lambda_1 \geq 1$, $D/O > \lambda_2 \geq 1$ *and* $t \in [0, T]$.

(iii) Suppose $S < \max\{K, (r/\delta)K\}$. *Then* $(S, O, t) \notin \mathcal{E}^{cdoc}$, *for all* $t \in [0, T)$ *and* $O \in [0, t \wedge D]$.

(iv) $(S, O, T) \in \mathcal{E}^{cdoc}$ *if and only if* $S \geq K$.

(v) $\mathcal{E} \subseteq \mathcal{E}^{cdoc} \subseteq \mathcal{E}^{doc}$.

Properties (i)-(iv) are standard and parallel those for standard options stated in Proposition 31. The main difference pertains to the fact that the exercise region is defined in terms of a triplet (S, O, t), reflecting the dependence on the occupation time O. Property (ii) identifies the basic nature of this relation. It shows that the exercise region is connected, up to the boundary of the domain, in the O-direction. The reason for this feature is the negative impact of the occupation time on the option value. When the occupation time increases the probability of an early knock-out increases, thereby reducing value. Immediate exercise is therefore optimal if it is optimal at the outset, i.e., at (S, O, t).

Property (v) shows the relations between the exercise regions associated with the one-touch down-and-out call, the cumulative barrier call and the standard call option. The inclusions reported reflect the ordering of prices

$$C^{do}(S, t; K, T, L) \leq C^{cdo}(S, O, t; K, T, D, L) \leq C(S, t; K, T).$$

American down-and-out calls are naturally worth less than standard call options that are not subject to knock-out prior to maturity. The American cumulative down-and-out call is also worth more than the one-touch down-and-out call because of its resilience to breaches of the barrier: the option remains alive at the first time at which the underlying price reaches the barrier. The ordering of prices implies the ordering of the exercise regions. If immediate exercise is optimal for the plain vanilla call, it must also be optimal for the two barrier options. Similarly, if it is optimal to exercise the cumulative down-and-out call it is also optimal to exercise the one-touch barrier call.

The results of Proposition 101 enable us to define the boundary of the exercise region as $B = \{B(t, O) : t \in [0, T], O \in [0, t \wedge D]\}$ where

$$B(t, O) = \inf\left\{S : S \in \mathcal{E}^{cdoc}(t, O)\right\}$$

is the smallest asset price in the exercise region at date t when the occupation time is O. Moreover, standard arguments can be invoked to show that $B(t, O)$ is a right-continuous, non-increasing function with respect to time. From property (ii) above it also follows that $B(t, O)$ is non-increasing with respect to the occupation time. Limiting values can be also be easily inferred. As the maturity date approaches the likelihood of a knock-out goes to zero and the exercise region converges to that of a vanilla option. If $t \geq D$ and the occupation time approaches the window the likelihood of a knock-out increases. In the limit (i.e., when O is very close to D, but strictly less) the cumulative barrier option becomes similar to a newly issued one-touch option that is knocked out at the next touch of the barrier. The difference between the exercise regions of the two options vanishes.

The definition of $B(t,O)$ shows that immediate exercise is optimal when $S \geq B(t,O)$. The exercise region can be written as

$$\mathcal{E}^{cdoc} \equiv \{(S,O,t) \in \mathbf{D} : S \geq B(t,O)\}$$

and the event $\{\tau_s = s\}$ is the same as $\{S_s \geq B(s,O_s)\}$. The EEP formula in Theorem 100 takes the more explicit form

Theorem 102 *Suppose that the underlying asset price follows the geometric Brownian motion process (4.1) and that the interest rate is constant. The value of an American-style cumulative down-and out call has the early exercise premium representation*

$$C^{cdo}(S,O,t) = c^{cdo}(S,O,t) + \pi^{cdo}(S,O,t,B(\cdot)) \tag{7.26}$$

for $O \in [0, t \wedge D]$ and $t \in [0,T]$, where $c^{cdo}(S,O,t)$ represents the European cumulative call option value and $\pi^{cdo}(S,O,t,B(\cdot))$ is the early exercise premium. These functions are given by

$$c^{cdo} = R^{\gamma}_{t,T} \int_{d_o(S,K)}^{+\infty} \int_0^{D-O} e^{mx} \left(Se^{\sigma x} - K\right) k_l^- \left(T-t, dv, dx\right) dvdx \tag{7.27}$$

$$\begin{aligned} \pi^{cdo} &= \int_t^T R^{\gamma}_{t,s} \int_0^{D-O} \int_{d_o(S,B(s,O+v))}^{+\infty} e^{mx} \left(\delta Se^{\sigma x} - rK\right) \\ &\quad \times k_l^- \left(s-t, dv, dx\right) dxdvds \end{aligned} \tag{7.28}$$

with $\gamma = r + \frac{1}{2}m^2$, $m \equiv \left(r - \delta - \frac{1}{2}\sigma^2\right)/\sigma$, $l = \log(L/S)/\sigma$ and $d_o(S,y) = \log(y/S)/\sigma$. The immediate exercise boundary B^{cdoc} solves the recursive non-linear integral equation

$$B^{cdoc}(t,O) - K = c^{cdo}(B^{cdoc}(t,O),O,t) + \pi^{cdo}(B^{cdoc}(t,O),O,t,B(\cdot)) \tag{7.29}$$

for $O \in [0, t \wedge D]$ and $t \in [0,T)$, subject to the boundary conditions

$$B^{cdoc}(T_-,O) = \max\{K, (r/\delta)K\}, \quad \text{for } O < D \tag{7.30}$$

$$B^{cdoc}(t,D_-) = B^{doc}(t), \quad \text{for } t \geq D. \tag{7.31}$$

At maturity $B_T = K \leq B_{T_-}$.

The EEP formula (7.26)-(7.28) identifies two value components embedded in the price of an American cumulative down-and-out call. The first is the value of the European cumulative down-and-out call with identical characteristics (K,T,D,L). The second is the value accruing from early exercise, prior to the minimum of the knock-out time and the maturity date. Both value components are written in terms of the joint density function, defined in section 7.6.2, for the occupation time and the cumulative continuously compounded asset return per unit volatility.

The exercise premium component shows that value is created by collecting the local benefits when immediate exercise is optimal and the knock-out provision is inactive. The innermost integral in (7.28) adds up those benefits for prices above the exercise boundary $B(s, O+v)$, conditional on a fixed time s and a fixed occupation time v. The middle integral sums up the cumulative benefits obtained over possible values of the occupation time, conditional on a fixed time s. The outer integral adds up the twice cumulated benefits over times up to the maturity date. The order of integration is dictated by the dependence of the exercise boundary on the occupation time. This can be contrasted with the European price (7.27) where the order of integration is immaterial.

As usual the EEP formula is parametrized by the exercise boundary B^{cdoc}, which, in this case, is a two dimensional surface depending on time and on the value of the occupation time. Given that the formula holds for all values of $(S, O, t) \in \mathbf{D}$, it holds on the boundary of the exercise region. This gives the integral equation (7.29). This equation, along with the boundary conditions (7.30)-(7.31), can in principle be used as the basis for a numerical scheme designed to compute the option value.

The transformations employed to derive the pricing expression in Theorem 102 can be carried out for several other occupation time derivatives. Cases requiring modifications of the results in Theorems 100 and 102 are those where the dependence on the occupation time takes a more complex form than (7.22).

7.9 Multiasset Claims

Occupation time derivatives can be easily generalized to the multiasset case. For a progressively measurable, closed set A and an n-dimensional vector of asset price processes S, a multiasset occupation time F-claim has payoff $F(S, K, O^{S,A})$ where

$$O_t^{S,A} = \int_0^t 1_{\{S_v \in A_v\}} dv, \quad t \in [0, T].$$

As before K represents a parameter such as a strike or a cap, depending on the specific context.

7.9.1 Symmetry Properties

Multiasset occupation time derivatives also satisfy certain symmetry properties. To state the main result, recall the scaling operator in Appendix A of chapter 6,

$$\lambda \circ_j S = (S^1, ..., S^{j-1}, \lambda S^j, S^{j+1}, ..., S^n)$$

defined for a constant λ and a vector of prices S (the jth component of the price vector is rescaled by the factor λ). Also for a given occupation time F-

claim with parameter K and for any $j = 1, ..., n$ define the associated F^j-claim obtained by permutation of the jth argument and the parameter

$$F^j(S, K, O^{S,A}) = F\left(\lambda^j \circ_j S, S^j, O^{S,A}\right)$$

with $\lambda^j = K/S^j$. With this notation the following natural generalization of Theorem 95 is obtained.

Theorem 103 *Consider an American-style occupation time F-claim with maturity date T and a payoff function $F(S, K, O^{S,A})$ which is homogeneous of degree one in (S, K). Let $V(S, K, O^{S,A}, r, \delta; \mathcal{F}_t)$ denote the value of the claim in the financial market with filtration $\mathcal{F}_{(\cdot)}$, asset prices S satisfying (2.26) and progressively measurable interest rate r. Pick some arbitrary index j and define*

$$\lambda^j(\delta) \equiv \frac{r}{\delta^j}.$$

Prior to exercise the value of the multiasset occupation time F-claim is

$$V(S_t, K, O^{S,A}, r, \delta; \mathcal{F}_t) = V^j(S_t^*, S^j, O^{S^*,A^*}, \delta^j, \lambda^j(\delta) \circ_j \delta; \mathcal{F}_t)$$

where $A^* = \{A^*(v, \omega), v \in [t, T]\}$ with

$$A^*(v,\omega) = \left\{x \in \mathbb{R}^n_+ : x_i = \frac{y_i S^j}{y_j}, \text{ for } i \neq j, x_j = \frac{KS^j}{y_j} \right. \\ \left. \text{and } y = (y_1, ..., y_n) \in A(v, \omega)\right\}$$

and $O_t^{S^*,A^*} \equiv O_t^{S,A}$. Here $V^j(S_t^*, S^j, O_t^{S^*,A^*}, \delta^j, \lambda^j(\delta) \circ_j \delta; \mathcal{F}_t)$ is the value of the F^j-claim with parameter $S^j = S_t^j$, maturity date T and occupation time $O_t^{S^*,A^*}$ in an auxiliary financial market with interest rate δ^j and in which the underlying asset prices follow the Ito processes

$$\begin{cases} dS_v^{i*} = S_v^{i*}[(\delta_v^j - \delta_v^i)dv + (\sigma_v^j - \sigma_v^i)dz_v^{j*}]; \text{ for } i \neq j \text{ and } v \geq t \\ dS_v^{j*} = S_v^{j*}[(\delta_v^j - r_v)dv + \sigma_v^j dz_v^{j*}]; \text{ for } i = j \text{ and } v \geq t \end{cases}$$

with respective initial conditions S^i for $i \neq j$ and K for $i = j$. The n-dimensional process z^{j*} is defined by

$$dz_v^{j*} = -d\widetilde{z}_v + \sigma_v^{j'} dv$$

for all $v \in [0, T]$, $z_0^{j*} = 0$. The optimal exercise time for the occupation time F-claim is the same as the optimal exercise time for the occupation time F^j-claim in the auxiliary financial market.

The major difference relative to the single asset case is the choice of numeraire associated with the multidimensionality of the relevant variables. Selecting a specific asset as the numeraire leads to one particular symmetric contract. Another choice can produce a very different symmetric payoff. In the end, all these contracts are related as they are derived from a common specification.

A simple illustration of this property is provided by the cumulative barrier exchange options with payoff function $\left(S^1 - S^2\right)^+ 1_{\{O_t^{S,A} \geq D\}}$, with $D \in [0, T]$. By taking asset 2 as the numeraire it can be verified that this exchange option is symmetric to a cumulative barrier call with suitably adjusted occupation time. If asset 1 functions as the numeraire then the exchange option is symmetric to a cumulative barrier put with adjusted occupation time. Other examples are provided by cumulative barrier max- and min-options. When there are two underlying assets, knock-in call options in this category have payoff functions of the form $(S_t^1 \vee S_t^2 - K)^+ 1_{\{O_t^{S,A} \geq D\}}$ (max-option) or $(S_t^1 \wedge S_t^2 - K)^+ 1_{\{O_t^{S,A} \geq D\}}$ (min-option), where $D \in [0, T]$. Put options have similar structures substituting put payoffs in place of call payoffs. Knock-out contracts are obtained by conditioning the payment on $\{O_t^{S,A} < D\}$ instead of $\{O_t^{S,A} \geq D\}$. Taking the cumulative barrier call max-option with knock-in window D as an illustrative example it can be verified that this contract is symmetric to cumulative barrier options to exchange the maximum of an asset and cash against another asset, for which the occupation time has been suitably adjusted. There are two symmetric contracts, both of the same type, depending on the asset chosen as the numeraire.

7.9.2 Valuation

Multiasset occupation time derivatives can be valued using the same principles as for the univariate case. Present value formulas such as (7.1) and (7.2) naturally apply. A multivariate version of the early exercise premium representation (7.24)-(7.25) also holds. In this expression the appreciation component of the payoff function, described in (7.25), is obtained by an application of the multidimensional version of Ito's lemma and reflects the impact of all prices. More explicit formulas can be obtained, in the standard model with constant coefficients, for certain contractual forms. Cumulative exchange options, for instance, can be easily valued in that context. For many multiasset contractual forms, however, explicit formulas may prove elusive. With that in mind, it is worth recalling that the duality method applies to OTDs. As mentioned before this approach is useful for constructing upper bounds for derivatives' prices. Finding tight bounds and approximations may be of particular interest in this context.

7.10 Appendix: Proofs

Proof of Theorem 95: Fix $t \in [0, T]$ and suppose that $O_t(S, A) = O_t(S^*, A^*)$. For any stopping time $\tau \in \mathcal{S}_{t,T}$ the occupation time $O_{0,\tau}(S, A)$ can be written as

$$\begin{aligned}
O_{0,\tau}(S, A) &= O_{0,t}(S, A) + \int_t^\tau 1_{\{S_v \in A(v,\omega)\}} dv \\
&= O_{0,t}(S^*, A^*) + \int_t^\tau 1_{\{S_v^* \in A^*(v,\omega)\}} dv = O_{0,\tau}(S^*, A^*)
\end{aligned}$$

where $S_v^* = KS/S_v, v \in [t,T]$ and $O_{0,\tau}(S^*, A^*)$ is the occupation time of the random set A^* by the process S^*. Performing the usual change of measure leads to the result stated. ■

Proof of Theorem 96: The proof of the theorem relies on the next 3 lemmas. Throughout it is assumed that the uncertainty is generated by a Brownian motion $\widetilde{z}$. A progressively measurable process is understood to be adapted to the Brownian filtration.

Lemma 104 *Let Z be a progressively measurable, continuous process with α-quantile $M^Z(\alpha,t)$ and define the transformed process $S = f(Z)$ where $f(\cdot) : \mathbb{R} \to \mathbb{R}_+$ is a continuous and increasing function. The α-quantile of S is the transform of the α-quantile of Z, that is $M(\alpha,t) = f\left(M^Z(\alpha,t)\right)$.*

Proof of Lemma 104: The definition of an α-quantile and the properties of the map f give

$$\begin{aligned}
M(\alpha,T) &= \inf\left\{x : \int_0^T 1_{\{S_v \leq x\}} dv > \alpha T\right\} \\
&= \inf\left\{x : \int_0^T 1_{\{Z_v \leq f^{-1}(x)\}} dv > \alpha T\right\}. \\
&= \inf\left\{f(y) : \int_0^T 1_{\{Z_v \leq y\}} dv > \alpha T\right\} \qquad \text{(defining } y = f^{-1}(x)) \\
&= f\left(\inf\left\{y : \int_0^T 1_{\{Z_v \leq y\}} dv > \alpha T\right\}\right) \qquad (\inf f(y) = f(\inf y)) \\
&= f\left(M^Z(\alpha,T)\right).
\end{aligned}$$

The equality on the second line uses the fact that f is increasing and continuous. The third line follows from the definition $x = f(y)$. The expression obtained represents a constrained minimization problem (of the function $f(y)$) subject to a constraint on y. For the fourth line note that the function $f(\cdot)$ is strictly increasing: the minimization can then be performed directly over the argument y. The last line is the definition of the α-quantile of Z. ■

Lemma 105 *Let Z be a progressively measurable, continuous process with $Z_0 = 0$ and α-quantile $M^Z(\alpha,T)$ over the interval $[0,T]$. Then*

$$M^Z(\alpha,T) = M^{\widehat{Z}}(\widehat{\alpha}(t,\widehat{y}_t^*), T-t) + Z_t$$

where $\widehat{y}_t^* = M^{\widehat{Z}}(\widehat{\alpha}(t,\widehat{y}_t^*), T-t)$ with $\widehat{Z}_v = Z_v - Z_t$,

$$\widehat{\alpha}(t,y) \equiv \frac{\alpha T - \int_0^t 1_{\{\widehat{Z}_v \leq y\}} dv}{T-t}$$

and $M^{\widehat{Z}}(\widehat{\alpha}(t,\widehat{y}_t^*), T-t)$ is the $\widehat{\alpha}(t,\widehat{y}_t^*)$-quantile of the process $\widehat{Z}$ in the interval $[t,T]$.

Proof of Lemma 105: Let Z be an adapted stochastic process with initial value $Z_0 = 0$. Then

$$\begin{aligned} M^Z(\alpha, T) &= \inf\left\{ y : \int_0^T 1_{\{Z_v \leq y\}} dv > \alpha T \right\} \\ &= \inf\left\{ y : \int_t^T 1_{\{Z_v \leq y\}} dv > \alpha T - \int_0^t 1_{\{Z_v \leq y\}} dv \right\} \\ &= \inf\left\{ y : \int_t^T 1_{\{Z_v - Z_t \leq y - Z_t\}} dv > \alpha T - \int_0^t 1_{\{Z_v - Z_t \leq y - Z_t\}} dv \right\} \\ &= \inf\left\{ \widehat{y}_t + Z_t : \int_t^T 1_{\{\widehat{Z}_v \leq \widehat{y}_t\}} dv > \widehat{\alpha}(t, \widehat{y}_t)(T - t) \right\} \end{aligned} \tag{7.32}$$

where $\widehat{y}_t = y - Z_t$ and

$$\widehat{\alpha}(t, \widehat{y}_t) \equiv \frac{\alpha T - \int_0^t 1_{\{\widehat{Z}_v \leq \widehat{y}_t\}} dv}{T - t} = \frac{\alpha T - \int_0^t 1_{\{Z_v - Z_t \leq y - Z_t\}} dv}{T - t}.$$

Let $\widehat{y}_t^*$ denote the solution of the minimization problem in (7.32). Then, by Lemma 104, the last line of (7.32) shows that $M^Z(\alpha, T) = \widehat{y}_t^* + Z_t$ where $\widehat{y}_t^* = M^{\widehat{Z}}(\widehat{\alpha}(t, \widehat{y}_t^*), T - t)$ with $\widehat{Z}_v = Z_v - Z_t$ and $M^{\widehat{Z}}(\widehat{\alpha}(t, \widehat{y}_t^*), T - t)$ the $\widehat{\alpha}(t, \widehat{y}_t^*)$-quantile of the process $\widehat{Z}$ in the interval $[t, T]$. ■

Lemma 106 *Suppose that $\widehat{Z}$ is a progressively measurable, continuous process with independent increments and let $\widehat{y}_t^*$ and $M^{\widehat{Z}}(\widehat{\alpha}(t, \widehat{y}_t^*), T - t)$ be as defined in Lemma 105. Then,*

$$\begin{aligned} & P\left[M^{\widehat{Z}}(\widehat{\alpha}(t, \widehat{y}_t^*), T - t) > z \middle| \mathcal{F}_t \right] \\ &= \widehat{P}\left[M^{\widehat{Z}}(\widehat{\alpha}(t, z), T - t) > z \right] 1_{\{0 < \widehat{\alpha}(t,z) \leq 1\}} + 1_{\{1 < \widehat{\alpha}(t,z)\}} \end{aligned}$$

where $P[\cdot | \mathcal{F}_t]$ is the conditional probability of $M^{\widehat{Z}}(\widehat{\alpha}(t, \widehat{y}_t^*), T - t)$ and $\widehat{P}$ represents the probability measure over the increments $\widehat{Z}_v = Z_v - Z_t$, $v \geq t$. The quantity $\widehat{\alpha}(t, z)$ is $\mathcal{F}_t$-measurable.

Proof of Lemma 106: Using the definition of $M^{\widehat{Z}}(\widehat{\alpha}(t, \widehat{y}_t^*), T - t)$ and the independence of increments gives

$$\begin{aligned}
& P\left[M^{\widehat{Z}}\left(\widehat{\alpha}\left(t,\widehat{y}_t^*\right),T-t\right)>z\middle|\mathcal{F}_t\right] \\
&= P\left[\left[\int_t^T 1_{\{\widehat{Z}_v\le z\}}dv\le\widehat{\alpha}\left(t,z\right)\left(T-t\right)\right|\mathcal{F}_t\right] \\
&= \widehat{P}\left[\int_t^T 1_{\{\widehat{Z}_v\le z\}}dv\le\widehat{\alpha}\left(t,z\right)\left(T-t\right)\right] \\
&= \widehat{P}\left[M^{\widehat{Z}}\left(\widehat{\alpha}\left(t,z\right),T-t\right)>z\right]1_{\{0<\widehat{\alpha}(t,z)\le1\}} \\
&\quad +1_{\{1<\widehat{\alpha}(t,z)\}}
\end{aligned}$$

where the probability measure $\widehat{P}$ is over the increments $\widehat{Z}_v = Z_v - Z_t$, for $v \geq t$, and $M^{\widehat{Z}}\left(\widehat{\alpha}\left(t,z\right),T-t\right)$ is the $\widehat{\alpha}\left(t,z\right)$-quantile of $\widehat{Z}$ in the interval $[t,T]$. Note that the equality on the third line uses the fact that $\widehat{\alpha}\left(t,z\right)$ is known at date t. ■

The proof of Theorem 96 proceeds as follows. Using the geometric Brownian motion assumption write the asset price as $S_t = S_0 \exp\left(\sigma Z_t\right)$ where

$$Z_t \equiv \widetilde{z}_t + mt \quad \text{with} \quad m = \frac{1}{\sigma}\left(r-\delta-\frac{1}{2}\sigma^2\right)$$

is a Brownian motion with drift m under the risk neutral measure. Lemma 104 can be invoked to conclude that $M\left(\alpha,T\right) = S_0\exp\left(\sigma M^Z\left(\alpha,T\right)\right)$. Moreover, Lemma 105 shows that $M^Z\left(\alpha,T\right) = M^{\widehat{Z}}\left(\widehat{\alpha}\left(t,\widehat{y}_t^*\right),T-t\right) + Z_t$ where $\widehat{y}_t^* = M^{\widehat{Z}}\left(\widehat{\alpha}\left(t,\widehat{y}_t^*\right),T-t\right)$ is the $\widehat{\alpha}\left(t,\widehat{y}_t^*\right)$-quantile of the process $\widehat{Z}$ in the interval $[t,T]$.

The price of the α-quantile option price can then be simplified as follows

$$\begin{aligned}
C^{\alpha}\left(S_t,t\right) &= R_{t,T}\widetilde{E}_t\left[\left(S_0\exp\left(\sigma M^Z\left(\alpha,T\right)\right)-K\right)^+\right] \\
&= R_{t,T}\widetilde{E}_t\left[\left(S_0\exp\left(\sigma M^{\widehat{Z}}\left(\widehat{\alpha}\left(t,y_t^*\right),T-t\right)+\sigma Z_t\right)-K\right)^+\right] \\
&= R_{t,T}S_0e^{\sigma Z_t}\widetilde{E}_t\left[\left(\exp\left(\sigma M^{\widehat{Z}}\left(\widehat{\alpha}\left(t,y_t^*\right),T-t\right)\right)-\frac{K}{S_0e^{\sigma Z_t}}\right)^+\right] \\
&= R_{t,T}S_t\widetilde{E}_t\left[\left(\exp\left(\sigma M^{\widehat{Z}}\left(\widehat{\alpha}\left(t,y_t^*\right),T-t\right)\right)-\frac{K}{S_t}\right)^+\right] \\
&= R_{t,T}S_t\int_{\log(K/S_t)/\sigma}^{\infty}\left(\exp\left(\sigma x\right)-\frac{K}{S_t}\right)\widehat{g}\left(x,t;\widehat{\alpha}\left(t,y_t^*\right),m,T\right)dx
\end{aligned}$$

where $\widehat{g}\left(x,t;\alpha,m,T\right)$ is the conditional density at date t of $M^{\widehat{Z}}\left(\widehat{\alpha}\left(t,y_t^*\right),T-t\right)$.

Integrating by parts and simplifying the resulting expression gives

$$\begin{aligned}
C^{\alpha}(S_t,t) &= R_{t,T}S_t\int_{\log(K/S_t)/\sigma}^{\infty}\sigma e^{\sigma x}\int_x^{\infty}\widehat{g}\left(v,t;\widehat{\alpha}\left(t,y_t^*\right),m,T\right)dvdx \\
&= R_{t,T}S_t\int_{K/S_t}^{\infty}\int_{\log(z)/\sigma}^{\infty}\widehat{g}\left(v,t;\widehat{\alpha}\left(t,y_t^*\right),m,T\right)dvdz \\
&= R_{t,T}S_t\int_{K/S_t}^{\infty}P\left[M^{\widehat{Z}}\left(\widehat{\alpha}\left(t,\widehat{y}_t^*\right),T-t\right)>\log(z)/\sigma\Big|\mathcal{F}_t\right]dz \\
&= R_{t,T}S_t\int_{K/S_t}^{\infty}\widehat{P}\left[M^{\widehat{Z}}\left(\widehat{\alpha}\left(t,\log(z)/\sigma\right),T-t\right)>\log(z)/\sigma\right] \\
&\quad\times 1_{\{0<\widehat{\alpha}(t,\log(z)/\sigma)\leq 1\}}dz+R_{t,T}S_t\int_{K/S_t}^{\infty}1_{\{1<\widehat{\alpha}(t,\log(z)/\sigma)\}}dz.
\end{aligned}$$

The second line, in this derivation, is obtained by applying the change of variables $z=e^{\sigma x}, dz=\sigma e^{\sigma x}dx$. The third line uses the definition of the conditional probability. The last line is based on the result in Lemma 106.

Using the density $g(z;\widehat{\alpha},m,T-t)$ of the $\widehat{\alpha}$-quantile of a Brownian motion over $[0,T-t]$ enables us to write

$$\begin{aligned}
&\widehat{P}\left[M^{\widehat{Z}}\left(\widehat{\alpha}\left(t,\log(z)/\sigma\right),T-t\right)>\log(z)/\sigma\right] \\
&= \int_{\log(z)/\sigma}^{\infty}g\left(x;\widehat{\alpha}\left(t,\log(z)/\sigma\right),m,T-t\right)dx
\end{aligned}$$

and

$$\begin{aligned}
C^{\alpha}(S_t,t) &= R_{t,T}S_t\int_{K/S_t}^{\infty}\left(\int_{\log(z)/\sigma}^{\infty}g\left(x;\widehat{\alpha}\left(t,\log(z)/\sigma\right),m,T-t\right)dx\right) \\
&\quad\times 1_{\{0<\widehat{\alpha}(t,\log(z)/\sigma)\leq 1\}}dz+R_{t,T}S_t\int_{K/S_t}^{\infty}1_{\{1<\widehat{\alpha}(t,\log(z)/\sigma)\}}dz.
\end{aligned}$$

The transformation

$$\begin{aligned}
\widehat{\alpha}(t,\log(z)/\sigma) &\equiv \frac{\alpha T-\int_0^t 1_{\{\widehat{Z}_v\leq\log(z)/\sigma\}}dv}{T-t} \\
&= \frac{\alpha T-\int_0^t 1_{\{Z_v\leq\log(z)/\sigma+Z_t\}}dv}{T-t} \\
&= \frac{\alpha T-\int_0^t 1_{\{S_v\leq S_t z\}}dv}{T-t} \\
&= \frac{\alpha T-O_t\left(S,A^{-}\left(S_t z\right)\right)}{T-t}
\end{aligned}$$

leads to the formula in the theorem. ■

Proof of Theorem 98: The arguments following the theorem show that

$$C^{pdi} = R_T^{\gamma} \int_{-\infty}^{+\infty} e^{my} \left(Se^{\sigma y} - K\right)^+ h_l\left(T, y\right) dy$$

where

$$h_l\left(T, y\right) \equiv \int_{-\infty}^{+\infty} E^s \left[1_{\{H_z^-(0,l,D) \leq T\}} n\left(y; z, T - H_z^-\left(0, l, D\right)\right)\right] \phi\left(dz\right).$$

To complete the proof it remains to characterize the function $h_l\left(T, y\right)$. This is accomplished in the next lemma which gives the Laplace transform of $h_l\left(T, y\right)$. ■

Lemma 107 *(Chesney, Jeanblanc-Picque and Yor [1997]) The function $h_l\left(T, y\right)$ has Laplace transform $\widehat{h}_l\left(\lambda, y\right) \equiv \int_0^{\infty} e^{-\lambda t} h_l\left(t, y\right) dt$ given by*

$$\widehat{h}_l\left(\lambda, y\right) = \begin{cases} \frac{e^{l\sqrt{2\lambda}}}{J(\lambda, D)} K_{\lambda,D}\left(y - l\right) & \text{if } l < 0 \\ \frac{1}{J(\lambda, D)} K_{\lambda,D}\left(y - l\right) \int_0^D e^{-\lambda x} \mu_l\left(dx\right) + \frac{e^{-\lambda D}}{2\sqrt{\lambda \pi D}} G_{\lambda,D}\left(l, y\right) & \text{if } l > 0 \end{cases}$$

where $\mu_l\left(dx\right) = \left(l \exp\left(-l^2/2x\right)/\sqrt{2\pi x^3}\right) dx$ is the density of the first hitting time of the level l by a Brownian motion and

$$J\left(\lambda, D\right) = D\sqrt{2\lambda}\Psi\left(\sqrt{2\lambda D}\right)$$

$$\Psi(z) = 1 + z\sqrt{2\pi} \exp\left(\frac{1}{2}z^2\right) N(z)$$

$$K_{\lambda,D}(a) = \int_0^{\infty} z \exp\left(-\frac{z^2}{2D} - |z + a| \sqrt{2\lambda}\right) dz$$

$$G_{\lambda,D}(l, y) = \int_{-\infty}^{l} \left[\exp\left(-\frac{z^2}{2D}\right) - \exp\left(-\frac{(z - 2l)^2}{2D}\right)\right] \exp\left(-|y - z| \sqrt{2\lambda}\right) dz.$$

In the special case $y > l$ the Laplace transform simplifies to

$$\widehat{h}_l\left(\lambda, y\right) = \begin{cases} \frac{\Psi\left(-\sqrt{2\lambda D}\right)}{\Psi\left(\sqrt{2\lambda D}\right)} \frac{e^{(2l-y)\sqrt{2\lambda}}}{\sqrt{2\lambda}} & \text{if } l < 0 \\ \frac{\Psi\left(-\sqrt{2\lambda D}\right)}{\Psi\left(\sqrt{2\lambda D}\right)} \frac{e^{(l-y)\sqrt{2\lambda}}}{\sqrt{2\lambda}} \int_0^D e^{-\lambda x} \mu_l\left(dx\right) + \frac{e^{-\lambda D}}{2\sqrt{\lambda \pi D}} G_{\lambda,D}\left(l, y\right) & \text{if } l > 0. \end{cases}$$

Proof of Lemma 107: See Chesney, Jeanblanc-Picque and Yor [1997], section 5 and Appendix. ■

Remark 7 *A more explicit expression for $K_{\lambda,D}(a)$ can be obtained by performing the integration. This gives*

$$K_{\lambda,D}(a) = \begin{cases} D \exp\left(-a\sqrt{2\lambda}\right) \Psi\left(-\sqrt{2\lambda D}\right) & \text{if } a > 0 \\ \\ D \exp\left(a\sqrt{2\lambda}\right) & \\ +2D\sqrt{\pi \lambda D} e^{\lambda D + a\sqrt{2\lambda}} \left[N\left(-e^+\left(a, \lambda, D\right)\right) \right. & \\ \qquad \left. -N\left(-\sqrt{2\lambda D}\right)\right] & \\ -2D\sqrt{\pi \lambda D} e^{\lambda D - a\sqrt{2\lambda}} N\left(e^-\left(a, \lambda, D\right)\right) & \text{if } a < 0. \end{cases}$$

with $e^{\pm}(a,\lambda,D) \equiv a/\sqrt{D} \pm \sqrt{2\lambda D}$.

The following expression for the function $\Psi_{\nu}^{+}(T,d,D,l)$ in section 7.6.3 is provided by Hugonnier [1999] and Moraux [2002].

Lemma 108 *(Hugonnier [1999], Moraux [2002]) For* $S < L$ *(or* $l > 0$*),*

$$\begin{aligned}\Psi_{\nu}^{+}(T,d,D,l) &= \int_{D}^{T}\int_{d\vee l}^{+\infty} e^{\nu x}\Gamma(l,x-l,s,T-s)\,dxds \\ &\quad + \int_{D}^{T}\int_{d\wedge l}^{l} e^{\nu x}\Gamma(2l-x,0,s,T-s)\,dxds. \qquad (7.33)\end{aligned}$$

For $S > L$ *(or* $l < 0$*) it holds that*

$$\begin{aligned}\Psi_{\nu}^{+}(T,d,D,l) &= \int_{D}^{T}\int_{d\wedge l}^{l} e^{\nu x}\Gamma(l-x,-l,s,T-s)\,dxds \\ &\quad + \int_{D}^{T}\int_{d\vee l}^{+\infty} e^{\nu x}\Gamma(0,x-2l,s,T-s)\,dxds \\ &\quad + \int_{d\vee l}^{+\infty} e^{\nu x}\Lambda_{-l}(T,-x)\,dx \qquad (7.34)\end{aligned}$$

with

$$\begin{aligned}&\int_{d\vee l}^{+\infty} e^{\nu x}\Lambda_{-l}(T,-x)\,dx \\ &= \exp\left(\frac{1}{2}\nu^{2}T\right)N\left(d_{1}\left(d\vee l,\nu,T\right)\right) \\ &\quad - \exp\left(\frac{1}{2}\nu^{2}T+2\nu l\right)N\left(d_{2}\left(d\vee l,l,\nu,T\right)\right) \qquad (7.35)\end{aligned}$$

and

$$d_{1}(d\vee l,\nu,T) = -\frac{d\vee l-\nu T}{\sqrt{T}} \qquad (7.36)$$

$$d_{2}(d\vee l,l,\nu,T) = -\frac{(d\vee l-2l-\nu T)}{\sqrt{T}}. \qquad (7.37)$$

Similar formulas can be established for $\Psi_{\nu}^{-}(T,d,D,l)$ *using* $k_{l}^{-}(T,ds,dx) = k_{l}^{+}(T,T-ds,dx)$.

Proof of Lemma 108: Assume $0 < D < T$ and recall that

$$\begin{aligned}\Psi_{\nu}^{+}(T,d,D,l) &= \int_{D}^{T}\int_{d\vee l}^{+\infty} e^{\nu x}\Gamma(l,x-l,s,T-s)\,dxds \\ &\quad + \int_{D}^{T}\int_{d\wedge l}^{l} e^{\nu x}\Gamma(2l-x,0,s,T-s)\,dxds \\ &\quad - \int_{D}^{T}\int_{d\wedge l}^{l} e^{\nu x}\delta_{0}(ds)\Lambda_{l}(t,x)\,dx.\end{aligned}$$

If $S < L$ (or $l > 0$) use

$$\int_D^T \int_{d \wedge l}^l e^{\nu x} \delta_0 (ds) \Lambda_l (T, x) \, dx = 0$$

(recall that $D > 0$) to show that (7.33) holds.

If $S > L$ (or $l < 0$) use $k_l^+ (T, ds, dx) = k_{|l|}^+ (T, T - ds, -dx)$ where

$$\begin{aligned} k_{|l|}^+ (T, T - ds, -dx) &= 1_{\{-x \geq -l\}} \Gamma (-l, -x + l, T - s, s) \, dsdx \\ &\quad + 1_{\{-x < -l\}} \Gamma (-2l + x, 0, T - s, s) \, dsdx \\ &\quad + 1_{\{-x < -l\}} \delta_0 (T - ds) \Lambda_{-l} (T, -x) \, dx \end{aligned}$$

and the symmetry relation $\Gamma (a, b, u, v) = \Gamma (b, a, v, u)$ to write

$$\Gamma (-l, -x + l, T - s, s) = \Gamma (l - x, -l, s, T - s)$$

$$\Gamma (-2l + x, 0, T - s, s) = \Gamma (0, x - 2l, s, T - s)$$

$$\begin{aligned} k_{|l|}^+ (T, T - ds, -dx) &= 1_{\{x \leq l\}} \Gamma (l - x, -l, s, T - s) \, dsdx \\ &\quad + 1_{\{x > l\}} \Gamma (0, x - 2l, s, T - s) \, dsdx \\ &\quad + 1_{\{x > l\}} \delta_0 (T - ds) \Lambda_{-l} (T, -x) \, dx \end{aligned}$$

and

$$\begin{aligned} \Psi_\nu^+ (T, d, D, l) &= \int_d^{+\infty} \int_D^T e^{\nu x} k_l^+ (T, ds, dx) \, dsdx \\ &= \int_D^T \int_{d \wedge l}^l e^{\nu x} \Gamma (l - x, -l, s, T - s) \, dxds \\ &\quad + \int_D^T \int_{d \vee l}^{+\infty} e^{\nu x} \Gamma (0, x - 2l, s, T - s) \, dxds \\ &\quad + \int_D^T \int_{d \vee l}^{+\infty} e^{\nu x} \delta_0 (T - ds) \Lambda_{-l} (T, -x) \, dx. \end{aligned}$$

To complete the proof of (7.34)-(7.37) note that $\Lambda_{-l} (T, -x) = \Lambda_l (T, x)$ and

$$\begin{aligned} \int_D^T \int_{d \vee l}^{+\infty} e^{\nu x} \delta_0 (T - ds) \Lambda_{-l} (T, -x) \, dx &= \int_{d \vee l}^{+\infty} e^{\nu x} \Lambda_{-l} (T, -x) \, dx \\ &= \int_{d \vee l}^{+\infty} e^{\nu x} \Lambda_l (T, x) \, dx. \end{aligned}$$

Following Moraux [2002] one can then show that

$$\begin{aligned}
&\int_{d\vee l}^{+\infty} e^{\nu x}\Lambda_l(T,x)\,dx \\
&= \int_{d\vee l}^{+\infty} e^{\nu x}\frac{1}{\sqrt{2\pi T}}\left[\exp\left(-\frac{x^2}{2T}\right)-\exp\left(-\frac{(2l-x)^2}{2T}\right)\right]dx \\
&= \exp\left(\frac{1}{2}\nu^2 T\right)\int_{(d\vee l-\nu T)/\sqrt{T}}^{+\infty}\frac{1}{\sqrt{2\pi}}\exp\left(-\frac{x^2}{2}\right)dx \\
&\quad -\exp\left(\frac{1}{2}\nu^2 T+2\nu l\right)\int_{(d\vee l-2l-\nu T)/\sqrt{T}}^{+\infty}\frac{1}{\sqrt{2\pi}}\exp\left(-\frac{x^2}{2}\right)dx \\
&= \exp\left(\frac{1}{2}\nu^2 T\right)N\left(d_1\left(d\vee l,\nu,T\right)\right) \\
&\quad -\exp\left(\frac{1}{2}\nu^2 T+2\nu l\right)N\left(d_2\left(d\vee l,l,\nu,T\right)\right)
\end{aligned}$$

with $d_1(d\vee l,\nu,T)$ and $d_2(d\vee l,l,\nu,T)$ as defined in (7.36)-(7.37). ■

Remark 8 *The following explicit formula for the function* Γ *in (7.19) and Lemma 108 is provided by Hugonnier [1999],*

$$\begin{aligned}
\Gamma(a,b,u,v) &= \frac{1}{\pi}\left[\frac{av+bu}{(u+v)^2\sqrt{uv}}\right]\exp\left(-\frac{a^2}{2v}-\frac{b^2}{2u}\right) \\
&\quad +\sqrt{\frac{2}{\pi}}\left(\frac{1}{u+v}\right)^{3/2}\left[1-\frac{(b-a)^2}{u+v}\right] \\
&\quad \times\exp\left(-\frac{(b-a)^2}{2(u+v)}\right)N\left(-\frac{au+bv}{\sqrt{uv(u+v)}}\right).
\end{aligned}$$

This expression, obtained by collecting terms and performing the integration in (7.19), proves useful for the implementation of the valuation formulas for cumulative barrier claims.

Proof of Theorem 100: The EEP representation in (7.24) follows immediately from Theorem 21. The appreciation component (7.25) of the payoff $Y = F(S,O)$ is obtained by an application of Ito's lemma. This derivation uses the evolution equation for the occupation time, $dO_t = 1_{\{S_t\in A_t\}}dt$, $t\in[0,T]$. ■

Proof of Proposition 101: Properties (i), (iii) and (iv) are natural extensions of similar properties for vanilla options, and are proved along the same lines. For property (ii) note that

$$C^{cdo}(\lambda_1 S,\lambda_2 O,t)\le C^{cdo}(\lambda_1 S,O,t)\le C^{cdo}(S,O,t)+(\lambda_1-1)S.$$

A standard argument can then be invoked to prove the claim. Property (v) follows from the bounds

$$C^{do}(S,t;K,T,L) \leq C^{cdo}(S,O,t;K,T,D,L) \leq C(S,t;K,T)$$

where the contractual characteristics associated with the options involved have been specified. Assume that the down-and-out options are alive. It is then clear that the optimality of immediate exercise of the vanilla call implies the optimality of immediate exercise of the two knock-out options (otherwise immediate exercise cannot be optimal for the vanilla option). Likewise, optimality of immediate exercise for the cumulative knock-out call implies optimality of immediate exercise for the standard knock-out call. The ordering of exercise regions follows. ■

Proof of Theorem 102: For a cumulative down-and-out call the EEP formula in Theorem 100 becomes

$$C^{cdo}(S,O,t) = c^{cdo}(S,O,t) + \pi^{cdo}(S,O,t,B(\cdot))$$

with

$$c^{cdo}(S,O,t) = \widetilde{E}_t\left[R_{t,T}1_{\{O_T(S,A^-(L))<D\}}(S_T-K)^+\right]$$

$$\pi^{cdo}(S,O,t,B(\cdot)) = \widetilde{E}_t\left[\int_{\tau_0}^{\tau_e\wedge T} R_{t,s}1_{\{\tau_s=s\}}(\delta S_s - rK)\,ds\right]$$

and $\tau_e = \inf\{v \in [0,T] : O_v \geq D\}$.

The European option value can be simplified using the additivity of the occupation time and the change of measure

$$dQ^s = \exp\left(-mW_T - \frac{1}{2}m^2T\right)dQ = \exp\left(-mZ_T + \frac{1}{2}m^2T\right)dQ$$

described in section 7.5. This gives

$$\begin{aligned}
& c^{cdo}(S,O,t) \\
&= E_t^s\left[R_{t,T}^{\gamma}1_{\{O+O_{t,T}<D\}}e^{m(Z_T-Z_t)}\left(Se^{\sigma(Z_T-Z_t)}-K\right)^+\right] \\
&= R_{t,T}^{\gamma}\int_{-\infty}^{+\infty}\int_0^T 1_{\{v<D-O\}}e^{mx}(Se^{\sigma x}-K)^+ k_l^-(T-t,dv,dx)\,dvdx \\
&= R_{t,T}^{\gamma}\int_{-\infty}^{+\infty}\int_0^{D-O} e^{mx}(Se^{\sigma x}-K)^+ k_l^-(T-t,dv,dx)\,dvdx \\
&= R_{t,T}^{\gamma}\int_{d_o(S,K)}^{+\infty}\int_0^{D-O} e^{mx}(Se^{\sigma x}-K)\, k_l^-(T-t,dv,dx)\,dvdx.
\end{aligned}$$

In the equality on the first line $O_{t,T}$ is the occupation time of the set $A^-(L)$ during the period $[t,T]$ and the conditional expectation $E_t^s[\cdot]$ is over the joint distribution of $Z_T - Z_t$ and $O_{t,T}$. The equality on the second line integrates over the bivariate law of $Z_T - Z_t$ and $O_{t,T}$. The next expression is obtained by restricting the second integral to the domain over which the indicator function is one. The final formula restricts the first integral to the domain over which the option payoff is positive.

Similar transformations can be performed to simplify the exercise premium component. Using the equality of events $\{\tau_s = s\} = \{S_s \geq B(s, O_s)\}$ where $B(s, O_s)$ is the exercise boundary at date s and simplifying the resulting expression gives

$$\begin{aligned}
&\pi^{cdo}(S, O, t, B(\cdot)) \\
&= \widetilde{E}_t\left[\int_{\tau_0}^{\tau_e \wedge T} R_{t,s} 1_{\{S_s \geq B(s,O_s)\}} (\delta S_s - rK)\, ds\right] \\
&= E_t^s\left[\int_t^T R_{t,s}^{\gamma} 1_{\{O+O_{t,s}<D\}} 1_{\{Se^{\sigma(Z_s - Z_t)} \geq B(s, O+O_{t,s})\}}\right. \\
&\quad \left. \times e^{m(Z_s - Z_t)}\left(\delta S e^{\sigma(Z_s - Z_t)} - rK\right) ds\right] \\
&\equiv \int_t^T R_{t,s}^{\gamma} \phi(S, O, s-t)\, ds
\end{aligned}$$

where

$$\begin{aligned}
&\phi(S, O, s-t) \\
&\equiv E_t^s\left[1_{\{O+O_{t,s}<D\}} 1_{\{Se^{\sigma(Z_s - Z_t)} \geq B(s, O+O_{t,s})\}} e^{m(Z_s - Z_t)}\left(\delta S e^{\sigma(Z_s - Z_t)} - rK\right)\right] \\
&= \int_{-\infty}^{+\infty}\int_0^T 1_{\{O+v<D\}} 1_{\{Se^{\sigma x} \geq B(s,O+v)\}} e^{mx}(\delta S e^{\sigma x} - rK) \\
&\quad \times k_l^-(s-t, dv, dx)\, dv dx \\
&= \int_0^{D-O}\int_{d_o(S, B(s,O+v))}^{+\infty} e^{mx}(\delta S e^{\sigma x} - rK)\, k_l^-(s-t, dv, dx)\, dx dv.
\end{aligned}$$

The steps in this derivation are, by now, standard. The only noteworthy aspect is the fact that the outside integral in the final expression for $\phi(S, O, s-t)$ is over the occupation time v during the period $[t, s]$, whereas the inside integral is over values of x for which immediate exercise at s is optimal. This order of integration is dictated by the dependence of the exercise boundary $B(s, O+v)$ on the occupation time v.

The recursive equation for the exercise boundary follows immediately from the EEP representation formula. The boundary conditions are discussed in the text preceding the theorem. ■

Proof of Theorem 103: Fix a time t and suppose that asset j is selected as the numeraire. By the homogeneity property the payoff function can be rescaled and rewritten as

$$
\begin{aligned}
F(S_\tau, K, O_\tau^{S,A}) &= \left(S_\tau^j/S_t^j\right) F\left(\frac{S_\tau}{\left(S_\tau^j/S_t^j\right)}, \frac{K}{\left(S_\tau^j/S_t^j\right)}, O_\tau^{S,A}\right) \\
&\equiv \left(S_\tau^j/S_t^j\right) F\left(\widehat{S}_\tau, S_\tau^{j,*}, O_\tau^{S,A}\right)
\end{aligned}
$$

for any stopping time $\tau \in \mathcal{S}_{t,T}$, where

$$
S_\tau^{j,*} \equiv \frac{K}{\left(S_\tau^j/S_t^j\right)} = \frac{KS_t^j}{S_\tau^j}
$$

$$
\begin{aligned}
\widehat{S}_\tau &= \frac{1}{\left(S_\tau^j/S_t^j\right)} \left(S_\tau^1, ..., S_\tau^{j-1}, S_\tau^j, S_\tau^{j+1}, ..., S_\tau^n\right) \\
&= \left(S_\tau^{1,*}, ..., S_\tau^{j-1,*}, S_t^j, S_\tau^{j+1,*}, ..., S_\tau^{n,*}\right)
\end{aligned}
$$

and where, for $i \neq j$, $S_\tau^{i,*} \equiv \left(S_\tau^i/S_\tau^j\right) S_t^j$. With the definitions

$$
F^j(S, K, O^{S,A}) \equiv F\left(\lambda^j \circ_j S, S^j, O^{S,A}\right)
$$

$$
\lambda \circ_j S = (S^1, ..., S^{j-1}, \lambda S^j, S^{j+1}, ..., S^n) \text{ and } \lambda^j = K/S^j
$$

(i.e., F^j is the same as F but where arguments j and $n+1$ have been exchanged) it follows that

$$
F(S_\tau, K, O_\tau^{S,A}) = \left(S_\tau^j/S_t^j\right) F\left(\widehat{S}_\tau, S_\tau^{j,*}, O_\tau^{S,A}\right) = \left(S_\tau^j/S_t^j\right) F^j\left(S_\tau^*, S_t^j, O_\tau^{S,A}\right).
$$

The same rescaling can be used to rewrite the occupation time during the period $[t, T]$, as

$$
\begin{aligned}
O_{t,\tau}^{S,A} &= \int_t^\tau 1_{\{S_v \in A_v\}} dv \\
&= \int_t^\tau 1_{\{S_v^* \in A_v^*\}} dv = O_{t,\tau}^{S^*,A^*}, \quad \text{for } \tau \in [t, T]
\end{aligned}
$$

where $A^* = \{A^*(t, v, \omega), v \in [t, T]\}$ with

$$
\begin{aligned}
A^*(t, v, \omega) &= \Big\{x \in \mathbb{R}_+^n : x_i = y_i S_t^j / y_j, \text{ for } i \neq j, x_j = KS_t^j/y_j \\
&\qquad \text{and } y = (y_1, ..., y_n) \in A(v, \omega)\Big\}.
\end{aligned}
$$

Assuming that occupation times have the same starting value at t, $O_t^{S^*,A^*} \equiv O_t^{S,A}$, one has $O_\tau^{S,A} = O_\tau^{S^*,A^*}$, for $\tau \in [t, T]$, and $F(S_\tau, K, O_\tau^{S,A}) = \left(S_\tau^j/S_t^j\right) \times F^j\left(S_\tau^*, S_t^j, O_\tau^{S^*,A^*}\right)$.

With this notation the price of the American multiasset OTD is

$$\begin{aligned}
V(S_t, K, O^{S,A}, r, \delta; \mathcal{F}_t) &= \sup_{\tau \in \mathcal{S}_{t,T}} \widetilde{E}_t \left[R_{t,\tau} F(S_\tau, K, O^{S,A}_\tau) \right] \\
&= \sup_{\tau \in \mathcal{S}_{t,T}} \widetilde{E}_t \left[R_{t,\tau} \left(S^j_\tau / S^j_t \right) F^j \left(S^*_\tau, S^j_t, O^{S^*,A^*}_\tau \right) \right] \\
&= \sup_{\tau \in \mathcal{S}_{t,T}} E^j_t \left[R^\delta_{t,\tau} F^j \left(S^*_\tau, S^j_t, O^{S^*,A^*}_\tau \right) \right]
\end{aligned}$$

where $R^\delta_{t,\tau} = \exp\left(-\int_t^\tau \delta_v dv \right)$ is the discount factor based on the rate δ and $E^j_t [\cdot]$ is the conditional expectation at t based on the measure

$$dQ^j = \left(S^j_T / S^j_0 \right) \exp\left(-\int_0^T (r_v - \delta_v)\, dv \right) dQ.$$

The last expression is the value at time t of an American OTD with payoff function F^j, written on the price vector S^*, with fixed parameter S^j_t and occupation time O^{S^*,A^*}, in an economy with interest rate equal to δ. It can be verified that the evolution of S^* in the modified economy is described by the equations reported in the theorem. Asset j, therefore, has dividend yield r while asset i's dividend yield remains the same. The vector of dividend rates can be written as $\lambda^j(\delta) \circ_j \delta$ with $\lambda^j(\delta) \equiv r/\delta^j$, as stated. ■

Chapter 8

Numerical Methods

This chapter presents numerical methods developed to compute the optimal exercise boundary and the value of an American option. Particular emphasis is placed on recent approaches. Our presentation starts with a brief summary of popular methods proposed during the last three decades (section 8.1). Approaches based on the integral equation for the boundary (section 8.2) and on simple approximations of the optimal exercise policy (section 8.3) are reviewed next. Both of these sections assume that the underlying asset price is a geometric Brownian motion process and that the interest rate is constant. Extensions to more general diffusions (section 8.4) and a summary of additional recent methodologies follow (section 8.5). A numerical study is carried out to compare the efficiencies of some of the approaches described (section 8.6). A presentation of methods for multiasset options (section 8.7) and for occupation time derivatives (section 8.8) concludes the chapter.

8.1 Numerical Methods for American Options

As shown in the previous chapters there is no known simple analytic formula for pricing an American option. As a result one must eventually resort to a numerical method to compute the value of a given contract. A number of approaches have been proposed for that purpose. Among the major ones we find computational methods based on lattices, partial differential equations, variational inequalities, integral equations, stopping time approximations and Monte Carlo simulation. In this section we briefly survey the first three approaches and the related literature. Other methods will be presented in more details in the following sections.

Lattice methods arc among the oldest approaches pursued and perhaps the simplest ones. They rely on the fact that a geometric Brownian motion process can be approximated by a binomial lattice and that the resulting option price constitutes an approximation of the true price. The advantage of a lattice-based approach is that it reduces an infinite dimensional problem to a finite dimensional one. This is accomplished by discretizing the time interval into a finite number of subintervals, n, and assuming that the underlying asset price has only a finite number of outcomes at each node of the tree. The most common lattice approach, the binomial method, postulates an asset price with two possible outcomes at each node (Su, Sd) where u (resp. d) is the gross return if the price goes up (resp. down). Thus, each node of the tree has two follower nodes. If u, d are constants the tree is said to be recombining. In this case the total number of nodes at step j is $j+1$, $j = 1, ..., n$. The American option price can then be easily computed in a backward manner starting from the last step, n (the maturity date). At that point the option value is the exercise value. So, for a call option $C(n,i) = (Su^i d^{n-i} - K)^+$, for $i = 0, ..., n$. At step $n-1$ there are n nodes, and therefore n option values. These are computed using risk neutral valuation. Thus,

$$C(n-1,i) = \max\{V(n-1,i), (Su^i d^{n-1-i} - K)^+\}$$

for $i = 0, ..., n-1$, where $V(n-1,i) = e^{-r(T/n)}(qC(n,i+1) + (1-q)C(n,i))$ is the continuation value, $(Su^i d^{n-1-i} - K)^+$ is the immediate exercise value and q the risk neutral probability. At step j there are $j+1$ prices to compute, that are obtained using the same formula (replacing $n-1$ by j). The option price at the initial date (with one node) is the unique value obtained at that point. The optimal exercise policy is the first time at which immediate exercise is optimal $\tau = \inf\{j \in \{0, ..., n\} : C(j,i) = (Su^i d^{j-i} - K)^+\}$. The optimal exercise boundary is $B(j) = Su^{i^*} d^{j-i^*}$ where $i^* = \inf\{i : C(j,i) = (Su^i d^{j-i} - K)^+\}$.

This simple approach was initiated by Cox, Ross and Rubinstein [1979]. A variant of the method uses a trinomial lattice, with three follower nodes issued from each node of the tree, to compute prices. This modification was proposed by Parkinson [1977] and Boyle [1988] and is analyzed by Omberg [1988]. Additional refinements of the binomial approach and improvements in its implementation can be found in Rendleman and Bartter [1979], Jarrow and Rudd [1983], Hull and White [1988], Amin [1991], Trigeorgis [1991], Tian [1993], Kim and Byun [1994], Leisen and Reimer [1995] and Curran [1995]. A proof of convergence for lattice methods appears in Amin and Khanna [1994]. The convergence of the optimal exercise boundary is proved by Lamberton [1993]. Convergence rates for European-style options are derived in Leisen and Reimer [1995]. For those contracts Diener and Diener [2004] carry out a detailed asymptotic analysis of the behavior of the binomial approximation. Numerical estimates of convergence rates for American-style option prices appear in Broadie and Detemple [1996]. Analytical bounds for convergence rates of American option prices in binomial lattices are derived by Lamberton [1998].[1]

Partial differential equations methods rely on the fundamental valuation equation derived by Black and Scholes [1993] and Merton [1993]. This equation

[1]The price of an American contingent claim can also be estimated by using a random walk approximation of the underlying Brownian motion. Error estimates for these types of approximations are provided by Lamberton [2002].

states that an American call option price $C(S,t)$ satisfies

$$C_t(S,t) + C_s(S,t)S(r-\delta) + \frac{1}{2}C_{ss}(S,t)S^2\sigma^2 - rC(S,t) = 0$$

subject to the boundary conditions $\lim_{t\to T} C(S,t) = (S-K)^+$, $\lim_{S\to 0}C(S,t) = 0$, $C(B_t,t) = (B_t - K)^+$ and $\lim_{S\to B_t} C_s(S,t) = 1$. Here C_t, C_s and C_{ss} are the partial derivatives with respect to time t and the price S. Finite difference methods to solve this equation were introduced by Schwartz [1977] and Brennan and Schwartz [1977, 1978]. Convergence of the Brennan and Schwartz algorithm is proved in Jaillet, Lamberton and Lapeyre [1990] and Zhang [1995]. Related numerical methods can be found in Courtadon [1982] and Hull and White [1990]. Some authors have also proposed approximations of the fundamental valuation PDE that admit closed form solutions. The quadratic method of MacMillan [1986] and Barone-Adesi and Whaley [1987], as well as the method of lines of Carr and Faguet [1995], fall in that category.

An approximation that echoes some features of the binomial lattice construction is the one proposed by Geske and Johnson [1984]. This approach discretizes the time interval into n subintervals and evaluates the American option price as a multidimensional integral involving the relevant payoffs at each discretization point. In effect the method uses an approximation of the true price corresponding to an exercise policy restricted to the n discretization points selected. Implementation of the formula requires numerical approximations of the multidimensional integrals and of the exercise boundary. Geske and Johnson [1984] also introduce Richardson extrapolation in the context of option pricing. This extrapolation technique has proved popular and is also used by Breen [1991], Bunch and Johnson [1992], Ho, Stapleton and Subrahmanyam [1994], Huang, Subrahmanyam and Yu [1996] and Carr and Faguet [1995]. For a full treatment of Richardson extrapolation see Marchuk and Shaidurov [1983].

The variational inequality approach to option pricing was introduced by Jaillet, Lamberton and Lapeyre [1990]. A discretization of the relevant variational inequality leads to a linear complementarity problem that can be solved by linear programming methods (see Cottle, Pang and Stone [1992]). An overview of PDE and variational inequality approaches is provided in the textbook of Wilmott, Dewyne and Howison [1993]. A short survey can be found in Broadie and Detemple [2004].

8.2 Integral Equation Methods

Integral equation methods are based on the EEP representation formula for the option price and the associated integral equation for the exercise boundary. The numerical implementation procedure suggested by this formula involves two steps. The first step consists in computing the optimal exercise boundary B, the second step in computing the option price taking the curve B as an input.

Implementation of this procedure starts with the equation for the optimal exercise boundary. Let us focus on the case of a call option with strike K and maturity date T. For the standard market model we know, from chapter 4, that the optimal exercise boundary B of this option solves the equation

$$B_t - K = c(B_t, t) + \int_t^T \phi(B_t, B_s, s-t)ds \tag{8.1}$$

where $c(B_t, t)$ is the European option value evaluated at $S_t = B_t$ and

$$\begin{aligned} \phi(S_t, B_s, s-t) &= \delta S_t e^{-\delta(s-t)} N\left(d\left(S_t, B_s, s-t\right)\right) \\ &\quad - rKe^{-r(s-t)} N\left(d_1\left(S_t, B_s, s-t\right)\right) \end{aligned}$$

with

$$\begin{aligned} d\left(S_t, B_s, s-t\right) &= \frac{\log(S_t/B_s) + \left(r - \delta + \frac{1}{2}\sigma^2\right)(s-t)}{\sigma\sqrt{s-t}} \\ d_1\left(S_t, B_s, s-t\right) &= d\left(S_t, B_s, s-t\right) - \sigma\sqrt{s-t}, \end{aligned}$$

subject to the terminal condition $B_{T_-} = K \max\{1, r/\delta\}$. This is a backward equation, which naturally suggests a recursive algorithm to compute its solution.

The algorithm works as follows. Divide the period $[0, T]$ into n equal subintervals and let $\Delta t = T/n$. For $j = 0, ..., n$ set $t(j) = j\Delta t$. We seek a step function approximation $\left\{B^{(n)}(j) \equiv B_{t(j)}^{(n)},\ j = 0, ..., n\right\}$ that can be computed recursively. Step function approximations have been suggested by Huang, Subrahmanyam and Yu [1996].

From the terminal condition we have $B^{(n)}(n) = K \max(1, r/\delta)$. Suppose now that $B^{(n)}(l)$ is known for all $l > j$. A non-linear equation for $B^{(n)}(j)$, parametrized by $\left\{B^{(n)}(l),\ l = j+1, ..., n\right\}$, is obtained by discretizing the integral on the right hand side of (8.1). This can be done in several ways by using different quadrature approximations of the integral. If we use the trapezoidal rule (see Press, Teukolsky, Vetterling and Flannery [1992]) we obtain the equation

$$\begin{aligned} & B^{(n)}(j) - K \\ = \;& c\left(B^{(n)}(j), t(j)\right) \\ & + \left[\phi\left(B^{(n)}(j), B^{(n)}(n), t(n-j)\right) + \phi\left(B^{(n)}(j), B^{(n)}(j), t(0)\right)\right] \frac{\Delta t}{2} \\ & + \sum_{q=j+1}^{n-1} \phi\left(B^{(n)}(j), B^{(n)}(q), t(q-j)\right) \Delta t \end{aligned} \tag{8.2}$$

with unknown $B^{(n)}(j)$. Its solution $B^{(n)}(j)$ can be computed using an iterative method. This produces an approximation of the boundary at time $t(j)$. Repeating this computation recursively from $j = n-1$ to 0 yields a step function approximation of the exercise boundary, $B^{(n)} = \left\{B^{(n)}(j),\ j = 0, ..., n\right\}$.

Several iterative methods are available to compute the solution of (8.2) at each step j. Kallast and Kivinukk [2003] have recently proposed to use the

Newton-Raphson procedure and shown that it has good efficiency properties.[2] Let $F\left(B^{(n)}(j)\right)$ denote the difference between the left and the right hand sides of (8.2). The Newton-Raphson scheme produces a sequence of approximations $\left\{B^{(n,k)}(j) : k = 0, 1, 2...\right\}$ at step j according to the map

$$B^{(n,k+1)}(j) = B^{(n,k)}(j) - \frac{F\left(B^{(n,k)}(j)\right)}{F'\left(B^{(n,k)}(j)\right)}$$

where $F'(x)$ is the derivative of $F(x)$. The initial value of the sequence can be selected as $B^{(n,0)}(j) = B^{(n)}(j+1)$. The scheme is run until the difference between successive iterates falls below some selected tolerance level ϵ. When $F\left(B^{(n,k)}(j)\right) > 0$ the next iterate decreases (i.e., $B^{(n,k+1)}(j) < B^{(n,k)}(j)$) because $F'\left(B^{(n,k)}(j)\right) > 0$. When $F\left(B^{(n,k)}(j)\right) < 0$ it increases. In numerical experiments this scheme converges, and does so quickly.

The algorithm described above produces an approximate exercise boundary, hence an approximate option value, for any fixed number of time steps n. The true price is the limit as n goes to infinity. As computations are based on finite discretizations of the time interval one may want to improve efficiency for a given selection n. This can be achieved by using the standard Richardson extrapolation scheme that gives the approximation $\widehat{C}(S,t) = 2C^{(n)}(S,t) - C^{(n-1)}(S,t)$ where $C^{(n)}(S,t)$ is the approximate price with n discretization points. Alternative extrapolation schemes, such as the one proposed by Geske and Johnson [1984], stating that $\widehat{C}(S,t) = 4.5C^{(n)}(S,t) - 4C^{(n-1)}(S,t) + 0.5C^{(n-2)}(S,t)$, can also be employed to that effect.

8.3 Exercise Time Approximations: LBA-LUBA

The approximations presented in the previous section are constructed directly from the recursive equation characterizing the optimal exercise boundary. An alternative approach is to restrict attention to a class of exercise policies that are easy to evaluate and to choose the best one within this class. This is the key idea behind the procedure proposed by Broadie and Detemple [1996]. Their approach produces a lower bound for the exercise boundary, lower and upper bounds for the option price and approximate option values that are called *lower bound approximation* (LBA) and *lower-upper bound approximation* (LUBA).

8.3.1 A Lower Bound for the Option Price

A particularly simple exercise policy consists in exercising at the first hitting time of a prespecified constant level L, or at maturity if this event does not materialize before the expiration date of the contract. This policy is represented by the stopping time

$$\tau_L = \inf\{v \in [t,T] : S_v = L\} \text{ or } \tau_L = T \text{ if no such time exists in } [t,T].$$

[2] Ju [1998] uses a two-dimensional Newton-Raphson scheme to solve for the free parameters in an exponential approximation of the boundary (see section 8.5.2).

This policy is certainly feasible for the holder of an American option that allows for unrestricted exercise time. Thus, our class of simple stopping times is $\mathcal{S}_{t,T}^{c} = \{\tau_L : L \geq 0\}$. Implementation of a stopping time $\tau_L \in \mathcal{S}_{t,T}^{c}$ gives the payoff of a capped option with automatic exercise at the cap L. Suppose that we hold an American call with strike K and maturity date T. In this case the exercise policy τ_L has a value equal to $C(S,t;L)$, the price of a capped call option with strike K, maturity date T and automatic exercise at the cap L. As τ_L is feasible for the American call the lower bound $C(S,t) \geq C(S,t,L)$ holds, and this for any $L \geq 0$. Given that the capped option is automatically exercised at $L = S$ we can restrict attention to constants $L \geq S$. Let $\mathcal{S}_{t,T}^{c}(S) = \{\tau_L : L \geq S\}$ be the corresponding class of stopping times. Maximization with respect to these exercise policies produces the highest lower bound[3]

$$C(S,t) \geq \max_{L \geq S} C(S,t;L) = \sup_{\tau_L \in \mathcal{S}_{t,T}^{c}(S)} \widetilde{E}_t \left[R_{t,\tau_L} (S_{\tau_L} - K)^+ \right] \equiv C^l(S,t).$$

As the price $C(S,t;L)$ is available in closed form this optimization can be performed using standard iterative methods (such as Newton-Raphson).

8.3.2 A Lower Bound for the Exercise Boundary

The lower bound for the American option price derived above can be used to construct a lower bound for the optimal exercise boundary. To derive this bound define

$$D(S,t) \equiv \left. \frac{\partial C(S,t;L)}{\partial L} \right|_{L=S} \tag{8.3}$$

which represents the derivative of the capped call with respect to the cap, evaluated at the point where the cap equals the asset price, i.e., $L = S$. When $D(S,t) > 0$ at a point S immediate exercise is suboptimal. Let $L_t^* \equiv \inf\{S : D(S,t) \leq 0\}$. This quantity is the smallest price at which immediate exercise is optimal when the option holder is restricted to policies of the form τ_L. Solving this optimization problem for $t \in [0,T)$ produces a curve $L^* = \{L_t^* : t \in [0,T)\}$. Basic intuition might suggest that L^* is a lower bound for the optimal exercise boundary B. This is indeed the case.

Proposition 109 *Suppose that the underlying asset price satisfies (2.2). Let B be the optimal exercise boundary for the American call option, and L^* the curve solving $L_t^* \equiv \inf\{S : D(S,t) \leq 0\}$, for all $t \in [0,T)$. Then $(r/\delta)K \vee K \leq L_t^* \leq B_t$, for all $t \in [0,T)$. Moreover, $\lim_{t \to T} L_t^* = K \max\{1, r/\delta\}$ and $\lim_{t \to -\infty} L_t^* = K \frac{b+f}{b+f-\sigma^2}$ where $b = \delta - r + \frac{1}{2}\sigma^2$ and $f = \sqrt{b^2 + 2r\sigma^2}$.*

The intuition underlying the lower bound is simple. Given that barrier policies τ_L are always feasible for the American call option it must be that immediate exercise is optimal for the capped option whenever it is optimal for the uncapped option. This implies that the derivative $D(S,t) \leq 0$ when

[3] Similar ideas can also be found in Bjerksund and Stensland [1992] and Omberg [1987].

$(S, t) \in \mathcal{E}$. Taking the lowest price at which this slope is non-positive (i.e., $L_t^* = \inf\{S : D(S,t) \leq 0\}$) will then give a lower bound for the exercise boundary B_t (i.e., $L_t^* \leq B_t$).

The lower bound L^* obtained via this procedure is very tight. As stated in the proposition, it coincides with the immediate exercise boundary B at the maturity date T_- and at very long maturities $T - t \to \infty$. As shown in Figure 8.1 the difference between the two curves is small at intermediate times.

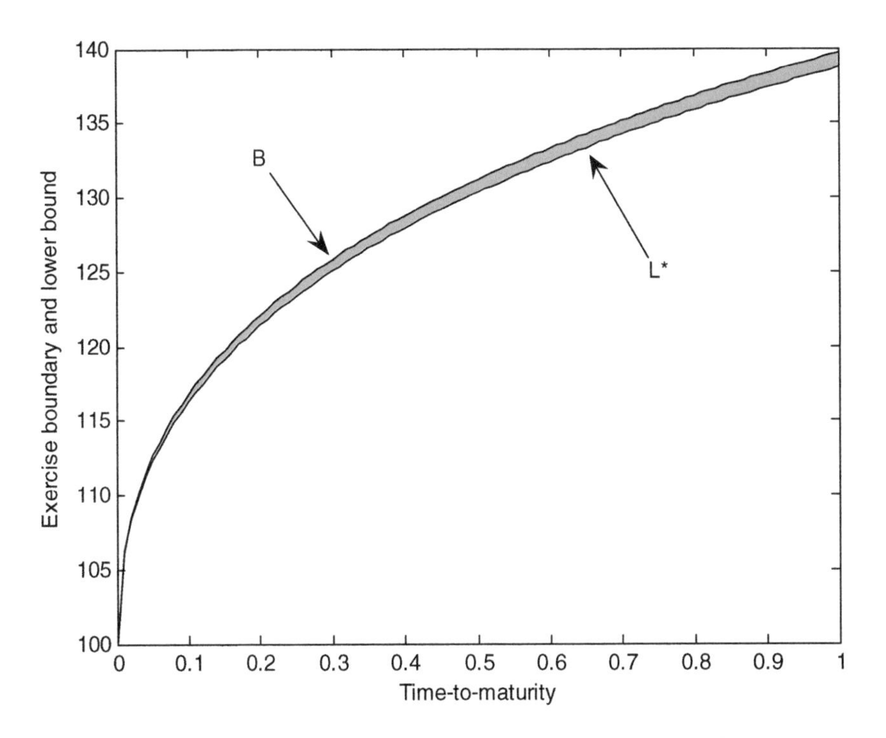

Figure 8.1: This figure illustrates the difference between the immediate exercise boundary B and its lower bound approximation L^*. The boundary B is computed using the integral equation. Parameter values are $K = 100$, $r = \delta = 0.06$, $\sigma = 0.2$. Both curves are calculated using 100 time steps.

8.3.3 An Upper Bound for the Option Price

In turn, the lower bound for the boundary can be used to produce an upper bound for the American option price. Substituting the lower bound L_t^* in place of B_t in the EEP formula gives the function

$$C^u(S,t) \equiv c(S,t) + \int_t^T \phi(S, L_s^*, s-t)\, ds.$$

This function depends parametrically on the curve L^*. It is easy to compute once L^* is known as it only depends on the cumulative normal distributions appearing in $c(S,t)$ and $\phi(S, L_s^*, s-t)$. Our next proposition shows that it is an upper bound for the price.

Proposition 110 *The American call option price is bounded above by the function $C^u(S,t)$. Hence $C^l(S,t) \leq C(S,t) \leq C^u(S,t)$ for all $(S,t) \in \mathbb{R}_+ \times [0,T]$.*

Intuition for the upper bound can also be gathered by recalling that the early exercise premium is given by the present value formula

$$\pi(S_t, t, B(\cdot)) = \widetilde{E}_t\left[\int_t^T e^{-r(s-t)}(\delta S_s - rK)\, 1_{\{S_s \geq B_s\}} ds\right].$$

Given that $B_s \geq L_s^* \geq (r/\delta)K$ (from Proposition 109) we see that the net benefits of exercising, $\delta S_s - rK$, are positive over the event $\{S_s \geq L_s^*\}$, and this for all $s \in [0,T]$. Substituting this event in place of $\{S_s \geq B_s\}$ in the premium and using the inclusion $\{S_s \geq B_s\} \subseteq \{S_s \geq L_s^*\}$ produces an upper bound. This upper bound corresponds to the function $C^u(S,t)$ defined above.

8.3.4 Price Approximations

The bounds C^u, C^l can be used to construct various approximations of the American call option. For example the weighted sums

$$C^1(S,t) = \lambda_1 C^l(S,t);\ \ \lambda_1 \geq 1$$

$$C^2(S,t) = \lambda_2 C^l(S,t) + (1-\lambda_2) C^l(S,t);\ \ \ \lambda_2 \in [0,1]$$

are suggested by Broadie and Detemple [1996]. The estimate $C^1(S,t)$ is a lower bound approximation (LBA) of the American option price; $C^2(S,t)$ is a lower-upper bound approximation (LUBA). They report that simple choices of the weights (e.g., $\lambda_1 = 1$ and $\lambda_2 = 0.5$) already provide good results in typical situations. Better weights can be obtained by using a regression approach based on a sample of parameter configurations. Nonparametric methods could also be employed to capture non-linearities in these relationships.

8.4 Diffusion Processes

The computational procedures presented in the previous sections can also be applied in more general diffusion settings. Specifically, suppose that the underlying asset price follows the risk neutralized process

$$dS_t/S_t = (r(S_t,t) - \delta(S_t,t))\, dt + \sigma(S_t,t)\, d\widetilde{z}_t \tag{8.4}$$

where $r(S,t)$ is the interest rate, $\sigma(S,t)$ the return volatility, $\delta(S,t)$ the dividend yield and $\widetilde{z}$ a Brownian motion under the risk neutral measure (the setting of section 4.6). As before assume that the coefficients $r(S,t), \sigma(S,t), \delta(S,t)$ are continuously differentiable and satisfy suitable conditions for the existence of a solution to this stochastic differential equation, that $\sigma(S,t)$ is positive and that $r(S,t)$ is a decreasing function of S and an increasing function of t, while $\delta(S,t)S$ is an increasing function of S. Also assume decreasing time-monotonicity. In the remainder of this section we show how to adapt integral equation and stopping time approximation methods to handle this setting.

8.4.1 Integral Equation Methods

Under the conditions of the model we know that the optimal exercise boundary is unique and satisfies the recursive integral equation (4.11) subject to the appropriate boundary condition. This equation can be simplified if the truncated moments in the integral equation can be written more explicitly, in terms of known density and cumulative distribution functions. In this case the exercise boundary B of a call option satisfies

$$B(t) - K = c(B(t), t) + \int_t^T \phi(B(t), t; B(v), v)\, dv$$

where the functions $c(B(t), t)$ and $\phi(B(t), t; B(v), v)$ involve the relevant density and distribution functions. This integral equation is subject to the terminal condition $B(T_-) = K \vee B^*$ where B^* solves the nonlinear equation

$$B = \frac{r(B,T)}{\delta(B,T)} K,$$

if an interior solution exists. If $\delta(S,T)S - r(S,T)K < 0$ for all possible realizations S of the process (8.4) at T set $B^* = \overline{S}$ (the highest possible value of S). If $\delta(S,T)S - r(S,T)K > 0$ for all possible realizations S of the process (8.4) at T set $B^* = \underline{S}$ (the lowest value of S). If an interior solution exists it is unique because $r(S,T)$ is decreasing in S while $\delta(S,T)S$ is increasing.

This recursive integral equation has the same structure as (8.1), albeit with a different integrand $\phi(B(t), t; B(v), v)$ and a boundary condition that involves the solution of a nonlinear equation. It is then clear that the numerical algorithms described in section 8.2 can also be used in this case.

8.4.2 Stopping Time Approximations: LBA and LUBA

Likewise, the stopping time approximations in section 8.3 can be adapted to handle the case of diffusion processes such as (8.4). In fact, the construction of the bounds is valid in the generality of this model, as it does not require the existence of closed form expressions for distributions or truncated moments.

The construction of the bounds proceeds as in section 8.3. A lower price bound is given by $C^l(S,t) = \max_{L \geq S} C(S,t;L)$ where $C(S,t;L)$ is the price of a capped call option with automatic exercise at the cap L, and same strike K and maturity date T as the American call option to be priced. A lower bound for the exercise boundary is given by the time-dependent function $L_t^* \equiv \inf\{S : D(S,t) \leq 0\}$ where $D(S,t)$ is defined as in (8.3). An upper price bound is then obtained by substituting L^* in place of B in the EEP formula. Proofs of these results can be found in Detemple and Tian [2002].

Therefore, all the elements of the LBA and LUBA algorithms extend to general diffusion processes satisfying the assumptions stated above. Although the construction does not require explicit distributions, this approach becomes particularly attractive when the capped call option price $C(S,t;L)$ has a closed form expression. In this case the computation of L^* and of the price bounds is fast and easy to perform.

8.5 Other Recent Approaches

8.5.1 Lattice Methods: Binomial Black-Scholes Algorithm

A modification to the binomial lattice method was proposed by Broadie and Detemple [1996]. This approach is based on the usual binomial recursive algorithm except that it replaces the continuation value at the last step prior to maturity (i.e., step $n-1$) by the Black-Sholes value. The approach is termed BBS (i.e., *B*inomial with *B*lack-*S*choles modification). The interest of this modification stems from its reduced error and smoother convergence to the true value. Combining the procedure with Richardson extrapolation gives another approximation called BBSR (i.e., *B*inomial with *B*lack-*S*choles modification and *R*ichardson extrapolation). Broadie and Detemple [1996, 1997b] show that BBS and BBSR perform significantly better than the standard binomial lattice methods when efficiency is measured by the trade-off between computation speed and error. They also show that LBA and LUBA give even better results than BBS and BBSR. Their numerical results are generated using a sample of about 2,500 randomly selected configurations for the parameters of the option pricing model.

8.5.2 Integral Equation: Non-linear Approximations

Instead of using step functions in order to approximate the solution of the integral equation one could employ piecewise non-linear functions. Ju [1998] proposes a multipiece exponential function. In his method the exercise boundary is approximated, within each subinterval, by an exponential function of time (e.g., $B_s = B_{t_i} \exp(b_i(s-t_i))$ for $s \in [t_i, t_{i+1})$). Integration by parts can then be used to express the integral of the discounted gains from exercise in terms of cumulative normal distributions. This gives an approximation of the recursive part of the integral equation that involves a single integral, the one in the cumulative normal distributions. A by-product is an approximation formula for the option value in terms of cumulative normals. Given that the exponential function has a free parameter b_i, that must be estimated along with the boundary point B_{t_i}, the equation for the boundary is solved in conjunction with the smooth pasting condition. Performing this computation recursively gives a multipiece exponential estimate of the boundary. An estimate of the option price is then obtained from the approximate option pricing formula.

This scheme can be combined with Richardson extrapolation to improve efficiency. Ju shows that a 3-point extrapolation scheme (where the points correspond to one, two and three subintervals) performs about as well as LUBA in terms of root mean square error and speed of computation. This result is obtained using a sample of $1,250$ randomly generated option parameter configurations.

8.5.3 Monte Carlo Simulation

Monte Carlo simulation has long been used for pricing European-style derivative securities. The method was introduced in the field by Boyle [1977]; a detailed survey can be found in Boyle, Broadie and Glasserman [1997]. Attempts at applications to American-style securities are more recent and were initiated by Tilley [1993], in the context of single asset models. The main difficulty in using simulation to price American claims is the need to derive the optimal stopping rule in conjunction with the claim's price. The recursive nature of this optimization problem conflicts with the inherent forward looking structure of the simulation approach. Various approaches have been proposed to deal with this difficulty. Some of the recent contributions include Barraquand and Martineau [1995], Carriere [1996], Broadie and Glasserman [1997a], [1997b], Broadie, Glasserman and Jain [1997], Raymar and Zwecher [1997], Carr [1998] and Longstaff and Schwartz [1999]. Further developments, based on duality (see section 3.4), can be found in Rogers [2002], Haugh and Kogan [2004] and Andersen and Broadie [2004].

Monte Carlo simulation is especially useful for dealing with contingent claims written on multiple underlying assets. Much of the recent literature cited above, in fact, attempts to develop the simulation-based approach for multiasset derivatives. A description of the some of these developments is provided in the last section of this chapter.

8.6 Performance Evaluation

To provide perspective on the numerical methods for univariate pricing problems we conduct a numerical study designed to evaluate the relative efficiencies of the methods presented. For parsimony we focus on the Binomial (B) algorithm, the Binomial-Black-Scholes (BBS) approach, the integral equation (IE) method and the lower and upper bound approximations (LBA and LUBA). Complementary results for additional methods can be found in Broadie and Detemple [1996], [1997b].

8.6.1 Experiment Design

The experimental design follows Broadie and Detemple [1996]. Suppose that we wish to compare the efficiency of J different approaches. To do so we consider a large set of N parameter configurations $\theta_i = [S_i, T_i, \delta_i, r_i, \sigma_i]$, $i = 1, ..., N$, drawn from a prespecified multivariate distribution. For each vector of parameter values θ_i we estimate the corresponding option price using each of the candidate methods. This gives a set of price estimates $\widehat{C}(\theta_i, j)$, where j represents the method used, $j = 1, ..., J$. The sample of values obtained $\{\widehat{C}(\theta_i, j) : i = 1, ..., N\}$ is then used to calculate speed and accuracy measures. Speed $s(j)$ is measured by the number of options prices calculated per second. Accuracy is measured by the root-mean-squared relative error criterion

$$RMSRE(j) = \sqrt{\frac{1}{N}\sum_{i=1}^{N}\left(\frac{\widehat{C}(\theta_i, j) - C(\theta_i)}{C(\theta_i)}\right)^2}$$

where $C(\theta_i)$ is the true price of the option. The true price is calculated using the binomial model with 15,000 steps. Accuracy is, of course, inversely related to root-mean-squared error. A plot of speed $s(j)$ versus accuracy $RMSRE(j)$, for each of the methods involved, provides perspective on the relative efficiencies of the approaches.

The sample considered in this study consists of $N' = 2,000$ initial parameter values. From this set, draws resulting in option prices less than 0.50 were eliminated. The resulting size of our parameter set is $N = 1,841$. Speed and error measures are calculated from this data set.

The multivariate parameter distribution is selected as follows. Parameters are independent of each other. The initial asset price S is uniformly distributed between 70 and 130, time to maturity T is uniform $[0.1, 1]$ with probability 0.75 and uniform $[1, 5]$ with probability 0.25, the dividend yield δ is uniform $[0, 0.10]$, the interest rate r is uniform $[0, 0.1]$ with probability 0.8 and null with probability 0.2, volatility σ is uniform $[0.1, 0.6]$. The strike price K is set equal to 100. This distribution is the same as in Broadie and Detemple [1996]. The coefficients of the regressions converting lower and upper bounds into price approximations (LBA and LUBA) are then the same as in their study (see Appendices B.6 and B.7 in Broadie and Detemple [1996]).

As mentioned above, the methods tested are B, BBS, IE, LBA and LUBA. The experiment is run in MATLAB, on a PC with a 2.4GHz Pentium 4 processor. As the built-in function *normcdf*$(\cdot)$, calculating the cumulative standard normal distribution, is computationally time consuming, it was replaced by a less accurate but faster approximation. This approximation intervenes in the programs for IE, LBA and LUBA.

8.6.2 Results and Discussion

Figure 8.2 illustrates the results of our study. The speed-accuracy trade-off is graphed in log-log scale for the 5 methods considered.

The version of the binomial algorithm tested uses the parametrization proposed by Hull and White [1988]. The plot is for time discretizations of 25, 50, 100, 200, 300, 400, 500, 600, 800, 1,000 and 1,200. The trade-off is clear: an increase in the number of time steps increases accuracy but also slows down computations. For $n = 400$ root-mean-squared relative error is about 6×10^{-4} and computation speed about 2.4×10. BBS is tested using the same parametrization of the tree (of course with Black-Scholes correction at the last step). The plot is for time discretizations of $25, 50, 100, 200, 300, 400, 500, 600$ and 800 . The graph shows uniform dominance over the binomial: for any fixed accuracy level BBS calculates more options per second than B.

The integral equation method is tested for discretizations $n = 4, 8$ and 16. The computation of the boundary approximation at each time step (i.e., the solution of the integral equation) is performed using a Newton-Raphson procedure. Strikingly, performance does not improve relative to the binomial method. For the sample drawn B dominates IE when the time discretization is 4 or 8, and performs about the same as IE with a discretization of 16. Both B and IE are dominated by BBS over the range of discretizations studied. The lower slope of the trade-off for IE, though, indicates that its relative performance should improve as the number of discretization points increases.

LBA and LUBA are also implemented using the Newton-Ralphson algorithm to solve the non-linear equations for the optimal barrier policies. LBA has an accuracy which is roughly of the order of 10^{-3} for a speed of computation about equal to 3.4×10^2. LUBA reaches an accuracy of about 2×10^{-4} for a speed in the neighborhood of 10^2. This represents an improvement over BBS by a factor of 10, and over B by a factor of 50.

These results are in line with those reported in Broadie and Detemple [1996], [1997b]. Broadly speaking they document the dominance of LBA and LUBA over the other candidate methods tested. BBS also exhibits good performance properties. Given the simple and intuitive nature of the method it may be the approach of choice for non-specialists. One notable difference is the slightly worse performance of IE. The small variation recorded here could be due to differences in precision parameters selected in the program and/or to differences in the implementation of built-in functions across software platforms.

8.7 Methods for Multiasset Options

This section provides a brief overview of numerical methods that have been developed to deal with multiasset options. We first survey lattice methods, then present more recent developments based on Monte Carlo simulation.

8.7.1 Lattice Methods

Multinomial lattice methods naturally extend the principles underlying the binomial approach to option pricing. As for the latter, the central idea is to discretize

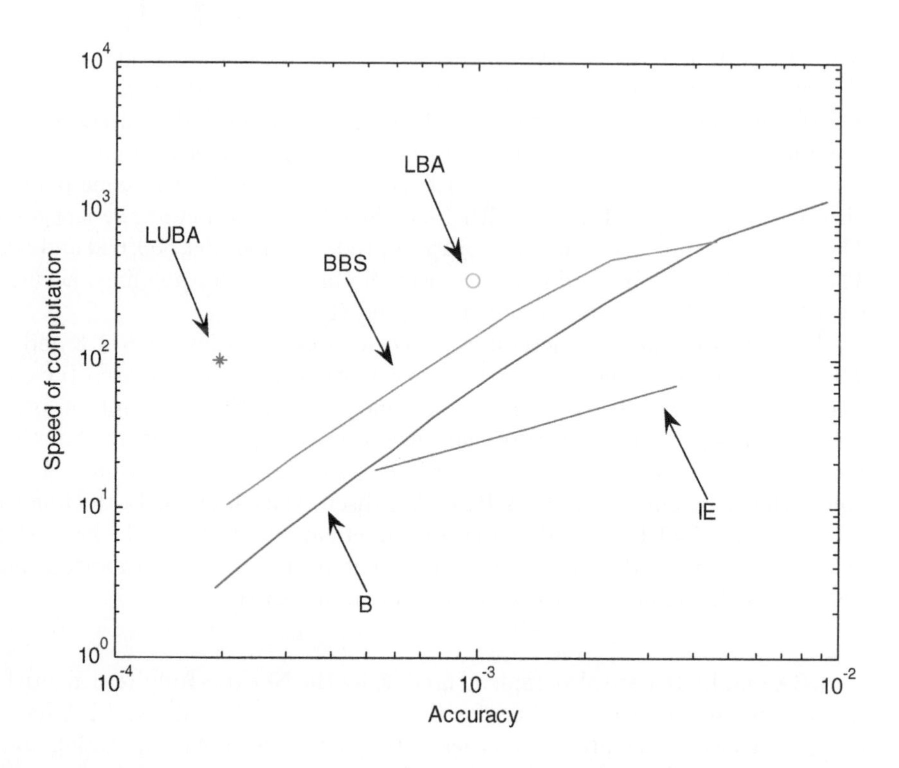

Figure 8.2: This figure shows the efficiency of the Binomial method (B), the Binomial Black-Scholes approach (BBS), the Integral Equation method (IE), the Lower Bound Approximation (LBA) and the Lower-Upper Bound Approximation (LUBA). Speed is measured in options per second (y-axis). Accuracy is measured by the root-mean-squared relative error (x-axis).

the prices of the underlying assets and then to use a backward dynamic programming algorithm to compute the option value. One of the new challenges is to find a discretization that provides not only a good description of the individual characteristics of assets but also of the relationships across prices. Various multinomial procedures have been proposed to that effect. Some of the most prominent ones include Boyle [1988], Boyle, Evnine and Gibbs [1989], Madan, Milne and Shefrin [1989], Cheyette [1990], He [1990], Kamrad and Ritchken [1991] and Rubinstein [1994].

The approach proposed by Boyle, Evnine and Gibbs [1989] illustrates the basic ingredients needed to deal with the multivariate case. Suppose that the objective is to price an American claim written on k underlying assets. This can be done by constructing an approximation of the price based on a lattice with 2^k branches issued from each node. For the two-asset case this gives 4 branches leading to 4 pairs of outcomes $\left(S^1u_1, S^2u_2\right)$, $\left(S^1u_1, S^2d_2\right)$, $\left(S^1d_1, S^2u_2\right)$ and

(S^1d_1, S^2d_2) with corresponding "risk neutral" probabilities q_{uu}, q_{ud}, q_{du} and q_{dd}. These are given by

$$q_{uu} = \frac{1}{4}\left(1+\rho+\sqrt{h}\left(\frac{\alpha_1}{\sigma_1}+\frac{\alpha_2}{\sigma_2}\right)\right)$$

$$q_{ud} = \frac{1}{4}\left(1-\rho+\sqrt{h}\left(\frac{\alpha_1}{\sigma_1}-\frac{\alpha_2}{\sigma_2}\right)\right)$$

$$q_{du} = \frac{1}{4}\left(1-\rho+\sqrt{h}\left(-\frac{\alpha_1}{\sigma_1}+\frac{\alpha_2}{\sigma_2}\right)\right)$$

$$q_{dd} = \frac{1}{4}\left(1+\rho-\sqrt{h}\left(\frac{\alpha_1}{\sigma_1}+\frac{\alpha_2}{\sigma_2}\right)\right)$$

where $\alpha_1 = r - \delta_i - 0.5\sigma_i^2$ for $i = 1, 2$, and $h = T/n$, with n representing the number of time steps.[4] The up and down returns are $u_i = \exp\left(\sigma_i\sqrt{h}\right)$ and $d_i = 1/u_i$, $i = 1, 2$. The (approximate) value of the claim, at any given node, is the maximum of the exercise payoff and the continuation value. The continuation value is calculated using the probability q described above, to calculate expected values. A similar construction applies for the k-asset case.

This procedure is quite natural and closely parallels the single asset case. The only subtlety involves the determination of the probabilities q and the price movements (u_i, d_i), $i = 1, 2$, at each node. Most variations of the approach select alternative parametrizations of these coefficients and/or modify the number of branches in the tree. For instance Boyle [1988] and Kamrad and Ritchken [1991] suggest a 5-branch lattice structure, for the two asset case, where the additional branch can be selected to represent the absence of a price movement. In He [1990] the price approximation is based on a 3-branch lattice.

The computational effort, for a k-dimensional lattice such as the one described above, is of the order $O\left(n^{k+1}\right)$ where n is the number of time steps. Thus, computation quickly becomes prohibitive as the dimensionality of the problem grows. In practice lattice methods can handle low dimensional problems where k is limited to four or less variables.

8.7.2 Monte Carlo Simulation

Monte Carlo simulation is effective in dealing with the curse of dimensionality that plagues the lattice approach. For high dimensional problems it usually becomes the only approach feasible. This is indeed the case for pricing problems involving European-style derivatives (see Boyle [1977]). The interest of the method for pricing American-style claims is much less apparent. As mentioned earlier (see section 8.5.3) the forward nature of the simulation method seems to be at odds with the backward structure of the pricing algorithm.

[4] The probabilities are selected so as to provide a good approximation of the characteristic function of the bivariate distribution of returns.

To overcome this difficulty Broadie and Glasserman [1997a] use simulation to construct price approximations in a recursive manner. The starting point is the simulation of a tree, with a fixed number of branches issued from each node, and a given number of time steps corresponding to exercise decision points. Each branch issued from a given node is drawn independently from the others. Once the tree has been set, backward induction is employed to compute, at each node, two estimates of the price, one biased high, the other biased low. The high estimate is obtained by using the familiar recursion from the binomial model. It is the maximum of the immediate exercise value and the continuation value. If $h(S)$ is the exercise payoff and $v(t)$ the continuation value, then the high estimate is $C^u(t) = \max\{h(S_t), v(t)\}$. This estimate is biased high because the estimated continuation value either overestimates the true continuation value, in which case C^u is automatically biased high (whether it is optimal to exercise or not) or it underestimates the true value, in which case it will be replaced by a larger quantity, the exercise payoff, in the event that it falls below this payoff. The result is an upward bias that propagates through the tree, from the terminal nodes back to the initial date. A formal argument relies on Jensen's inequality applied to the convex function $\max\{h(S_t), v(t)\}$.

In order to construct the low estimate Broadie and Glasserman dissociate the exercise decision from the reward collected in the event of exercise. Suppose that there are b branches, $j = 1, ..., b$, issued from a node and let h be the time step. First, select the first branch $j = 1$ and use branches $j = 2, ..., b$ to calculate the continuation value $v(t, 1)$ from the low estimates at the next date $C^l(t + 1, j)$ for $j = 2, ..., b$. The exercise decision is based on the maximum of $v(t, 1)$ and the exercise payoff $h(S_t)$. The exercise reward, on the other hand, is taken to be

$$Y(t,1) = \begin{cases} h(S_t) & \text{if } h(S_t) \geq v(t,1) \\ e^{-rh}C^l(t+1,1) & \text{if } h(S_t) < v(t,1) \end{cases}$$

where $C^l(t + 1, 1)$ is the low estimate at time $t + 1$ at the end of branch $j = 1$. Repeating this operation for each branch leads to a collection of rewards $Y(t, j), j = 1, ..., b$. The low estimate is then obtained by taking the average over this sample,

$$C^l(t) = \frac{1}{b}\sum_{j=1}^{b} Y(t,j). \tag{8.5}$$

Understanding the reasons for a low bias in this estimate is more delicate. Intuition can be gained by noticing that each reward function $Y(t, j)$ is a random variable with expected value, at time t, given by $E_t[Y(t, j)] = h(S_t)p(t, j) + E_t\left[e^{-rh}C^l(t + 1, j)\right](1 - p(t, j))$ where $p(t, j)$ is the conditional probability at date t of immediate exercise based on $v(t, j)$.[5] The upper bound,

$$E_t[Y(t,j)] \leq \max\left\{h(S_t), E_t\left[e^{-rh}C^l(t+1,j)\right]\right\}$$

[5] The quantity $p(t, j)$ is the conditional probability at time t of the event $\{h(S_t) \geq v(t, j)\}$. Given the independence of branches, the event $\{h(S_t) \geq v(t, j)\}$ is independent of the estimate $C^l(t + 1, j)$. The expression for $E_t[Y(t, j)]$ follows.

follows immediately. If the estimate $C^l(t+1,j)$ is biased low, at time $t+1$, the expectation on the right hand side is bounded above by the true continuation value. But then $E_t[Y(t,j)] \leq C(t)$. Because branches are drawn independently we can now conclude, from (8.5), that the estimate $C^l(t)$ is also biased low, i.e., $E_t\left[C^l(t)\right] = \frac{1}{b}\sum_{j=1}^{b} E_t[Y(t,j)] \leq C(t)$.

This approach is powerful to the extent that it does not limit the number of assets underlying the contract (the variable S is a vector) and produces an interval of values for the true price of the derivative. Practical implementation, however, is limited by the fact that the amount of work is exponential in the number of exercise decisions. This has led to various modifications of the algorithm designed to improve performance and applicability (e.g., Broadie and Glasserman [1997b], Broadie, Glasserman and Jain [1997]).

8.7.3 Monte Carlo Simulation and Duality

The duality formula, described in Theorem 24, provides a systematic approach for the construction of an upper bound approximation for the price of an American-style derivative. The method suggested by the formula is straightforward. It can be decomposed in three steps:

1. select a set of martingales that are null at the initial date $M^i, i = 1, ..., m$,
2. for each i compute $E\left[\sup_{t\in[0,T]}(D_t - M_t^i)\right]$, where D is the discounted payoff, and
3. choose the minimum value obtained (i.e., minimize over i).

This procedure is implemented by Rogers [2002] who also suggests the following variation. Once a collection of martingales has been identified construct the linear combination $M^w = \sum_i w_i M_i$. Then use Monte Carlo simulation, with a small number of trajectories, in order to calculate the optimal minimizing weights w_i^*. Finally, construct an upper price bound by running a new simulation with a large number of paths to reestimate the value of the net discounted payoff based on the optimal weights.

In the examples studied by Rogers this approach proves to be accurate, often producing very tight upper bounds. Its main drawback is the computation time associated with the optimization over the set of weights. In the case of exotic options, in which a set of interesting martingales is not a priori obvious, this minimization can be time consuming.

One possible remedy is a careful choice of martingales in the first step of the procedure. After all the duality result states that the price of the American claim is attained if the martingale in the Doob-Meyer decomposition of the Snell envelope is selected. Of course identifying this martingale is typically not possible. But clever choices can be made that will often speed up computations and produce accurate results, even for exotic options (see Andersen and Broadie [2004] and Haugh and Kogan [2004]).

8.8 Methods for Occupation Time Derivatives

To conclude this chapter we provide a survey of numerical methods for occupation time derivatives. The path-dependent nature of these contracts complicates their evaluation. Standard approaches nevertheless apply, modulo suitable modifications. The methods investigated include those based on the numerical inversion of Laplace transforms, PDE-based methods and Binomial/Trinomial trees.

8.8.1 Laplace Transforms

Valuation formulas for occupation time derivatives are often expressed in terms of inverse Laplace transforms of density functions. In specific cases, such as the cumulative barrier claims (see section 7.6), the inverse Laplace transform admits an explicit solution. The calculation of prices is then reduced to the evaluation of the single or multiple integrals associated with the pricing operator and the claim's payoff. Quadrature methods can be used to carry out these integral evaluations.

In other cases, such as the Parisian option (see section 7.5), the inverse of the Laplace transform does not have an obvious analytical solution. In these situations the density function has an integral representation, resulting in additional integrals in the pricing formula. Quadrature methods can again be used to handle the computation of these multiple integrals. Numerical schemes based on approximations of inverse Laplace transforms can also be employed. Approximation-based approaches that have been used in the context of OTD pricing include the Weeks, Padé and Fourier-series methods.

Calculation of prices based on Laplace transforms and their inverses are carried out by Chesney, Cornwall, Jeanblanc-Picqué, Kentwell and Yor [1997], Hugonnier [1999], Linetsky [1999] and Fusai and Tagliani [2001]. Numerical schemes for inverting Laplace transforms can be found in Singhal, Vlach and Vlach [1975], Davies and Martin [1979], Abate and Whitt [1992], Chaudhury, Lucantoni and Whitt [1994] and Craddock, Heath and Platen [2000].

8.8.2 PDE-based Methods

Consider a European-style occupation time derivative written on an asset price S and depending on the time O spent by S in a constant set A. Let $\mathbb{D} \subset \mathbb{R}_+ \times [0,T]$ be an open set with boundary $\partial\mathbb{D}$. The derivative's payoff is $F(S,O,T)$ at the maturity date T in the event $(S,O) \in \mathbb{D}$ and $G(S,D,t)$ if $(S,O) \in \partial\mathbb{D}$ and $t \in [0,T]$. Under standard differentiability assumptions the derivative's price $V(S,O,t)$ satisfies the fundamental valuation PDE

$$V_t + V_s S(r-\delta) + \frac{1}{2} V_{ss} S^s \sigma^2 + V_o 1_{\{S\in A\}} = rV \tag{8.6}$$

for $(S, O, t) \in \mathbb{D} \times [0, T)$, subject to the boundary conditions

$$\begin{cases} V(S,O,T) = F(S,O,T) & \text{for } (S,O) \in \mathbb{D} \\ V(S,O,t) = G(S,O,t) & \text{for } (S,O) \in \partial\mathbb{D}. \end{cases} \tag{8.7}$$

Additional boundary conditions (such as limiting conditions) may apply depending on the exact nature of the contract. The dependence on the occupation time of a set amounts to an additional state variable to be accounted for in the valuation equations.

For American-style contracts the set of boundary conditions is augmented by the addition of early exercise provisions,

$$\begin{cases} V(S,O,t) \geq F(S,O,t) & \text{for } (S,O,t) \in \mathbb{D} \times [0,T] \\ V(B,O,t) = F(B,O,t) & \text{for } (B,O,t) \in \mathbb{D} \times [0,T) \\ V_s(B,O,t) = F_s(B,O,t) & \text{for } (B,O,t) \in \mathbb{D} \times [0,T). \end{cases} \tag{8.8}$$

The first condition in (8.8) stipulates that the contract value cannot fall below the immediate exercise value. The second condition mandates equality if the asset price lies on the boundary B of the exercise region. The last condition is the smooth-pasting condition, expressing the optimality of exercise along the boundary B.

Finite difference methods can be used for the numerical resolution of (8.6)-(8.7). Implicit or explicit methods can be applied. Numerical approximations of the derivatives are computed by discretizing over the time domain and the bivariate domain of the state variables. Various discretization schemes, such as forward, backward or central differencing, can be employed for that purpose.

Cumulative barrier claims are examples of contracts that fit in this framework. For instance, in the case of a knock-out claim with window D, the survival domain is $\mathbb{D} = \mathbb{R}_+ \times [0, D)$ and its boundary $\partial\mathbb{D} = \mathbb{R}_+ \times \{D\}$. In the absence of a rebate the payoff is null $(G(S,O,t) = 0)$ if the time spent in the set A reaches the threshold D. Derivatives with reset provisions mandate the addition of further conditions to capture the impact of resets. Parisian options, for instance, involve a reset of the occupation time if the underlying price reaches the barrier after an excursion of duration less than the window. The supplemental condition is $V(S,O,t) = V(S,0,t)$ if $S = L$, where L is the prespecified barrier.

PDE-based methods for OTDs are discussed in Haber, Schönbucher and Wilmott [1999] and Vetzal and Forsyth [1999].

8.8.3 Binomial/Trinomial Lattices

Binomial and trinomial lattices can also be amended so as to handle the valuation of OTDs. The main difficulty in implementing a lattice-based approach is to keep track of the occupation time of the relevant set as the asset price evolves along the tree. The *Forward Shooting Grid algorithm* (FSG) resolves this difficulty by adding a new dimension, measuring depth (or height), to the tree. In effect the two-dimensional lattice becomes a three-dimensional lattice where the third dimension captures the number k of occurrences of the event (the number of time intervals spent in the set A). The lattice can be visualized as a superposition of planes, each corresponding to a value of the occupation counter. The top plane corresponds to a count equal to $k = 0$. The next one, one step below, to $k = 1$, and so on. When the asset price moves to a follower node, it remains in the same plane as long as the counter stays the same. If the counter increases by one unit, the follower node is reached by taking a step down, i.e., moving down to the next plane. Construction of the lattice proceeds in this manner until maturity.

Valuation is then carried out along familiar lines. The value at a given node is obtained by taking the sum of the discounted values at follower nodes. Follower nodes lie in the same plane if the counter stays the same as the price moves to the follower node; they lie one step below if the counter increases by a unit. The algorithm proceeds recursively, starting with terminal values, at the maturity date, and moving backward in time toward the initial date. For a binomial lattice, with up and down steps (u, d), this recursion can be summarized by

$$\begin{aligned} V(S,k,j) &= e^{-rh}\{q_u[V(Su,k,j+1)1_{\{Su\notin A\}} \\ &\quad +V(Su,k+1,j+1)1_{\{Su\in A\}}] \\ &\quad +q_d[V(Sd,k,j+1)1_{\{Sd\notin A\}} \\ &\quad +V(Sd,k+1,j+1)1_{\{Sd\in A\}}]\} \end{aligned} \tag{8.9}$$

for $j = 1, ..., n$ and $k = 0, ..., d$. This backward recursion is subject to the relevant boundary conditions at the maturity date n and the upper bound of the occupation time domain d. American-style provisions can be readily incorporated by taking the maximum of the continuation value, given by the algorithm above, and the immediate exercise value. Claims with discrete monitoring dates can also be incorporated by selecting a time grid such that monitoring dates correspond to a multiple of the time step. Finally, claims with reset provisions can be handled by substituting the function describing the dynamics of the counter k in the second argument of the continuation value, on the right hand side of (8.9). For a Parisian knock-out claim with window count d, based on the occupation time of the set A, this function is given by $f(x,k) = (k+1)1_{\{x\in A\}}$ where $x = Su$ or Sd.

The adaptation of the Forward Shooting Grid algorithm to OTDs is discussed by Kwok and Lau [2001].[6] FSG algorithms were originally pioneered by Hull and White [1993] and Ritchken, Sankarasubramaniam and Vijh [1993] in order to price Asian and Lookback options. Further results about the method can be found Barraquand and Pudet [1996].

[6] Another lattice-based approach is proposed by Avellaneda and Wu [1999]. Their method uses the density function of the relevant hitting time in order to identify the appropriate boundary conditions in the tree.

8.9 Appendix: Proofs

Proof of Proposition 109: The definition of the capped call option $C(S,t;L)$ gives the lower bound for the American call price $C(S,t) \geq C(S,t;L)$. This bound implies that the constrained optimization problem

$$\max_{L} \{C(S,t;L) : L \geq S\}$$

has the solution $L = S$ when $(S,t) \in \mathcal{E}$ (as immediate exercise is optimal for the American call). Thus,

$$D(S,t) \equiv \left. \frac{\partial C(S,t;L)}{\partial L} \right|_{L=S} \leq 0$$

for all $(S,t) \in \mathcal{E}$ and

$$L_t^* \equiv \inf \{S : D(S,t) \leq 0\} \in \overline{\mathcal{C}}_t$$

where $\overline{\mathcal{C}}_t$ is the closure of the open set $\mathcal{C}_t$ (the time t-section of the continuation region). We conclude that $L_t^* \leq B(t)$.

To show that $L_t^* \geq K$ it suffices to note that exercise when $L < K$ has a negative payout. Suppose now that $(r/\delta)K > K$. The proof that $L_t^* \geq (r/\delta)K$ relies on the intuition that exercise at a point $L_1 < (r/\delta)K$ must be suboptimal because the local net benefits from exercising are negative. To formalize this idea let $\tau_i = \inf \{v \in [t,T] : S_v = L_i\}$ for $i = 1,2$, where $L_1 < (r/\delta)K$ and $L_2 = (r/\delta)K$, and suppose that $S_t \leq L_1$. Clearly $\tau_1 < \tau_2$. On the event $\{\tau_1 = t\}$ (where $S_{\tau_1} = L_1$) we have, by Ito's lemma,

$$\begin{aligned} L_1 - K &= \widetilde{E}_{\tau_1}\left[e^{-r(\tau_2-\tau_1)}(S_{\tau_2} - K)1_{\{\tau_2 \leq T\}}\right] \\ &\quad -\widetilde{E}_{\tau_1}\left[\int_{\tau_1}^{\tau_2} e^{-r(v-\tau_1)}(rK - \delta S_v)\,ds 1_{\{\tau_2 \leq T\}}\right] \\ &\quad +\widetilde{E}_{\tau_1}\left[e^{-r(T-\tau_1)}(S_T - K)1_{\{\tau_2 > T\}}\right] \\ &\quad -\widetilde{E}_{\tau_1}\left[\int_{\tau_1}^{T} e^{-r(v-\tau_1)}(rK - \delta S_v)\,ds 1_{\{\tau_2 > T\}}\right]. \end{aligned}$$

Given that the policy of exercising at time τ_2 if $\tau_2 \leq T$, or at maturity if $\tau_2 > T$, has value

$$\begin{aligned} V(\tau_2) &= \widetilde{E}_{\tau_1}\left[e^{-r(\tau_2-\tau_1)}(S_{\tau_2} - K)1_{\{\tau_2 \leq T\}}\right] \\ &\quad +\widetilde{E}_{\tau_1}\left[e^{-r(T-\tau_1)}(S_T - K)^+ 1_{\{\tau_2 > T\}}\right] \end{aligned}$$

we conclude that

$$\begin{aligned} L_1 - K &= V(\tau_2) - \widetilde{E}_{\tau_1}\left[\int_{\tau_1}^{\tau_2 \wedge T} e^{-r(v-\tau_1)}(rK - \delta S_v)\,ds\right] \\ &\quad -\widetilde{E}_{\tau_1}\left[e^{-r(T-\tau_1)}(S_T - K)^- 1_{\{\tau_2 > T\}}\right], \end{aligned}$$

which implies $V(\tau_2) > L_1 - K$ (since $rK - \delta S_v > 0$ for $t \leq \tau_2$). We conclude that immediate exercise at $S_t = L_1 < (r/\delta)K$ is dominated by the policy of waiting until L_2 is reached.

For the limiting values of L^* note that when $t \to T$ the optimal exercise boundary for the call option converges to the point, $B_{T\_} = K \max\{1, r/\delta\}$. Given that $(r/\delta)K \vee K \leq L_t^* \leq B_t$ for $t \in [0, T)$ we obtain $\lim_{t \to T} L_t^* = B_{T\_}$. When $T - t \to \infty$ the American call has a constant exercise boundary equal to $B_{-\infty} = (b + f)/(b + f - \sigma^2)$ (see Corollary 35). This policy lies in the class of constant barrier policies. It follows immediately that $L_t^* \equiv \inf\{S : D(S, t) \leq 0\}$ converges to $B_{-\infty}$. ■

Proof of Proposition 110: The proof is outlined in the text following the proposition. ■

Bibliography

[1] Abate, J. and W. Whitt, "The Fourier-Series Method for Inverting Transforms of Probability Distributions," *Queueing Systems*, 10, 1992, 5-88.

[2] Ait-Sahlia, F., "Optimal Stopping and Weak Convergence Methods for Some Problems in Financial Economics," Ph.D. Dissertation, Stanford University, 1995.

[3] Akahori, J., "Some Formulae for a New Type of Path-Dependent Option," *Annals of Applied Probability*, 5, 1995, 383-388.

[4] Amin, K., "On the Computation of Continuous Time Option Prices using Discrete Approximations," *Journal of Financial and Quantitative Analysis*, 26, 1991, 477-496.

[5] Amin, K. and A. Khanna, "Convergence of American Option Values from Discrete- to Continuous-Time Financial Models," *Mathematical Finance*, 4, 1994, 289-304.

[6] Andersen, L. and M. Broadie, "A Primal-Dual Simulation Algorithm for Pricing Multi-Dimensional American Options," *Management Science*, 50, 2004, 1222-1234.

[7] Avellaneda, M. and L. Wu, "Pricing Parisian-Style Options with a Lattice Method," *International Journal of Theoretical and Applied Finance*, 2, 1999, 1-16.

[8] Bachelier, L., "Théorie de la Spéculation," *Annales de l'École Normale Supérieure*, 17, 1900, 21–86. In *The Random Character of Stock Market Prices*, Paul H. Cottner ed., The MIT Press, Cambridge, Mass., 1964.

[9] Barles, G., J. Burdeau, M. Romano and N. Samsoen, "Critical Stock Price Near Expiration," *Mathematical Finance*, 5, 1995, 77-95.

[10] Barone-Adesi, G. and R. Whaley, "Efficient Analytical Approximation of American Option Values," *Journal of Finance*, 42, 1987, 301-320.

[11] Barraquand, J. and D. Martineau, "Numerical Valuation of High Dimensional Multivariate American Securities," *Journal of Financial and Quantitative Analysis*, 30, 1995, 383-405.

[12] Barraquand, J. and T. Pudet, "Pricing of American Path Dependent Contingent Claims," *Mathematical Finance*, 6, 1996, 17-51.

[13] Bensoussan, A., "On the Theory of Option Pricing," *Acta Applicandae Mathematicae*, 2, 1984, 139-158.

[14] Bergman, Y., B. Grundy and Z. Wiener, "General Properties of Option Prices," *Journal of Finance*, 51, 1996, 1573-1610.

[15] Bermin, H. P., "A General Approach to Hedging Options: Applications to Barrier and Partial Barrier Options," *Mathematical Finance*, 12, 2002, 199-218.

[16] Bermin, H. P., "Hedging Options: The Malliavin Calculus Approach versus the Delta-Hedging Approach," *Mathematical Finance*, 13, 2003, 73-84.

[17] Bjerksund, P. and G. Stensland, "Closed Form Approximation of American Options," *Scandinavian Journal of Management*, 1992.

[18] Bjerksund, P. and G. Stensland, "American Exchange Options and a Put-Call Transformation: A Note," *Journal of Business, Finance and Accounting*, 20, 1993, 761-764.

[19] Black, F. and J. Cox, "Valuing Corporate Securities: Some Effects of Bond Indenture Provisions," *Journal of Finance,* 31, 1976, 351-367.

[20] Black, F. and M. Scholes, "The Pricing of Options and Corporate Liabilities," *Journal of Political Economy*, 81, 1973, 637-654.

[21] Boyle, P. P., "Options: A Monte Carlo Approach," *Journal of Financial Economics*, 4, 1977, 323-338.

[22] Boyle, P. P., "A Lattice Framework for Option Pricing with Two State Variables," *Journal of Financial and Quantitative Analysis*, 23, 1988, 1-12.

[23] Boyle, P. P., "Barrier Options," Working paper, University of Waterloo, 1992.

[24] Boyle, P. P., M. Broadie and P. Glasserman, "Monte Carlo Methods for Security Pricing," *Journal of Economic Dynamics and Control*, 21, 1997, 1276-1321.

[25] Boyle, P. P., J. Evnine and S. Gibbs, "Numerical Evaluation of Multivariate Contingent Claims," *Review of Financial Studies*, 2, 1989, 241-250.

[26] Boyle, P. P. and S. M. Turnbull, "Pricing and Hedging Capped Options," *Journal of Futures Markets*, 9, 1989, 41-54.

[27] Breen, R., "The Accelerated Binomial Option Pricing Model," *Journal of Financial and Quantitative Analysis*, 26, 1991, 153-164.

[28] Brennan, M. and E. Schwartz, "The Valuation of American Put Options," *Journal of Finance*, 32, 1977, 449-462.

[29] Brennan, M. and E. Schwartz, "Finite Difference Methods and Jump Processes Arising in the Pricing of Contingent Claims: A Synthesis," *Journal of Financial and Quantitative Analysis*, 13, 1978, 461-474.

[30] Broadie, M. and J. B. Detemple, "American Capped Call Options on Dividend-Paying Assets," *Review of Financial Studies*, 8, 1995, 161-191.

[31] Broadie, M. and J. B. Detemple, "American Option Valuation: New Bounds, Approximations, and a Comparison of Exisiting Methods," *Review of Financial Studies*, 9, 1996, 1211-1250.

[32] Broadie, M. and J. B. Detemple, "The Valuation of American Options on Multiple Assets," *Mathematical Finance*, 7, 1997a, 241-285.

[33] Broadie, M. and J. B. Detemple, "Recent Advances in Numerical Methods for Pricing Derivative Securities," *Numerical Methods in Finance*, L. C. G. Rogers and D. Talay, eds., Cambridge University Press, Cambridge, 1997b, 43-66.

[34] Broadie, M. and J. B. Detemple, "American Options on Dividend Paying Assets," in *Topology and Markets,* G. Chichilnisky, ed., *Fields Institute Communications,* 22, 1999, 69-97.

[35] Broadie, M. and J. B. Detemple, "Option Pricing: Valuation Models and Applications," *Management Science*, 50, 2004, 1145-1177.

[36] Broadie, M. and P. Glasserman, "Pricing American-Style Securities Using Simulation," *Journal of Economic Dynamics and Control*, 21, 1997a, 1323-1352.

[37] Broadie, M. and P. Glasserman, "A Stochastic Mesh Method for Pricing High-Dimensional American Options," Working Paper, Columbia University, 1997b.

[38] Broadie, M., P. Glasserman and G. Jain, "Enhanced Monte Carlo Estimates for American Option Prices," *Journal of Derivatives*, 5, 1997, 25-44.

[39] Bunch, D. S. and H. Johnson, " A Simple and Numerically Efficient Valuation Method for American Puts Using a Modified Geske-Johnson Approach," *Journal of Finance*, 47, 1992, 809-816.

[40] Carr, P., "Randomization and the American Put," *Review of Financial Studies*, 11, 1998, 597-626.

[41] Carr, P. and D. Faguet, "Fast Accurate Valuation of American Options," Working paper, Cornell University, 1995.

[42] Carr, P., R. Jarrow and R. Myneni, "Alternative Characterizations of American Put Options," *Mathematical Finance*, 2, 1992, 87-106.

[43] Carr, P. and R. Jarrow, "The Stop-Loss Start-Gain Paradox and Option Valuation: A New Decomposition into Intrinsic and Time Value," *Review of Financial Studies*, 3, 1990, 469-492.

[44] Carriere, J., "Valuation of the Early-Exercise Price for Options Using Simulations and Nonparametric Regression," *Insurance, Mathematics and Economics*, 19, 1996, 19-30.

[45] Chaudhury, G. L., Lucantoni, D. M. and W. Whitt, "Multidimensional Transform Inversion with Applications to the Transient M/G/1 Queue," *Annals of Applied Probability*, 5, 1994, 389-398.

[46] Chesney, M., J. Cornwall, M. Jeanblanc-Picqué, G.Kentwell and M. Yor, "Parisian Pricing," *Risk*, 10, 1997, 77-79.

[47] Chesney, M., M. Jeanblanc-Picqué and M. Yor, "Brownian Excursions and Parisian Barrier Options," *Advances in Applied Probability*, 29, 1997, 165-184.

[48] Cheyette, O., "Pricing Options on Multiple Assets," *Advances in Futures and Options Research*, 4, 1990, 69-81.

[49] Cottle, R., J.-S. Pang and R. Stone, *The Linear Complementarity Problem*, Academic Press, Boston, 1992.

[50] Courtadon, G., "A More Accurate Finite Difference Approximation for the Valuation of Options," *Journal of Financial and Quantitative Analysis*, 17, 1982, 697-705.

[51] Cox, J. C., "Notes on Options I: Constant Elasticity of Variance Diffusion," working paper, Stanford University, 1975.

[52] Cox, J. C., "The Constant Elasticity of Variance Option Pricing Model," *Journal of Portfolio Management*, 22, 1996, 15-17.

[53] Cox, J. and S. Ross, "The Valuation of Options for Alternative Stochastic Processes," *Journal of Financial Economics*, 3, 1976, 145-166.

[54] Cox, J. C., S. A. Ross and M. Rubinstein, "Option Pricing: A Simplified Approach," *Journal of Financial Economics*, 7, 1979, 229-263.

[55] Cox, J. and M. Rubinstein, *Option Markets*, Prentice Hall, Englewood Cliffs, New Jersey, 1985.

[56] Craddock, M., D. Heath and E. Platen, "Numerical Inversion of Laplace Transforms: A Survey of Techniques with Applications to Derivative Pricing," *Journal of Computational Finance*, 4, 2000, 57-81.

[57] Curran, M., "Accelerating American Option Pricing in Lattices," *Journal of Derivatives*, 3, 1995, 8-17.

[58] Dassios, A., "The Distribution of the Quantile of a Brownian Motion with Drift and the Pricing of Related Path-Dependent Options," *Annals of Applied Probability*, 5, 1995, 389-398.

[59] Davies, B. and B. Martin, "Numerical Inversion of the Laplace Transform: A Survey and Comparison of Methods," *Journal of Computational Physics*, 33, 1979, 1-32.

[60] Davis, M. H. A. and I. Karatzas, "A Deterministic Approach to Optimal Stopping, with Applications," in *Probability, Statistics and Optimization: A Tribute to Peter Whittle*, F. P. Kelly, ed., Wiley, New York, 1994, 455-466.

[61] Davydov, D. and V. Linetsky, "The Valuation and Hedging of Barrier and Lookback Options under the CEV Process," *Management Science*, 47, 2001, 949-965.

[62] Detemple, J. B., "American Options: Symmetry Properties," *Handbooks in Mathematical Finance: Topics in Option Pricing, Interest Rates and Risk Management*, J. Cvitanic, E. Jouini and M. Musiela, eds., Cambridge University Press, Cambridge, England, 2001, 67-104.

[63] Detemple, J. B., S. Feng and W. Tian, "The Valuation of American Call Options on the Minimum of Two Dividend-paying Assets," *Annals of Applied Probability*, 13, 2003, 953-983.

[64] Detemple, J. B., R. Garcia and M. Rindisbacher, "A Monte-Carlo Method for Optimal Portfolios," *Journal of Finance*, 58, 2003, 401-446.

[65] Detemple, J. B. and W. Tian, "The Valuation of American Options for a Class of Diffusion Processes," *Management Science*, 48, 2002, 917-937.

[66] Diener, F. and M. Diener, "Asymptotics of the Price Oscillations of a Vanilla Option in a Tree Model," *Mathematical Finance*, 13, 2004, 271-294.

[67] Dixit, A. and R. Pindyck, *Investment Under Uncertainty*, Princeton University Press, 1994.

[68] Duffie, D., "Stochastic Equilibria: Existence, Spanning Number, and The 'No Expected Financial Gain From Trade' Hypothesis," *Econometrica*, 54, 1986, 1161-1184.

[69] El Karoui, N., M. Jeanblanc-Picque and S. Shreve, "Robustness of the Black and Scholes Formula," *Mathematical Finance*, 8, 1998 93-126.

[70] El Karoui, N. and I. Karatzas, "A New Approach to the Skorohod Problem and its Applications," *Stochastics and Stochastics Reports*, 34, 1991, 57-82.

[71] El Karoui, N., *Les Aspects Probabilistes du Controle Stochastique*, Lecture Notes in Mathematics, 876, Springer-Verlag, Berlin, 1981, 73-238.

[72] Emanuel, D. C. and J. D. MacBeth, "Further Results on the Constant Elasticity of Variance Call Option Pricing Model," *Journal of Financial and Quantitative Analysis*, 7, 1984, 533-54.

[73] Embrecht, P., L.C.G. Rogers and M. Yor, "A Proof of Dassios' Representation of the α-Quantile of Brownian Motion with Drift," *Annals of Applied Probability*, 5, 1995, 757-767.

[74] Evans, J. D., R. Kuske and J. B. Keller, "American Options on Assets with Dividends near Expiry," *Mathematical Finance*, 12, 2002, 219-237.

[75] Flesaker, B., "The Design and Valuation of Capped Stock Index Options," Working Paper, University of Illinois at Urbana-Champaign, 1992.

[76] Fournié, E., J.-M. Lasry, J. Lebuchoux, P.-L Lions and N. Touzi, "Applications of Malliavin Calculus to Monte Carlo Methods in Finance," *Finance and Stochastics*, 3, 1999, 391-412.

[77] Francois, P. and E. Morellec, "Capital Structure and Asset Prices: Some Effects of Bankruptcy Procedures," *Journal of Business*, 77, 2004, 387-411.

[78] Fusai, G., "Corridor Options and Arc-Sine Law," *Annals of Applied Probability*, 10, 2000, 634-663.

[79] Fusai, G. and A. Tagliani, "Pricing of Occupation Time Derivatives: Continuous and Discrete Monitoring," *Journal of Computational Finance*, 5, 2001, 1-37.

[80] Galai, D., A. Raviv and Z. Wiener, "Liquidation Triggers and the Valuation of Equity and Debt," Working paper, Hebrew University of Jerusalem, 2005.

[81] Gao, B., J.-z. Huang and M. Subrahmanyam, "The Valuation of American Barrier Options Using the Decomposition Technique," *Journal of Economic Dynamics and Control*, 24, 2000, 1783-1827.

[82] Geltner, D., T. Riddiough and S. Stojanovic, "Insights on the Effect of Land Use Choice: The Perpetual Option on the Best of Two Underlying Assets," *Journal of Urban Economics*, 39, 1996, 20-50.

[83] Geman H., N. El Karoui and J.-C. Rochet, "Changes of Numeraire, Changes of Probability Measure and Option Pricing," *Journal of Applied Probability*, 32, 1995, 443-458.

[84] Geske, R., "A Note on an Analytical Valuation Formula for Unprotected American Options on Stocks with Known Dividends," *Journal of Financial Economics*, 7, 1979, 375-380.

[85] Geske, R. and H. Johnson, "The American Put Option Valued Analytically," *Journal of Finance*, 39, 1984, 1511-1524.

[86] Grabbe, O., "The Pricing of Call and Put Options on Foreign Exchange," *Journal of International Money and Finance*, 2, 1983, 239-253.

[87] Haber, R. J., P. J. Schönbucher and P. Wilmott, "Pricing Parisian Options," *Journal of Derivatives*, 6, 1999, 71-79.

[88] Harrison, M. and D. Kreps, "Martingales and Arbitrage in Multiperiod Security Markets," *Journal of Economic Theory*, 20, 1979, 381-408.

[89] Harrison, M. and S. Pliska, "Martingales and Stochastic Integrals in the Theory of Continuous Trading," *Stochastic Processes and Their Applications*," 11, 1981, 215-260.

[90] Haugh, M. and L. Kogan, "Approximating Pricing and Exercising of High-Dimensional American Options: A Duality Approach," *Operations Research*, 52, 2004, 258-270.

[91] He, H., "Convergence from Discrete- to Continuous-Time Contingent Claim Prices," *Review of Financial Studies*, 3, 1990, 523-546.

[92] Ho, T. S., S. Stapleton and M. G. Subrahmanyam, "A Simple Technique for the Valuation and Hedging of American Options," *Journal of Derivatives*, 2, 1994, 52-66.

[93] Hobson, D. G., "Robust Hedging via Coupling," *Finance and Stochastics*, 2, 1998, 329-347.

[94] Huang, J., M. Subrahmanyam and G.Yu, "Pricing and Hedging American Options: A Recursive Integration Method," *Review of Financial Studies*, 9, 1996, 277-330.

[95] Hugonnier, J., "The Feynman-Kac Formula and Pricing Occupation Times Derivatives," *International Journal of Theoretical and Applied Finance*, 2, 1999, 153-178.

[96] Hull, J., *Options, Futures and Other Derivatives*, 5th edition, Prentice Hall, New Jersey, 2002.

[97] Hull, J. and A. White, "The Use of Control Variate Techniques in Option Pricing," *Journal of Financial and Quantitative Analysis*, 23, 1988, 237-251.

[98] Hull, J. and A. White, "Valuing Derivative Securities Using the Explicit Finite Difference Method," *Journal of Financial and Quantitative Analysis,* 25, 1990, 87-100.

[99] Hull, J. and A. White, "Efficient Procedures for Valuing European and American Path-Dependent Options," *Journal of Derivatives*, 1, 1993, 21-32.

[100] Jacka, S. D., "Optimal Stopping and the American Put," *Mathematical Finance*, 1, 1991, 1-14.

[101] Jacka, S. D. and J. Lynn, "Finite-Horizon Optimal Stopping, Obstacle Problems and the Shape of the Continuation Region," *Stochastics and Stochastics Reports*, 39, 1992, 25-42.

[102] Jaillet, P., D. Lamberton and B. Lapeyre, "Variational Inequalities and the Pricing of American Options," *Acta Applicandae Mathematicae*, 21, 1990, 263-289.

[103] Jarrow, R. and A. Rudd, *Option Pricing*, Dow Jones - Irwin, Homewood, Illinois, 1983.

[104] Johnson, H., "The Pricing of Complex Options," Working paper, Louisiana State University, 1981.

[105] Johnson, H., "An Analytic Approximation for the American Put Price," *Journal of Financial and Quantitative Analysis*, 18, 1983, 141-148.

[106] Johnson, H., "Options on the Maximum or the Minimum of Several Assets," *Journal of Financial and Quantitative Analysis*, 22, 1987, 227-283.

[107] Ju, N., "Pricing an American Option by Approximating Its Early Exercise Boundary as a Multi-Piece Exponential Function," *Review of Financial Studies*, 11, 1998, 627-646.

[108] Kac, M., "On Distributions of Certain Wiener Functionals," Transactions of the American Mathematical Society, 65, 1949, 1-13.

[109] Kac, M., "On Some Connections Between Probability and Differential and Integral Equations," *Proceedings of the 2nd Berkeley Symposium on Mathematical Statistics and Probability*, 1951, 189-215.

[110] Kallast, S. and A. Kivinukk, "Pricing and Hedging American Options Using Approximations by Kim Integral Equations," *European Finance Review*, 7, 2003, 361-383.

[111] Kamrad, B. and P. Ritchken, "Multinomial Approximating Models for Options with k State Variables," *Management Science*, 37, 1991, 1640-1652.

[112] Karatzas, I., "On the Pricing of American Options," *Appl. Math. Optim.*, 17, 1988, 37-60.

[113] Karatzas, I. and S. Shreve, *Brownian Motion and Stochastic Calculus*, Springer-Verlag, New York, 1988.

[114] Karatzas, I. and S. Shreve, *Methods of Mathematical Finance*, Springer-Verlag, New York, 1998.

[115] Kholodnyi, V. A. and J. F. Price, *Foreign Exchange Option Symmetry*, World Scientific, London, 1998.

[116] Kim, I. J., "The Analytic Valuation of American Options," *Review of Financial Studies*, 3, 1990, 547-572.

[117] Kim, I. J. and S. Byun, "Optimal Exercise Boundary in a Binomial Option Pricing Model," *Journal of Financial Engineering*, 3, 1994, 137-158.

[118] Kim, I. J. and G. Yu, "An Alternative Approach to the Valuation of American Options and Applications," *Review of Derivatives Research*, 1, 1996, 61-86.

[119] Kwok, Y. K. and K. W. Lau, "Pricing Algorithms for Options with Exotic Path-Dependence," *Journal of Derivatives*, Fall 2001, 28-38.

[120] Lamberton, D., "Convergence of the Critical Price in the Approximation of American Options," *Mathematical Finance*, 3, 1993, 179-190.

[121] Lamberton, D., "Error Estimates for the Binomial Approximation of American Put Options," *Annals of Applied Probability*, 8, 1998, 206-233.

[122] Lamberton, D., "Brownian Optimal Stopping and Random Walks," *Applied Mathematics and Optimization*, 45, 2002, 283-324.

[123] Leisen, D. and M. Reimer, "Binomial Models for Option Valuation - Examining and Improving Convergence," Working paper B-309, University of Bonn, Bonn, 1995.

[124] Lintner, J., "The Valuation of Risky Assets and the Selection of Risky Investments in Stock Portfolios and Capital Budgets," *Review of Economics and Statistics*, 47, 1965, 13-37.

[125] Little, T., V. Pant and C. Hou, "A New Integral Representation of the Early Exercise Boundary for American Put Options," *Journal of Computational Finance*, 2000.

[126] Longstaff, F. A. and E. A. Schwartz, "Valuing American Options by Simulations: A Simple Least-Squares Approach," *Review of Financial Studies*, 14, 2001, 113-147.

[127] Madan, D., F. Milne and H. Shefrin, "The Multinomial Option Pricing Model and its Brownian and Poisson Limits," *Review of Financial Studies*, 4, 1989, 251-266.

[128] Marchuk, G. and V. Shaidurov, *Difference Methods and Their Extrapolations*, Springer-Verlag, New York, 1983.

[129] Margrabe, W., "The Value of an Option to Exchange One Asset for Another," *Journal of Finance*, 33, 1978, 177-186.

[130] McDonald, R. and M. Schroder, "A Parity Result for American Options," *Journal of Computational Finance*, 1998. Working paper, Northwestern University, 1990.

[131] McKean, H. P., "A Free Boundary Problem for the Heat Equation Arising from a Problem in Mathematical Economics," *Industrial Management Review*, 6, 1965, 32-39.

[132] Merton, R. C., "Theory of Rational Option Pricing," *Bell Journal of Economics and Management Science*, 4, 1973a, 141-183.

[133] Merton, R. C., "An Intertemporal Capital Asset Pricing Model," *Econometrica*, 41, 1973b, 867-888.

[134] Moraux, F., "On Cumulative Parisian Options," *Finance*, 23, 202, 127-132.

[135] Moraux, F., "Valuing Corporate Liabilities When the Default Threshold Is not an Absorbing Barrier," Working paper, Universite de Rennes 1, 2004.

[136] Mossin, J., "Security Pricing and Investment Criteria in Competitive Markets," *American Economic Review*, 59, 1969, 749-756.

[137] Miura, R, "A Note on Look-Back Options based on Order Statistics," *Hitosubashi Journal of Commerce and Management*, 27, 1992, 15-28.

[138] Myneni, R., "The Pricing of the American Option," *Annals of Applied Probability*, 2, 1992, 1-23.

[139] Ocone, D. and I. Karatzas, "A Generalized Clark Representation Formula, with Application to Optimal Portfolios," *Stochastics* 34, 187-220.

[140] Omberg, E., "The Valuation of American Put Options with Exponential Exercise Policies," *Advances in Futures and Options Research*, 2, 1987, 117-142.

[141] Omberg, E., "Efficient Discrete Time Jump Process Models in Option Pricing," *Journal of Financial and Quantitative Analysis*, 23, 1988, 161-174.

[142] Parkinson, M., "Option Pricing: The American Put," *Journal of Business*, 50, 1977, 21-36.

[143] Press, W., S. Teukolsky, W. Vetterling and B. Flannery, *Numerical Recipes in C: The Art of Scientific Computing*, 2nd edition, Cambridge University Press, Cambridge, England, 1992.

[144] Raymar, S. and M. Zwecher, "A Monte Carlo Valuation of American Call Options on the Maximum of Several Stocks," *Journal of Derivatives*, 5, 1997, 7-23.

[145] Rendleman, R. and B. Bartter, "Two-State Option Pricing," *Journal of Finance*, 34, 1979, 1093-1110.

[146] Ritchken, P., L. Sankarasubramaniam and A. M. Vijh, "The Valuation of Path Dependent Contracts on the Average," *Management Science*, 39, 1993, 1202-1213.

[147] Rogers, C., "Monte Carlo Valuation of American Options," *Mathematical Finance*, 12, 2002, 271-286.

[148] Roll, R., "An Analytic Valuation Formula for Unprotected American Call Options on Stocks with Known Dividends," *Journal of Financial Economics*, 5, 1977, 251-258.

[149] Rubinstein, M., "One for Another," *Risk*, 4, 1991, 30-32.

[150] Rubinstein, M., "Return to Oz," *Risk*, 7, 1994, 67-71.

[151] Rubinstein M. and E. Reiner, "Breaking Down the Barriers," *Risk*, 4, 1991, 28-35.

[152] Rutkowski, M., "The Early Exercise Premium Representation of Foreign Market American Options," *Mathematical Finance*, 4, 1994, 313-325.

[153] Samuelson, P. A., "Rational Theory of Warrant Pricing," *Industrial Management Review*, 6, 1965, 13-31.

[154] Schroder, M., "Computing the Constant Elasticity of Variance Option Pricing Formula," *Journal of Finance*, 44, 1989, 211-219.

[155] Schroder, M., "Changes of Numeraire for Pricing Futures, Forwards and Options," *Review of Financial Studies*, 12, 1999, 1143-1163.

[156] Schwartz, E. S., "The Valuation of Warrants: Implementing a New Approach," *Journal of Financial Economics*, 4, 1977, 79-93.

[157] Sharpe, W., "Capital Asset Prices: A Theory of Market Equilibrium under Conditions of Risk," *Journal of Finance*, 19, 1964, 425-442.

[158] Singhal, K., J. Vlach and M. Vlach, "Numerical Inversion of Multidimensional Laplace Transforms," *Proceedings of the IEEE*, 63, 1975, 1627-1628.

[159] Stulz, R. M., "Options on the Minimum or the Maximum of Two Risky Assets," *Journal of Financial Economics*, 10, 1982, 161-185.

[160] Takács, L., "On a Generalization of the Arc-Sine Law," *Annals of Applied Probability*, 6, 1996, 1035-1040.

[161] Tan, K. and K. Vetzal, "Early Exercise Regions for Exotic Options," unpublished manuscript, University of Waterloo, 1994.

[162] Tian, Y., "A Modified Lattice Approach to Option Pricing," *Journal of Futures Markets*, 13, 1993, 563-577.

[163] Tilley, J., "Valuing American Options in a Path Simulation Model," *Transactions of the Society of Actuaries*, 45, 1993, 83-104.

[164] Trigeorgis, L., "A Log-transformed Binomial Numerical Analysis Method for Valuing Complex Multi-Option Investments," *Journal of Financial and Quantitative Analysis*, 26, 1991, 309-326.

[165] Van Moerbeke, P., "On Optimal Stopping and Free Boundary Problems," *Arch. Rational Mech. Analysis*, 60, 1976, 101-148.

[166] Vetzal, K. R. and P. A. Forsyth, "Discrete Parisian and Delayed Barrier Options: A General Numerical Approach," *Advances in Futures and Options Research*, 10, 1999, 1-15.

[167] Wu, L., Y. K. Kwok and H. Yu, "Asian Options with the American Early Exercise Feature," *International Journal of Theoretical and Applied Finance*, 2, 1999, 101-111.

[168] Yor, M., "The Distribution of Brownian Quantiles," *Journal of Applied Probability*, 32, 1995, 405-416.

Index

For Product Safety Concerns and Information please contact our EU representative GPSR@taylorandfrancis.com Taylor & Francis Verlag GmbH, Kaufingerstraße 24, 80331 München, Germany

T - #0010 - 230425 - C34 - 234/156/15 [17] - CB - 9781584885672 - Gloss Lamination